Radiation Sensors with Three-Dimensional Electrodes

Series in Sensors

Principles of Electrical Measurement
Slawomir Tumanski
January 20, 2006

Novel Sensors and Sensing
Roger G. Jackson
September 30, 2004

Hall Effect Devices, Second Edition
R.S. Popovic
December 01, 2003

Sensors and Their Applications XII
S. J. Prosser, E. Lewis
September 01, 2003

Thin Film Magnetoresistive Sensors
S Tumanski
June 08, 2001

Electronic Noses and Olfaction 2000: Proceedings of the 7th International Symposium on Olfaction and Electronic Noses, Brighton, UK, July 2000
Julian W. Gardner, Krishna C. Persaud
January 01, 2001

Sensor Materials
P.T Moseley, J Crocker
January 01, 1996

Biosensors: Microelectrochemical Devices
M Lambrechts, W Sansen
January 01, 1992

Radiation Sensors with Three-Dimensional Electrodes

Cinzia Da Vià

The University of Manchester, UK,
and Stony Brook University, USA

Gian-Franco Dalla Betta

The University of Trento, Italy

Sherwood Parker

(formerly of) the University of Hawaii, USA

CRC Press
Taylor & Francis Group
Boca Raton London New York

CRC Press is an imprint of the
Taylor & Francis Group, an **informa** business

CRC Press
Taylor & Francis Group
6000 Broken Sound Parkway NW, Suite 300
Boca Raton, FL 33487-2742

Printed on acid-free paper

International Standard Book Number-13: 978-1-4987-8223-4 (Hardback)

Library of Congress Cataloging-in-Publication Data

Names: Viá, Cinzia da, author. | Betta, G. F. (Gian-Franco Dalla Betta), author. | Parker, Sherwood, author.
Title: Radiation sensors with 3D electrodes / Cinzia Da Viá (The University of Manchester, UK, and Stony Brook University, USA), Gian-Franco Dalla Betta (The University of Trento, Italy), Sherwood Parker (formerly of) the University of Hawaii, USA).
Description: Boca Raton, FL : CRC Press, [2019] | Series: Series in sensors | Includes bibliographical references and index.
Identifiers: LCCN 2018041973| ISBN 9781498782234 (hardback : alk. paper) | ISBN 9781498782241 (ebook : alk. paper)
Subjects: LCSH: Radiation—Measurement—Instruments. | Nuclear counters. | Electrodes. | Silicon diodes.
Classification: LCC QC795.5 .V53 2019 | DDC 539.7/7—dc23
LC record available at https://lccn.loc.gov/2018041973

Visit the Taylor & Francis Web site at
http://www.taylorandfrancis.com

and the CRC Press Web site at
http://www.crcpress.com

*We would like to dedicate this book to
Sherwood Parker: a friend, a colleague,
and a mentor who left us too soon for
complications from a devastating illness.*

---

*Meeting you and working with you have
been a privilege and an inspiration, and
your example will continue to be a reference
to both of us. Sadly you did not see the
completion of this opus. We missed your
comments and corrections, but we hope
you would be as proud of it as we are.*

Contents

About the Authors

Cinzia Da Vià is a Professor of Physics at the University of Manchester. UK, and is currently a Visiting Professor at Stony Brook University (USA). She received her PhD in Physics from the University of Glasgow, Scotland, in 1997 and is an expert in semiconductor detector development for high-energy physics and medical applications. She has authored more than 300 papers, of which several are on the evaluation of the first 3D sensor prototypes ever fabricated. Member of the ATLAS Experiment at the CERN Large Hadron Collider since 2007, she was the founder and leader of the 3D ATLAS Pixel R&D Collaboration (2007–2014), which successfully designed and industrialized the first 3D sensors to be installed in a collider. 3Ds have been successfully operating in the ATLAS experiment since 2014. She is currently involved in novel 3D sensor designs, 3D printed dosimetry, quantum imaging, and vertical integration of smart-systems. She is one of the founders of the ERDIT Network to promote radiation imaging technology research across different applications in Europe, and she is a member of the Independent Committee of the ATTRACT initiative, which funds innovative technologies in the field of radiation detection and imaging across Europe.

Gian-Franco Dalla Betta is a Full Professor of Electronics at the University of Trento, Italy. He received the M.S. degree in electronics engineering from the University of Bologna, Italy, in 1992 and the PhD in microelectronics from the University of Trento, Italy, in 1997. From 1997 to 2002, he was with the Institute for Scientific and Technological Research (ITC-IRST) of Trento, Italy, as a researcher and in November 2002, he moved to the University of Trento. His main research expertise is in design, simulation, fabrication, and experimental characterization of silicon integrated devices and circuits, with emphasis on radiation sensors and 3D sensors, of which he was among the first to design a double-sided layout. On these and related topics he has been the author or co-author of more than 350 papers published in international journals and conference proceedings. As a member of the 3D ATLAS R&D Collaboration from the start, he coordinated the design of the 3D pixel sensors that are now installed in the ATLAS Insertable B-Layer at the CERN Large

Hadron Collider (LHC), the first application of this technology in a high-energy physics experiment. His current activities include the development of new 3D sensor designs for neutron and fast-timing applications and the coordination of the Italian R&D effort aimed at a new generation of 3D sensors for the Phase 2 upgrades of the ATLAS and CMS projects at the high luminosity LHC.

Sherwood Ira Parker (1932–2018) was a pioneer in experimental physics. He developed the first scientific silicon readout integrated circuit (Microplex), the first monolithic charged particle sensors, and the first 3D silicon detectors used to prove the existence of the Higgs Boson particle. He collaborated with many leading research scientists and laboratories around the world, including CERN, FERMI, and SLAC. Dr. Parker also developed detectors for use in digital mammography and held seven patents. These achievements were recognized when he was awarded the Glenn Knoll Radiation Instrumentation Outstanding Achievement Award in 2015 from the IEEE Nuclear and Plasma Sciences Society. And despite severe mobility impairments caused by ALS, he continued contributing to key innovative work on high-speed signals with 3D radiation detectors.

Acknowledgments

The completion of this book was possible thanks to the contribution of many people over many years. We hope to remember everyone and apologize in advance if we have inadvertently overlooked any names.

We would like to start by thanking our friends and collaborators from many years, Chris Kenney, Jasmine Hasi, Angela Kok, Julie Segal, Stephen Watts, Kenway Smith (who regretted he did not process 3D in GaAs first), Bengt Svensson (who sadly is not with us any more), Thor-Erik Hansen, Ole Rohne, Marco Povoli, Roberto Mendicino, D.M.S. Sultan, Marcello Borri, Ching-Hung Lai, Clara Nellist, Iain Haugthon, Nicholas Dann, Sebastian Grinstein, Maurizio Boscardin, Giulio Pellegrini, Elisa Vianello, Per Hansson, Su Dong, Dmitri Tsibichev, Daniela Bortoletto, Norbert Wermes, Alessandro La Rosa, Philippe Grenier, Stanislav Pospisil, Tomas Slavicek, and Thomas Fritszch for their invaluable contribution through ideas, hard work, and insight over the years and for this work specifically. We would also like to thank Claudio Piemonte, Sabina Ronchin, Francesca Mattedi, Alberto Pozza, Gabriele Giacomini, Nicola Zorzi, Alvise Bagolini, Celeste Fleta, David Quirion, Manuel Lozano, Heidi Sandaker, Bjarne Stugu, Ozhan Koybasi, Andrea Zoboli, Jörn Lange, Stefano Terzo, Luciano Bosisio, Serena Mattiazzo, Hartmut Sadrozinski, Gregor Kramberger, Vladimir Cindro, Igor Mandic, Marko Mikuz, Juha Kalliopuska, Simo Eränen, Jaakko Härkönen, Richard Bates, Chris Parkes, David Pennicard, Val O'Shea, Victoria Wright, Sally Seidel, Martin Hoeferkamp, Jessica Metcalfe, Ulrich Parzefall, Michael Köhler, Riccardo Mori, Liv Wiik, Simon Eckert, Gregor Pähn, Cristopher Betancourt, Karl Jacobs, Enver Alagoz, Mayur Bubna, Fabio Ravera, Ada Solano, Margherita Obertino, Marta Ruspa, Andrea Micelli, Marco Meschini, Alberto Messineo, Claudia Gemme, Andrea Gaudiello, and David-Leon Pohl, and Christer Frojid for their work and support.

We also thank Yoshinobu Unno, Stanislav Pospisil, and Norbert Wermes for giving their positive feedback and comments during the editing of this book. Nothing in it would exist without the long hours spent by many students within the 3D ATLAS R&D and the IBL collaborations in test beams. Also, thanks to those who believed in our work and enthusiasm during the most challenging moments of 3D development, in particular, Marzio Nessi and Nanni Darbo: It really meant a lot to get your support when the majority was ready to bet a production would have never been possible in time!

We are also deeply indebted to our editorial contact, Rebecca Davies, for her constant patience, persistence, and kindness despite our multiple delayed submissions.

Finally, a huge thanks goes to our families (in particular Vania and Luis) for being so understanding when we could not always join them for holidays and celebrations during the past two years and for giving us continuous encouragement and support.

CHAPTER 1

Introduction

Silicon radiation detectors have been used extensively for many years in a large variety of industrial, medical, and scientific applications and are part of the core instrumentation in many areas of fundamental and applied research.

In the early 1980s, Joseph Kemmer achieved a major technological breakthrough in silicon sensors processing with the pioneering use of the planar fabrication process derived from microelectronics. The key innovations consisted of (1) exploiting the passivation properties of silicon dioxide, which allowed keeping the thermal budget to a minimum, and (2) using detectors with ion-implanted junctions making possible the design of fine pitch electrodes segmentation with very low leakage currents. This revolution, along with the development of low-noise, low-power microelectronic readout circuits, opened the field of position-sensitive radiation detection and imaging using planar silicon sensors. Silicon strip and silicon drift detectors became available first, followed by silicon pixel detectors a few years later.

Ever since this historical moment, silicon detector technologies have been continuously advancing. Now more complex and reliable detectors can be obtained, with outstanding performance in terms of energy, timing and position resolution, long-term stability, and radiation tolerance.

A new paradigm shift in silicon sensor technology became possible in the mid-1990s when Sherwood Parker and collaborators introduced bulk micromachining in combination with microelectronics' very-large-scale integration (VLSI) in the processing. By exploiting the third dimension within the silicon substrate, several interesting features could be obtained, related either to radiation-sensing properties or to some ancillary functions. Unlike planar detectors, where electrodes are confined to the wafer surfaces, in 3D, electrodes penetrate partially or entirely throughout the substrate, perpendicularly to the surface. This architecture offers a number of substantial advantages with respect to the planar one such as fast signals, reduced bias voltage and power dissipation, and extreme radiation tolerance, making 3D detectors ideal candidates for some critical applications, especially in high-energy or nuclear physics. These advantages, however, come at the expense of a more complicated and expensive fabrication process, which is only possible in facilities with microelectronic and micro-electro-mechanical-systems (MEMS) technologies in the same cleanroom.

For about a decade since their invention, 3D sensors have undergone a relatively slow research and development (R&D) phase, with only a small number of scientists working on few prototypes built at the Stanford nanofabrication facility by the original inventors. But starting in the mid-2000s, more devices inspired by the original design were processed at European facilities using modified technologies, and experimental results confirmed the great potential of these devices. The establishment of the ATLAS 3D Sensors Collaboration for the Large Hadron Collider (LHC) luminosity upgrade in 2007 led to an impressive boost in the development of 3D sensors, with experimental confirmation of their remarkable radiation tolerance with relatively low power dissipation, and the demonstration of medium-volume production, which culminated in their use for the ATLAS Insertable B-Layer (IBL) project in 2014. These accomplishments paved the way for the use of 3D sensors in other pixel detector systems in Phase 1 upgrades at the LHC such as ATLAS Forward Physics (AFP) and CMS-TOTEM Precision Proton Spectrometer (CT-PPS), and made them an appealing option for the innermost tracking layers at further high luminosity LHC (HL-LHC) phases.

This book presents a comprehensive and up-to-date review of 3D detectors, covering relevant aspects of device physics and simulation, fabrication technologies and design issues, selected experimental results, and application fields.

The book is organized as follows:

Chapter 2 provides a brief overview of some basic theoretical concepts necessary to understand 3D sensors. These concepts include interaction of radiation with silicon, the operation principle of silicon sensors, main sensor types, and signal formation and processing.

Chapter 3 reviews the effects of radiation damage on silicon (both surface and bulk), their macroscopic consequences for sensors performance, and some measures to counteract the effects in practice.

Chapter 4 provides a comprehensive description of 3D silicon sensors, from the operation principle to simulations and experimental results relevant to both the original Stanford device and all the alternative design variants reported thus far.

Chapter 5 describes the key fabrication technology steps used in 3D silicon sensors, ranging from general aspects of the processing to detailed descriptions of the different flavors made at facilities, which developed alternative 3D designs.

Chapter 6 addresses the radiation hardness of 3D sensors, with details on its dependence on electrodes' geometry and a comprehensive overview of experimental results supporting the theoretical model used to explain the signal formation of 3D sensors.

Chapter 7 describes the industrialization process for the ATLAS IBL project, explaining the strategy adopted to reach this crucial milestone, the quality assurance system used during the production, and the main achievements of the 3D ATLAS Collaboration, which led to the first application of 3D sensors in a high-energy physics (HEP) experiment.

Chapter 8 reports active edges and slim edges applied to planar sensors, which are one of the key side products of 3D technology. The different design solutions so far reported are reviewed, and selected experimental results are recalled.

Chapter 9 outlines the applications of 3D silicon sensors and related concepts, ranging from those that are well established such as HEP to emerging applications, including medical imaging, dosimetry, and neutron detection. The chapter ends with an outlook on 3D sensors made from different semiconductor substrates.

Finally, the Appendix presents excerpts from Sherwood Parker's recollections of his work on silicon detectors, including 3Ds.

CHAPTER 2

Silicon Radiation Sensors

This chapter presents an overview of silicon radiation sensors. Many of the concepts described here assume some familiarity with fundamentals of semiconductor physics, which may easily be found in several well-established textbooks [1,2]. Applications of semiconductors such as radiation detectors and related readout electronics are also fully described in the literature. Particular references [3–7] are the main source of information used for this chapter. Although the material covered in this chapter is intentionally limited to some basic aspects, it should be sufficient to help the reader understand the operation and performance of three-dimensional (3D) silicon sensors reported later in the text.

2.1 INTRODUCTION

Semiconductor have been used as radiation detectors in nuclear physics and X- and γ-ray spectroscopy since the early 1960s [4]. Nowadays they are extensively used in a variety of applications spanning from nondestructive diagnostics and process control in industry to biomedical imaging, from cultural heritage to security, as well as in many fields of fundamental and applied research, including high-energy physics, photon science, and astronomy.

Simply put, a semiconductor detector can be described as a solid-state ionization chamber, with an operating principle that is similar to a gaseous detector. When traversing a semiconductor detector, an impinging particle releases energy that creates electron-hole pairs along its path; in the presence of an electric field, electrons and holes are separated and start drifting, inducing a signal on the electrodes. However, in comparison to the gaseous counterpart, radiation absorption is higher in semiconductors because their density is larger. Moreover, because the ionization energy is rather low (just 3.6 eV), semiconductor detectors are characterized by an excellent intrinsic energy resolution. Other important advantages of semiconductors are their fast time response, a linearity of performance over a wide range, good stability, low noise, and the possibility of adjusting the effective detection volume by changing the junction depletion bias voltage. (This topic will be explained in more detail shortly.) With respect to gaseous detectors, semiconductor ones can be fabricated with a wide variety of geometries, yielding an excellent spatial resolution

TABLE 2.1 Main physical properties of some semiconductor materials.

Semiconductor	ρ (g/cm^3)	Z	E_g (eV)	E_{ion} (eV)	μ_e (cm^2/Vs)	μ_h (cm^2/Vs)
Silicon	2.33	14	1.12	3.6	1450	450
Germanium	5.33	32	0.67	2.9	3900	1900
GaAs	5.32	31–33	1.43	4.3	8500	450
CdTe	6.20	48–52	1.44	4.7	1050	100
4H-SiC	3.16	14–6	3.23	7.8	800–1000	50–115
Diamond	3.51	6	5.47	13	1800–2200	1200–1600

and with a very compact size, although this latter characteristic can sometimes be a limitation for some applications.

Important drawbacks of semiconductor detectors are the difficulty in achieving and controlling internal charge amplification (this is only feasible in avalanche-based detectors), which makes the signal processing more delicate, and the high sensitivity of these devices to radiation damage.

Various semiconductor materials, including elemental semiconductors and binary/ternary compounds, are suitable for radiation detector applications. Their physical properties are very important for the correct choice of the material that is most suitable for a particular application. Table 2.1 summarizes these properties for some semiconductors of practical interest [8]: density ρ, atomic number Z, band-gap energy E_g, ionization energy E_{ion}, electron mobility μ_e, and hole mobility μ_h.

Atomic number and material density influence the stopping power, so when photons or X-rays are interacting, the higher the material density, the higher the detection efficiency. Another important parameter is the mean ionization energy, defined as the average radiation energy loss per electron-hole pair production. For all semiconductors, this energy is related to the band gap and is normally two to three times larger because part of the energy is dissipated and used to create phonons [9]. In principle, low band-gap semiconductor detectors are better because they would give rise to a higher number of generated carriers, and therefore a higher signal for the same radiation, thus allowing for a good intrinsic energy resolution. As an example, germanium has unsurpassed characteristics for γ-ray spectroscopy applications because of high-density and low-ionization energy. However, the drawback is that a lower band gap also means a higher thermal generation of carriers, which in turn means higher leakage current, more noise, and, finally, lower energy resolution. For this reason, low band-gap semiconductors like germanium cannot be operated at room temperature and must be cooled at low temperatures (liquid nitrogen is commonly used) to limit the noise level of the spectroscopic system.

Among available semiconductor materials, GaAs and especially CdTe, CdZnTe are suited to room temperature X- and γ-ray detection, owing to their high Z, though their technology is not yet fully mature. Other semiconductors such as diamond and SiC are the object of increasing interest since they promise low or zero radiation-induced leakage currents, but several issues are still to be solved, including the availability of large-size wafers (diamond) and defect-free raw material (SiC).

Although its physical properties are not the best in some respects, silicon is by far the most studied, understood, manufactured, and used semiconductor for radiation detection. The main reason for this success is the same one that determined the success of silicon in microelectronics: Apart from the possibility of reaching high purity, silicon is the only semiconductor material that has a native oxide with good interface properties, which allows for reliable manufacturability and high-integration technology.

2.2 INTERACTION OF RADIATION WITH SILICON

Silicon detectors are most suited to applications that involve charged particles and low-energy X-rays. The operation of a radiation detector obviously depends on the way radiation interacts with the material composing the detector itself. Different interaction mechanisms can be encountered, depending on the detector material and on the type of radiation. A detailed analysis of all these interaction mechanisms would lie outside the purpose of this book; for this reason, only some basic concepts are addressed in the following.

2.2.1 Charged Particles

Charged particles can be divided into heavy charged particles, including all charged nuclei from protons to the heaviest fission fragments, and light charged particles, including electrons and positrons, that may be emitted through natural or induced nuclear decay or may be produced by the acceleration of electrons in an electromagnetic field [4].

When passing through matter, charged particles interact through Coulomb forces with atomic electrons and lose part of their energy. The latter, also known as ionizing energy loss (IEL), depends on the particle's initial energy and on material properties such as density and atomic number. Electrons, for instance, owing to their low mass, can be easily deflected, whereas protons or heavy ions proceed in a more straightforward manner within matter. As mentioned previously, protons, pions, muons, and electrons lose their energy primarily because of inelastic collisions with the matter's atomic electrons. A small amount of energy is lost at every collision, and part of this energy creates free charge in the material. The average energy loss of a charged particle per unit path length in a medium follows the Bethe–Block formula [10], which is described by Eqn. 2.1:

$$-\frac{dE}{dx} = 4\pi N_A r_e^2 m_e c^2 z^2 \frac{Z}{A}\frac{1}{\beta^2}\left[\frac{1}{2} ln\left(\frac{2m_2 c^2\beta^2\gamma^2 T_{max}}{I^2}\right) - \beta^2 - \frac{\delta(\gamma)}{2}\right]. \quad (2.1)$$

where N_A is Avogadro's number, r_e the classical electron radius, m_e the electron mass, c the speed of light, z the charge of the incident particle, Z the atomic number, and A the atomic mass of the considered material. $\beta = v/c$ represents the velocity, $\gamma = \frac{1}{\sqrt{1-\beta^2}}$, T_{max} the maximum kinetic energy that can be imparted to a free electron in a single collision, I the mean ionisation energy, and δ the density effect correction.

A schematic representation of a dE/dx versus $\beta\gamma$ plot is shown in Fig. 2.1a. Four regions can be distinguished: (1) a steep decrease proportional to $1/\beta^2$; (2) a minimum of energy

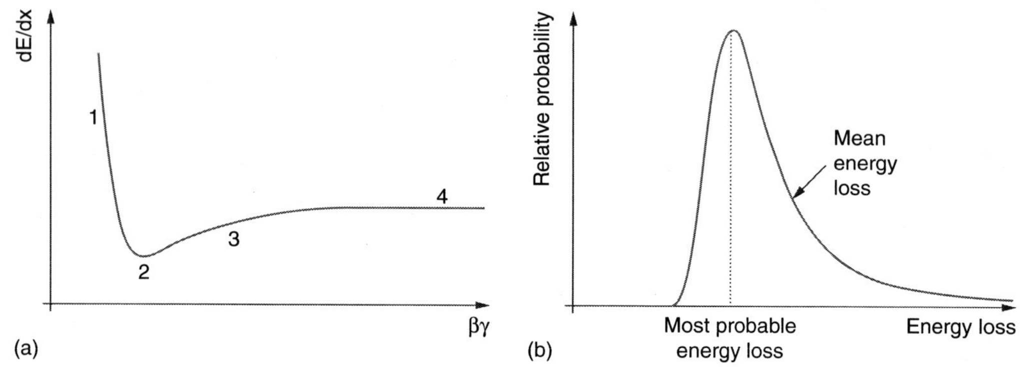

FIGURE 2.1 Schematic representations of (a) Bethe–Bloch formula, and (b) Landau distribution energy in Si due to a MIP.

loss (minimum ionization), which is reached at $\beta\gamma \cong 3$; (3) a slow "relativistic" increase proportional to $\ln(\gamma)$; and (4) a saturation region due to density effects.

For high-energy physics applications, particularly important is the minimum energy deposited in the medium. As can be seen in Fig. 2.1a, for high-energy particles, the mean energy transferred to matter reaches this minimum and then does not change significantly. Particles having energies high enough to reach this minimum are known as minimum ionizing particles (MIPs) and can be considered the "worst-case" signals for a detection system in terms of required sensitivity.

A complete analysis of these kinds of phenomena is not the object of this book, but some additional remarks are important for later discussions. When a particle interacts with matter, some statistical fluctuations in the number of collisions and energy transfer per scattering event are present. The first process is generally modeled with a Poisson distribution, while the second one is described with a "straggling function." In rare cases, high-energy recoil electrons (known as δ-electrons or δ-rays), produced by the interaction, have sufficient energy to become ionizing particles themselves and are responsible for an asymmetry in the collected spectrum, with a longer tail toward higher energies. The resulting energy deposition spectrum for an MIP crossing a silicon detector follows the Landau distribution [11], as is schematically shown in Fig. 2.1b. Due to the asymmetry, the most probable energy loss, corresponding to the distribution peak, differs from the mean energy loss, which is shifted at higher energies by about 30%. In silicon, as shown in Table 2.1, the average energy needed to create an electron-hole pair is 3.6 eV, which is about three times larger than the band gap. The most probable value (MPV) of the Landau distribution depends on the thickness of the detector [12]. For a standard 300 μm thickness, the most probable number of generated electron-hole pairs in one micron is 76, with an average value of 108. Hence, in a 300 μm thick silicon sensor, an, MIP will deposit a most probable charge of about $300 \times 76 = 22{,}800$ electrons, corresponding to about 3.6 fC. This is the tiny amount of charge that the readout system will therefore need to be able to detect.

Another important effect for charged particles is the so-called *multiple scattering;* that is, the trajectory of a particle traversing a medium is deviated (scattered) multiple times by small angles, mainly due to Coulomb interaction with the nuclei of the material. After multiple interactions, the scattering angle roughly follows a Gaussian distribution, with the root mean square (rms) value given by Eqn. 2.2 [13]:

$$\theta_{plane}^{rms} = \frac{13.6MeV}{\beta pc} z \sqrt{\frac{x}{X_0}} \left[1 + 0.038 \cdot ln\left(\frac{x}{X_0} \right) \right] \tag{2.2}$$

where z is the charge number of the considered particle and x/X_0 is the thickness of the absorption medium in units of radiation length. The angle θ is expressed in rad, the particle momentum p in MeV, and the velocity β in units of the speed of light c. The radiation length x_0 of silicon is 9.36 cm. As an example, a pixel detector built for Large Hadron Collider (LHC) experiments has a thickness of about 2% of a radiation length per layer, changing the trajectory of a 1 GeV particle by an angle of about 0.1 rad.

For lower energy particles, or in the presence of very thick absorbers, the velocity of the particle is strongly reduced by the energy loss in the material. This results in an increased ionization as shown in Fig. 2.1a. If the particle is completely stopped inside the medium, most of its energy is released near the end of the trajectory, an effect known as *Bragg peak* [3]. As an example, α particles have large kinetic energy (5MeV) and low speed (about 5% the speed of light), resulting in a penetration of just a few microns in silicon, with most of the energy released close to the stopping point, as shown in Fig. 2.2.

The main interaction mechanism for fast electrons (with both negative and positive charge) is still based on the Coulomb force; however, because their mass is equal to that of the orbital electrons with which they interact, these particles can be easily deflected from their original trajectory, resulting in a tortuous path in the absorbing material. For this reason, a projected range can be conveniently used that is a function of the energy. Moreover, electrons can also lose energy through radiative processes, thus resulting in *bremsstrahlung* or electromagnetic radiation, but this type of interaction is negligible for low Z elements like silicon [3].

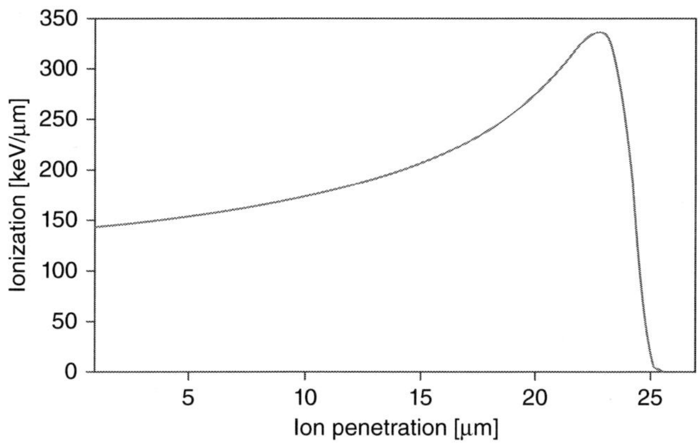

FIGURE 2.2 Bragg peak for a 5 MeV alpha particle in silicon calculated with SRIM [14].

2.2.2 Photons

Photons are the particles associated to the quantum description of the electromagnetic field, which are generated during the relaxation of an atom or a nucleus that have absorbed energy. Photons have zero rest mass, and they travel at the speed of light (c) in vacuum. They can be described either as particles carrying a certain energy (E_{ph}) or as waves of a certain frequency (ν) or wavelength (λ). These parameters are strictly related by Eqn. 2.3:

$$E_{ph} = h\upsilon = h\frac{c}{\lambda} \tag{2.3}$$

where h is the Planck constant. Electromagnetic radiation includes microwaves, far and near infrared light, visible light, ultraviolet light, soft X-rays, X-rays, and γ-rays. Visible light photons have wavelength ranging from approximately 0.4 to 0.7 μm.

Photon interactions with matter have been the object of extensive studies for a long time. As an example, Einstein was awarded the Nobel Prize in 1921 for his discovery of the photoelectric effect. Here we focus on the main interactions between photons and silicon that are useful for detector applications.

2.2.2.1 Photon Energy Close to the Energy Gap

When the energy of an incident photon exceeds the silicon band gap (E_g = 1.12 eV at room temperature), a photon can be absorbed by exciting a valence band electron to the conduction band, thus creating an electron-hole pair [1]. The maximum wavelength useful for absorption corresponds to the energy gap; so, in silicon it is λ_{max} = 1.12 μm, in the near infrared region. Silicon is therefore transparent to radiations whose wavelength exceeds λ_{max}, but absorption is possible if free carriers in the conduction band are excited to higher energy levels. This type of interaction is more likely to occur in extrinsic silicon (with high doping concentration) than in high-purity silicon [2].

The photon absorption process has to observe the laws of conservation for energy and momentum. The energy conservation law predicts that the photon energy will be completely transferred to the valence band electron. However, since silicon is an indirect gap semiconductor, additional momentum is required for the electron band-to-band transition, which cannot be provided by the quasi-particle photon itself. The conservation of momentum is satisfied by the presence of a third particle, known as a phonon, which is present in the lattice in the form of thermal vibration [1].

As they penetrate through silicon, interacting through either absorption or scattering mechanisms, photons are removed from their incident beam. By doing so, the intensity of the radiation, I, decreases exponentially with the depth, x, as described by Lambert–Beer's equation 2.4 [3]:

$$I(x) = I_0 e^{-\alpha x} \tag{2.4}$$

where I_0 is the beam initial intensity and α is the absorption (or attenuation) coefficient, which is a function of the photon energy and the absorbing material. The absorption coefficient for silicon at room temperature is shown in Fig. 2.3 for energies close to the energy gap. Note that if the energy of the incident photon is higher than the silicon zero momentum gap (3.4 eV, corresponding to a wavelength of about 0.4 μm), the absorption probability strongly

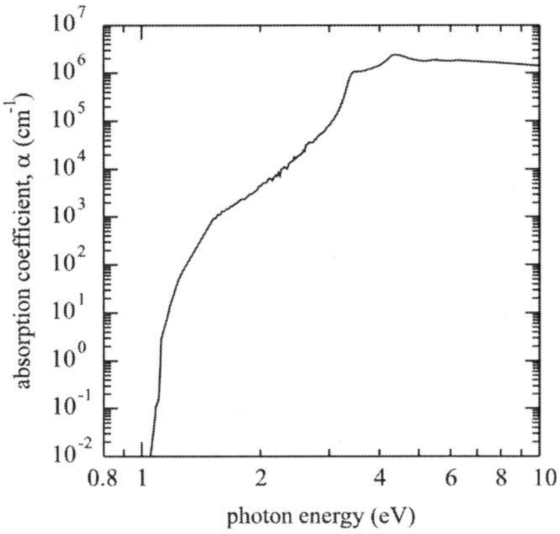

FIGURE 2.3 Absorption coefficient for Si at room temperature for energies close to the energy gap.

increases and then tends to saturate because direct transitions, not involving phonons, become possible. Hence, for low-wavelength photons, the detector dead layer, due to the presence of a highly doped region (typically in the order of several tenths of microns), which will be explained in chapter 5, limits the detection efficiency. In contrast, for photons in the near infrared, the absorption is much smaller, and a relatively thick detector layer would be required to obtain a high efficiency. Moreover, a silicon sensor used for visible light detection will be normally covered with an oxide layer. Since this layer is usually very thin and the attenuation coefficient of oxide is low, light is not significantly absorbed; yet reflection phenomena at the boundaries of this layer might influence the detector efficiency. When designing a photodetector, the thickness of the oxide layer can be optimized for efficiency improvement. As it will be explained later, additional layers of other materials, like silicon nitride, can be utilized, thus realizing a multilayer antireflecting coating [2].

2.2.2.2 Photon Energy Much Higher Than the Energy Gap

Unlike visible or IR (Infra Red) photons, X-rays and γ-rays have energies much higher than the energy gap. The interaction of these photons with silicon is based on three different mechanisms: (1) photoelectric effect, (2) Compton scattering, and (3) pair production. All the above-mentioned interaction processes contribute to the beam absorption, but each of them is dominant in a specific range of energies, as shown in Fig. 2.4.

These mechanisms can be briefly described as follows.

1. Photoelectric Effect

   This process is dominant for photon energies below than 50 keV. It results in the total absorption of a photon that, interacting with a silicon atom, makes it release an energetic electron from one of its bound shells [3]. Since the presence of a large mass is required to conserve momentum in the photoelectric process, the innermost (K-shell) electrons are more likely to be involved in this process. The released electron will gain a kinetic energy $E_k = E_{ph} - E_b$, where E_b is the binding energy of the electron. This kinetic energy is dissipated by secondary ionization while the electron travels inside the material for a small distance, thus creating electron-hole pairs. The excited atom will return to equilibrium through a cascade of electron transitions from the

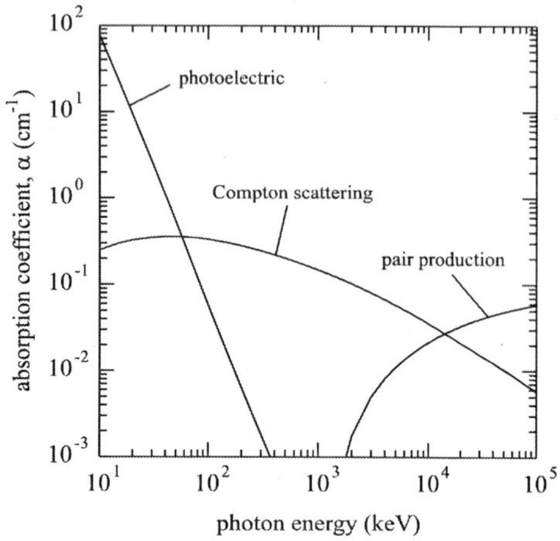

FIGURE 2.4 Absorption coefficient for Si at room temperature for energies much higher than the energy gap.

outer to the inner shells. This phenomenon is usually accompanied by the emission of characteristic X-rays, which, in most cases, are absorbed within a close distance from the position of interaction of the initial photon, so that the total charge signal corresponds to the total conversion of the photon energy. More rarely, the emitted X-rays might escape from the detector, thus influencing its response. Moreover, the emission of an Auger electron[1] may sometimes occur during the atom deexcitation process [3].

The absorption coefficient at 20 keV for silicon is 10.35 cm^{-1} which gives an interaction probability of only 26% for 300 μm thickness. At the same energy, 90% efficiency with silicon could only be achieved for 2.3 mm of active thickness, which is impractical. That's why silicon detectors are suitable only for soft X-rays, where the absorption length is comparable to the detector thickness. Other materials featuring high atomic number such as germanium, GaAs, or CdTe are more suitable for high-energy X-ray spectroscopy, thanks to their higher absorption coefficient (see Table 2.1).

2. Compton Scattering
   This process is dominant for photon energies ranging from 50 keV to 14 MeV. It consists of a collision between the incident photon and a weakly bound electron, resulting in the electron acquiring some of the photon's original energy and leaving a lower energy scattered photon. The electron, also known as "recoil" electron, will cause ionization in the detector, thus creating electron-hole pairs, while the scattered photon can either undergo further Compton scattering or can be absorbed by photoelectric process, depending on its residual energy. More rarely, the photon can escape from the detector. As the energy transferred by the photon to the electron can vary from event to event, the energy spectrum is characterized by the presence of the so-called Compton continuum, extending from zero to the maximum energy available for a head-on

[1] The **Auger effect** is a physical phenomenon in which the filling of an inner-shell vacancy of an atom is accompanied by the emission of an electron from the same atom. When a core electron is removed, leaving a vacancy, an electron from a higher energy level may fall into the vacancy, resulting in a release of energy. This energy is usually released in the form of an emitted photon but can also be transferred to another electron, which is ejected from the atom; this second ejected electron is called an **Auger electron**.

collision. However, the energy deposited by the scattered photon will also contribute to the energy spectrum, where energy peaks will be present due to photoelectric effect.

3. Pair Production

   This process can only take place when the incident photon energy exceeds the 1.022 MeV required to create an electron–positron pair, but it does not become dominant until the photon energy exceeds 14 MeV. In this interaction, which must take place in the Coulomb field of a nucleus, the original photon disappears, while its excess energy (over the 1.022 MeV) is transferred to the electron and the positron as kinetic energy [3]. Electron and positron will produce ionization by losing energy through collisions along their tracks. When the positron finally comes to a rest, it annihilates producing two photons, each having 511 keV energy emitted in opposite directions. These photons can be reabsorbed through the photoelectric effect or Compton scattering, or they can escape from the detector, thus affecting the energy spectrum. However, as can be seen in Fig. 2.4, the attenuation coefficient at these high energies is so low, that pair production in silicon is not important in practice.

2.2.3 Neutrons

Neutrons, like photons, do not have an electrical charge and consequently are not affected by the Coulomb force. They interact mainly with nuclei, and for this reason they can penetrate matter for relatively long distances (~ centimeters) before being entirely absorbed or scattered. Quantum mechanics governs the interactions between neutrons and nuclei, but a classical unit to describe the probability of interactions is the cross section σ. Its unit is the barn that corresponds to 10^{-24} cm^2.

Neutrons are classified according to their energy, which strongly affects the cross section. In slow neutrons, their kinetic energy is low, and the possibility to interact by nucleus scattering is very low. In this case, there is a high cross section for absorbing the neutron that induces nuclear reactions. The possible reactions are (n, γ), which can be difficult to detect because of the γ-ray background, and (n, α), (n, p), and (n, fission), which are the most convenient for detectors. On the contrary, fast neutrons are characterised by a very low absorption cross section, but their particle energy makes them suitable for a scattering. To be detected with silicon sensors, neutrons first have to interact with a high-neutron cross section converting material that can be deposited over the detector itself. Being heavy particles, neutrons primarily interact with the absorber by knocking on its nuclei, releasing their energy to electrons or heavy charged products of nuclear reactions, which can finally be detected by the previously cited interaction methods [3].

2.3 SEMICONDUCTOR PHYSICS

2.3.1 Silicon as a Detector Material

As stated at the beginning of this chapter, silicon detectors work in principle as an ionizing chamber. The simplest configuration one can imagine consists of an absorbing medium sandwiched between two metal electrodes, making either ohmic or Schottky contacts with the semiconductor. A signal is induced on the electrodes as the electron-hole pairs generated by radiation are separated by an electric field and drift towards the electrodes. To be clearly distinguished, signals should be large enough compared to noise; thus, a high signal-to-noise ratio (SNR) is necessary. This, however, can actually lead to two contradictory requirements: On one hand, to have a large signal, radiation should generate many electron-hole pairs, and this would require a low-ionization energy, hence a small band gap. On the other hand, to have low noise, there should be few intrinsic charge carriers, which is possible only with large band gaps. As an example, in intrinsic silicon substrates, with the size of a typical silicon sensor (1 cm^2 area, 300 μm thickness), there are ~4×10^8 electron-hole pairs at room temperature, whereas an MIP would normally generate only ~2×10^4 electron-hole pairs. The signal would be completely lost in this higher number of free-charge carriers! It is therefore essential to reduce the amount of free-charge carriers by several orders of magnitude. In principle, for silicon and other small band-gap semiconductors, this can be obtained by cooling, since the intrinsic concentration exponentially depends on the temperature [1]. However, the most effective solution is to create a p-n junction and to operate it in full depletion conditions, as will be extensively reviewed in the next paragraph. On the contrary, wide band-gap semiconductors have a much larger resistivity, that is, a much lower intrinsic concentration (e.g., 10^{-27} cm^{-3} in diamond). As a result, they can be operated at room temperature with ohmic or Schottky contacts.

2.3.2 The p-n Junction in Reverse Bias

Modern silicon radiation sensors operate on a reverse-biased p-n junction diode, which is the simplest structure one can build in silicon technology. Two silicon regions, respectively doped with acceptor atoms on one side and with donor atoms on the other side, are joined together. The acceptor-doped region is called p-side (anode), and the donor-doped region is called n-side (cathode). Due to the large concentration difference of charge carriers between the two regions, electrons (holes) will diffuse across the junction, leaving a positive (negative) charge in the n-side (p-side). As a result, a volume free of mobile charges is created across the junction, called space charge region (SCR) or depletion region (see Fig. 2.5).

Due to the distribution of fixed donor and acceptor ions, an electric field is established across the SCR, which counteracts further diffusion of holes and electrons through the junction. This electric field represents to the so-called built-in potential (V_{bi}), which depends on the doping level of the two regions:

$$V_{bi} = \frac{kT}{q} ln\left(\frac{N_a N_d}{n_i^2}\right) \tag{2.5}$$

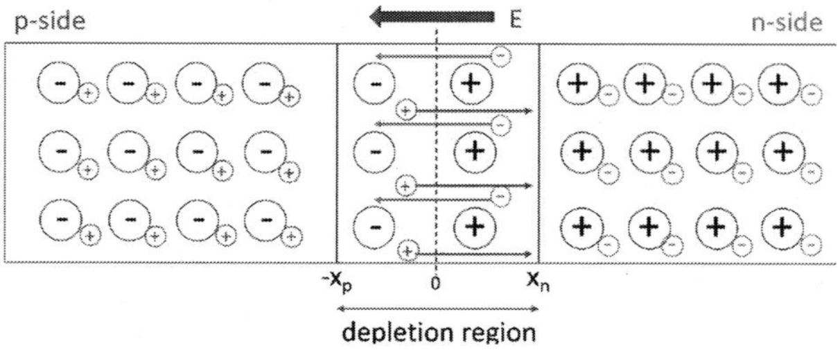

FIGURE 2.5 Schematic structure of a p-n junction.

where N_a is the acceptor concentration of the p-side, N_d the donor concentration in the n-side, and n_i is the intrinsic concentration in silicon.

Assuming a one-dimensional abrupt junction approximation, it is possible to extract the electric field and electrostatic potential distributions across the junction by successive integrations of Poisson's equation [1]:

$$\frac{d^2\varphi}{dx^2} = -\frac{\partial E}{\partial x} = -\frac{\rho(x)}{\varepsilon_0 \varepsilon_{Si}} \tag{2.6}$$

where φ is the electrostatic potential, E is the electric field, $\rho(x)$ is the charge density, and $\varepsilon_0 \varepsilon_{si}$ are the dielectric constant of the vacuum and silicon (~1 pF/cm) respectively. For solving the differential equation, boundary conditions depend on the continuity of the electric field and the electric potential, as well as from the charge neutrality of the p-n junction, as expressed by Eqn. 2.7:

$$N_a x_p = N_d x_n \tag{2.7}$$

where x_p and x_n are the widths of the depletion regions in the p- and n-side of the junction, respectively. The total depletion width (w_{depl}) is equal to:

$$w_{depl} = x_p + x_n = \sqrt{\frac{2\varepsilon_0 \varepsilon_{Si} (N_a + N_d)}{q N_a N_d} V_{bi}} \tag{2.8}$$

Since typically one side of the junction is much more heavily doped than the other, the depletion region will extend dominantly only on the side where the doping concentration is lower. As an example, assuming an n-type bulk material, Eqn. 2.8 can be simplified as:

$$w_{depl} \cong x_n = \sqrt{\frac{2\varepsilon_0 \varepsilon_{Si}}{q N_d} V_{bi}} \qquad (N_a \gg N_d) \tag{2.9}$$

As already stated, to be able to detect charges generated by radiation, it is necessary to fully deplete the bulk of the detector. Depletion is achieved by applying an external potential to the junction, in the same direction as the built-in potential, so as to further remove charge carriers and extend the SCR width to the entire device thickness. The width of the depleted layer, the electric field and electrostatic potential distributions as functions of the applied bias voltage (V_{bias}) can still be extracted from the integration of Poisson's equation. In particular, the width of the depletion layer can be calculated as:

$$w_{depl} = x_p + x_n = \sqrt{\frac{2\varepsilon_0 \varepsilon_{Si} (N_a + N_d)}{q N_a N_d} (V_{bi} + V_{bias})} \tag{2.10}$$

Again, taking into consideration the difference in the doping levels and the fact that the built-in potential is generally considerably lower than the externally applied voltage, assuming an n-type bulk material it is possible to simplify Eqn. 2.10 as:

$$w_{depl} \cong x_n = \sqrt{\frac{2\varepsilon_0\varepsilon_{Si}}{qN_d} V_{bias}} \tag{2.11}$$

The full depletion voltage (V_{fd}) is the bias voltage needed to extend the SCR to the entire thickness of the silicon substrate (d). Under the same simplifying assumptions used in 2.11, it can be expressed as:

$$V_{fd} \cong \frac{qN_d d^2}{2\varepsilon_0\varepsilon_{Si}} \tag{2.12}$$

It should be noted that the full depletion voltage is directly proportional to the doping concentration of the bulk. So, to maintain a reasonable value of V_{fd}, it is mandatory to use high-resistivity silicon as a starting material. Moreover, it should also be stressed that V_{fd} is directly proportional to the square of the sensor thickness, which should therefore be limited to a few hundred micrometers to keep V_{fd} reasonably small.

When the applied bias voltage exceeds the full depletion voltage, the device is said to be *overdepleted*. One of the most important parameters of a p-n junction in reverse bias is its capacitance, which can approximately be assumed to be the same as that of a parallel-plate capacitor with a silicon dielectric of thickness equal to the depletion region width:

$$C_d = A_d \frac{\varepsilon_0\varepsilon_{Si}}{w_{depl}} = A_d \sqrt{\frac{qN_d\varepsilon_0\varepsilon_{si}}{2(V_{bi} + V_{bias})}} \tag{2.13}$$

where A_d is the diode area. So, the capacitance is inversely proportional to the square root of the bias voltage (see Fig. 2.6a), until it saturates to a minimum (C_{fd}) as full depletion is reached:

$$C_{fd} = A_d \frac{\varepsilon_0\varepsilon_{Si}}{d} \tag{2.14}$$

Measuring the device capacitance is therefore an effective way to estimate the full depletion voltage. This is even more evident when we look at the knee in the $1/C^2$–V curve (see Fig. 2.6b).

Another fundamental parameter for the p-n junction in reverse bias is the leakage current (or dark current), which is very low as compared to the diffusion current in forward bias. The leakage current (I_{lk}) has multiple components, as highlighted in Fig. 2.7 and summarized in Eqn. 2.15.

$$I_{lk} = I_{gen} + I_{diff} + I_{surf} \tag{2.15}$$

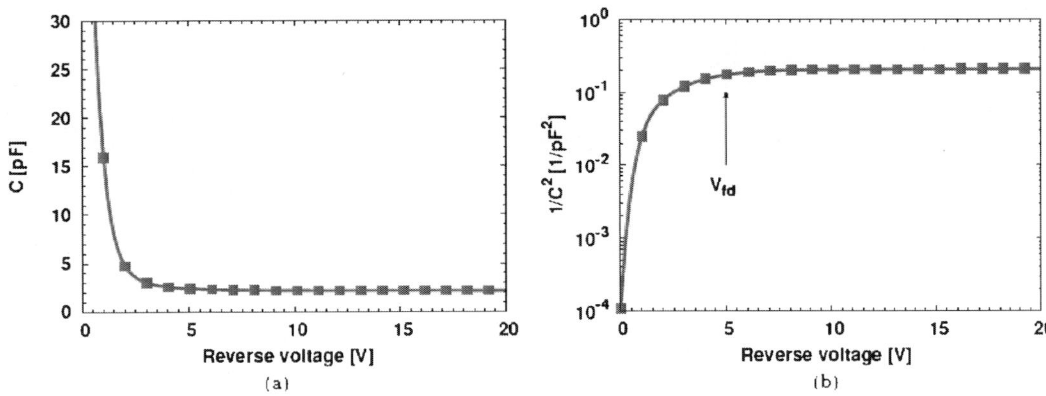

FIGURE 2.6 (a) Capacitance–voltage (C-V) curve of a diode in reverse bias, and (b) $1/C^2$–V curve of the same device.

The first contribution (I_{gen}) comes from the thermal generation of charge carriers within the depletion region, and can be written as:

$$I_{gen} = \frac{A_d q n_i w_{depl}}{\tau_g} \tag{2.16}$$

where q is the electron charge and τ_g is the generation lifetime. Note that the thermal generation current is directly proportional to the width of the depletion region, so it increases with the square root of the bias voltage, until saturation occurs at full depletion (see Fig. 2.8). It should be stressed that thermal generation and recombination in silicon are described by Shockley–Read–Hall (SRH) statistics [1], which is based on localized energy levels within the band gap, due to lattice defects or unwanted impurities (e.g., Au, Cu, Fe). The density and the energy of these deep-level states directly influence the carrier lifetimes, calling for a very high substrate purity and quality of the fabrication process in order to minimize thermal generation.

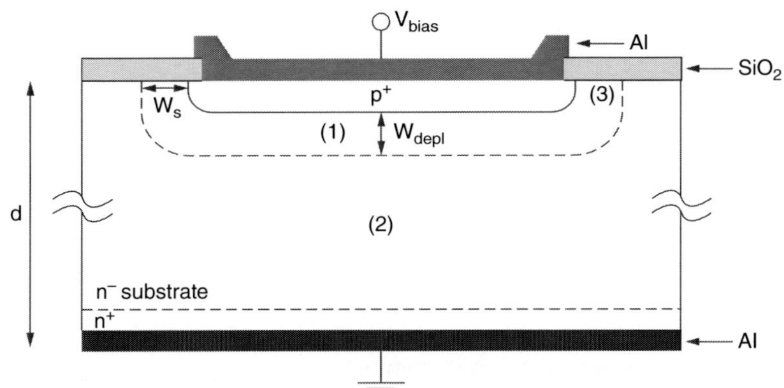

FIGURE 2.7 Schematic cross section of a reverse-biased diode (not to scale), showing different contributions to the leakage current: (1) generation current, (2) diffusion current, (3) surface current.

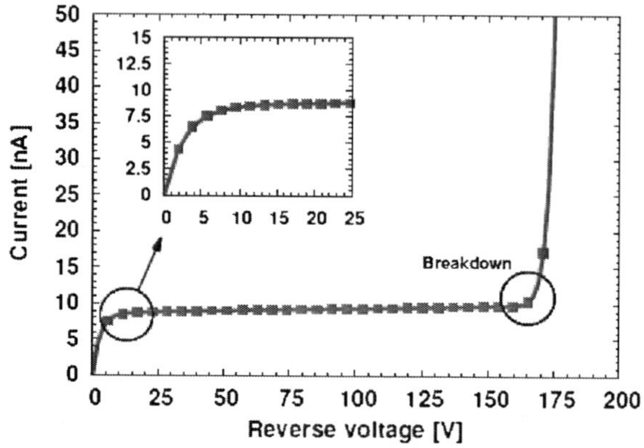

FIGURE 2.8 Current-voltage (I-V) curve of a diode in reverse bias.

The second contribution to the leakage current (I_{diff}) comes from diffusion of the minority carriers from the quasi-neutral region, elsewhere outside the SCR, to the depletion region. With respect to holes in an n-type bulk, it can be written as:

$$I_{diff} = \frac{A_d q n_i^2 D_p}{N_d L_p} \tag{2.17}$$

where D_p is the diffusion coefficient of holes and L_p is the diffusion length of holes, which in turn can be written as:

$$L_p = \sqrt{D_p \tau_p} \tag{2.18}$$

where τ_p is the recombination lifetime of holes. In the case of high-purity silicon, the diffusion length can be very high (~1 cm), resulting in a very low diffusion current. Nevertheless, diffusion current is sometimes non-negligible since the other components of the leakage currents are expected to be very low as well.

The third component of the leakage current (I_{surf}) is due to the generation of carriers at the depleted surface beneath the SiO_2 passivation layer. It can be written as:

$$I_{surf} = q n_i s_0 P_d w_s \tag{2.19}$$

where s_0 is the surface generation velocity, P_d is the diode perimeter, and w_s is the lateral extension of the depletion region at the surface. Note that, in most situations of practical interest, w_s is much smaller than w, due to the effect of oxide charges. As a result, apart from some device geometries where the perimeter is relatively long (e.g., in strip sensors), the leakage current is dominated by thermal generation in the depleted bulk. The latter is strongly dependent on the temperature through both n_i and τ_g. This phenomenon suggests operating the device in a cool environment in order to decrease the leakage current, especially after irradiation. The leakage current at temperature T_2 can be predicted with reference to leakage current at temperature T_1, using Eqn. 2.20 [15]:

$$\frac{I(T_2)}{I(T_1)} = \left(\frac{T_2}{T_1}\right)^2 \exp\left[-\frac{E_{eff}}{2k_B}\left(\frac{1}{T_2} - \frac{1}{T_1}\right)\right] \tag{2.20}$$

where k_B is the Boltzmann's constant and E_{eff} is the effective band gap (~1.21 eV). As a rule of thumb, it can be assumed that current doubles every 7 °K.

The p-n junction in reverse bias can be operated at large voltage up to the breakdown point, beyond which the current shows a steep increase (see Fig. 2.8). The breakdown mechanism can be driven by three effects: avalanche multiplication, thermal runaway, and tunneling. Avalanche multiplication is the most important mechanism for junction breakdown in p-n junctions used for radiation detection. Carrier multiplication occurs when the electric field is high enough for the kinetic energy of carriers to increase to such an extent that they can break the covalent bond of another electron in the silicon lattice by colliding with it. This phenomenon is called impact ionization, and it leads to the generation of a new electron-hole pair [1]. In silicon, the critical electric for avalanche multiplication at room temperature is $\approx 4 \times 10^5$ V/cm. If the electric field is high enough, the avalanche effect can self-sustain, causing a large increase in the leakage current of the device. If the bias voltage is also very high, the power developed in the sensor ($P = VI$) can be enough to cause an uncontrollable self-heating mechanism, also known as thermal runaway. To prevent this phenomenon, which could eventually destroy the device, an effective cooling system is necessary. At even larger fields, of the order of 10^6 V/cm, band-to-band tunneling effects could take place, leading to very high current, but this requires both sides of the junctions to be highly doped, which is not the case in silicon sensors.

2.4 POSITION-SENSITIVE SENSORS

The position of the incident radiation is required for many applications. This is especially the case for silicon detectors used for tracking elementary particles in high-energy colliders. p-n junctions can be arranged in many different ways to form simple diode or finely segmented structures, allowing position sensing of the incident particle, such as microstrip or pixel geometries, as schematically shown in Fig. 2.9 and outlined in the following paragraphs.

2.4.1 Pad (Diode)

The simplest sensor structure is a diode, also known as *pad* detector (see Fig. 2.9a). A pad is made on a high-resistivity substrate with a wide range of possible thicknesses (from some tens of micrometers to few millimeters). With reference to an n-type substrate, one side of the detector is p^+ doped (junction), and the other side is n^+ doped (ohmic contact), while the bulk is of (quasi) intrinsic type (hence the name p-i-n diode). Ultrapure silicon ingots obtained from floating-zone (FZ) refinement have bulk resistivity from 1.0 to 30 kΩ cm. As it happens for simple diodes, p-n junctions are generally reverse biased, so that

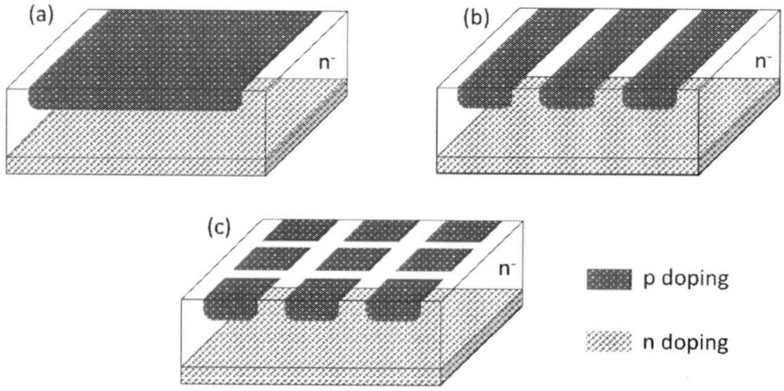

FIGURE 2.9 Schematic structures of different sensor geometries: a) pad, b) strip, and c) pixel.

electron-hole pairs generated inside the depletion region are stripped away from the SCR and are collected at the p^+ and n^+ contacts. When operated in full depletion mode, the drift velocity of charge carriers is limited only by the saturation drift velocity of electrons and holes, that is, about 10^7 cm/s in silicon. As a reference, the response time of a fully depleted silicon detector can be of the order of a few nanoseconds. In practice, not all the generated carriers are always collected at the p^+ and n^+ contacts. Some of them could be lost because of competitive processes such as charge trapping and recombination. Outside the SCR, electron-hole recombination is predominant, and carrier transport is based on diffusion. Therefore, ionization events occurring in the nondepleted regions can contribute to detector efficiency, but they increase the response time. According to Eqn. 2.12, depending on the substrate doping concentration and thickness, it turns out that large bias voltages might be required to fully deplete a p-i-n detector. As an example, a diode made on a standard 300 μm thick substrate with a doping concentration of 10^{12} cm^{-3} has a full depletion voltage of 70 V. The only ways to sustain these high voltages, while maintaining low reverse leakage currents and avoiding breakdown problems, are (1) to have a very good fabrication technology, able to ensure carrier lifetime of the order of milliseconds without degrading the ultrapure starting material, and (2) to optimize the detector layout.

2.4.2 Strip Sensors

p-n junctions can be arranged to form strip sensors, as shown in Fig. 2.9b. With respect to pad sensors, instead of having a single collecting electrode, the anode is segmented in many collecting electrodes a few tens of micrometers wide and up to a few centimeters long. All microstrips are biased at the same reverse voltage: Since the charge generated inside the detector by the incident radiation is not completely localized, the pulse signal can be spread out over two or more adjacent strips. In principle, each microstrip should be connected to an electronic readout channel. In this way, a spatial resolution of a few tens of micrometers or better can be achieved, depending on the strip pitch and readout mode. In more advanced designs, the backside of the wafers (ohmic side) is also segmented in microstrips at the expense of a significant complication in the technology. With respect to those on the junction side, the strips on the ohmic side can be either orthogonal or just

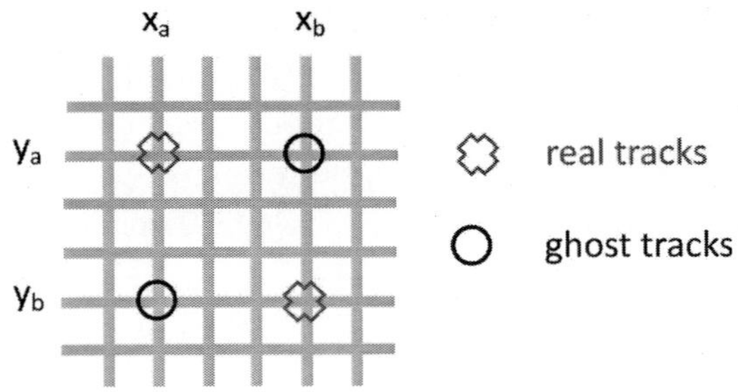

FIGURE 2.10 The problem of ghost tracks in double-sided strip sensors in case of high hit rates.

rotated by a small angle, in the so-called stereo configuration. By doing so, 2D devices are obtained, with a fine pixel granularity resolution in the *x*-*y* direction. However, it should be stressed that double-sided microstrip sensors are not suited to high-particle-rate environments. This problem is illustrated in Fig. 2.10: In case they impinge on the detector at the same time, two pairs of particles could fire the same pairs of strips, though their hit position is different.

Another disadvantage of microstrip sensors is their high capacitance, which imposes the use of very low noise preamplifiers (see Section 2.6). In fact, because of the proximity of electrodes to each other (a few tens of μm at most), the capacitance of each strip is much higher than that of a simple diode where the two electrodes are at a distance that corresponds to the substrate thickness. The capacitance of a strip electrode is made by two contributions: the interstrip capacitance, which depends mainly on the strip pitch (but also on other design parameters, like the field plate, and on the surface charge), and the strip-to-back capacitance, which depends on the substrate thickness. In planar microstrip sensors, the first contribution to the capacitance is by far the most important, owing to the peculiar strip aspect ratio. The primary advantage of strip sensors is the possibility of realizing detectors with very large active area (tens of cm^2) while preserving a very good position resolution, making this type of sensor a workhorse for tracking in high-energy physics and nuclear physics.

2.4.3 Pixel Sensors

Pixel geometries (see Fig. 2.9c) add further complexity in efforts to obtain a "true" two-dimensional position resolution of the incident particle without ambiguities in case of high particle rates. Depending on the specific configuration, pixel sensors are fabricated with either a single-sided (p-on-n, n-on-p) or a double-sided process (n-on-n, while p-on-p is rarely used). Electrodes are arranged as small implanted regions on the silicon area mimicking a checkerboard. In fact, the simplest pixel shape is a square, but rectangular pixels are often used, and hexagonal pixels can also be implemented.

The main issue of pixel systems is the readout of each single channel, which requires complicated and expensive interconnect techniques, like bump bonding and flip-chipping [6]. The pixel layout should be designed in order to perfectly match the layout of the readout chip. The resulting sensors are generally referred to as hybrid pixel detectors. The chip is typically connected to the sensor through small solder bumps. The main limitation to the pixel size usually comes from the electronics because a large amount of circuit functions must be fit in a single pixel, with complexity changing depending on the type of readout to be implemented. Hybrid pixel detectors are used in many applications ranging from high energy physics vertex detectors to medical and biological applications, homeland security, industrial control and more.

2.4.4 Drift Detector

The silicon drift detector (SDD), also called the semiconductor drift chamber, was first proposed by Gatti and Rehak in 1984 [16, 17]. It is a device made on high-resistivity n-type silicon wafers with rectifying p-n junctions implanted on both sides. The detector is fully depleted of mobile charges by applying a suitable reverse voltage to the p-n junctions on both sides of the wafer. An electrostatic potential parallel to the surface is then superimposed onto the depleting vertical potential by means of resistive voltage dividers. A schematic view of a drift detector is reported in Fig. 2.11, showing the basic operation mechanism: While holes are swept away by the p^+ electrodes close to the point of interaction, electrons generated inside the volume of the detector are drifted along the bottom of the potential valley toward one small collecting anode, where they finally induce a signal.

Depending on the application, the number and position of the anodes is different. In the cylindrical SDD, which is the most commonly used for spectroscopic applications, only one anode is placed at the center of the detector area, surrounded by several rings of cathodes. The drift time of electrons measures the radial coordinate of the interaction point, since the electric field and the drift velocity are known. The drift time for a detector that is a few millimeters wide is in the order of microseconds, and the radial position resolution that can be achieved is less than 10 μm. For tracking applications, SDDs of the "linear" type, having two-dimensional position resolution, can also be obtained and exploit the same basic mechanism (drift time) as cylindrical SDDs for the resolution in the first dimension, whereas the resolution on the second dimension is obtained by fragmenting the anode in several adjacent pixel anodes [18].

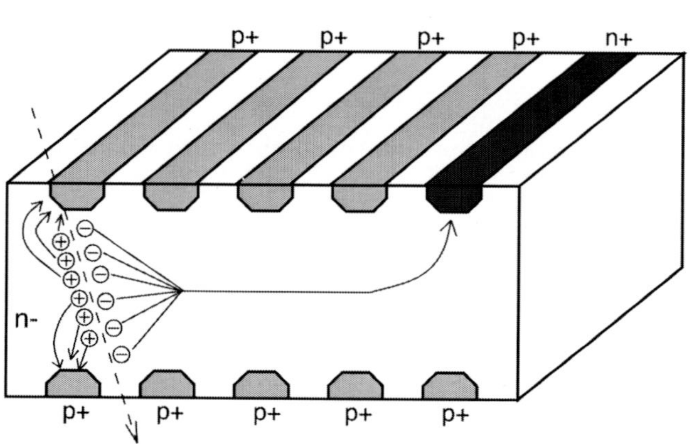

FIGURE 2.11 Schematic view of a silicon drift detector.

The main advantage of silicon drift detectors is that the anode capacitance is much lower than that of standard junction detectors of the same dimensions. As a consequence, as will be explained in Section 2.6, the electronic system noise can be reduced to a much lower level, and the detector can operate with higher energy resolution and at lower shaping times [19]. One disadvantage of SDDs is the rather complicated bias electronic circuitry necessary to distribute the longitudinal potential among all the p^+ electrodes. The external biasing circuitry can be avoided by using an integrated voltage divider, consisting of either implanted resistors or MOS transistors [20].

2.5 SIGNAL FORMATION

2.5.1 Charge Motion

The interaction of ionizing radiation with silicon ultimately leads to the creation of a certain amount of electron-hole pairs. The motion of charge carriers within the sensor volume is dominated by the drift length (L_{drift}) and by the diffusion length (L_{diff}) in the depleted and nondepleted regions, respectively. When these lengths are long enough, the charge signal becomes largely independent of the incident point of absorption within the detector-sensitive volume. The drift length is related to other physical parameters, as expressed by Eqn. 2.21:

$$L_{drift} = v\tau \tag{2.21}$$

where v is the drift velocity and τ is the lifetime. In turn, the drift velocity can be expressed as:

$$v = \mu E \tag{2.22}$$

where μ is the mobility and E the electric field magnitude. At high fields (~10^5 V/cm), the velocity saturates to a value of ~10^7 cm/s for both electron and holes.

The diffusion length is related to other physical parameters as expressed by Eqn. 2.23:

$$L_{diff} = \sqrt{D\tau} \tag{2.23}$$

where D is the diffusion coefficient, which is proportional to the mobility as stated by Einstein's relation:

$$D = \mu \frac{k_B T}{q} \tag{2.24}$$

These equations apply to both electrons and holes by using their respective parameters. As can be seen, if the mobility and lifetime values are high, the drift and diffusion lengths will be long. These high values can be obtained by using high-purity materials and very clean fabrication processes, which also minimize the impact of charge-trapping phenomena, thus enhancing the collection efficiency.

From previous equations, it can easily be estimated that the collection times of electrons and holes generated within a depleted region a few 100s μm width are of the order nanoseconds. On the contrary, since mobile charges generated within the neutral region will move by diffusion, their collection times are much longer, of the order of microseconds.

It should also be considered that during their motion towards the collecting electrode, electrons and holes also spread as a cloud from their point of origin, due to diffusion. The radius of the diffusive cloud (r_c) at the time carriers will reach the electrode can be written as:

$$r_c = \sqrt{2Dt_c} \tag{2.25}$$

where t_c is the collection time, which of course is longer for the carriers generated farther from the electrodes.

Another important aspect of silicon sensors used for particle tracking regards the presence of a magnetic field to allow for measurements of the particle momentum. However, the magnetic field, besides influencing the trajectory of the particle, also affects the motion of free charges within the sensor volume. As a result, electrons and holes drifting toward electrodes are deflected by the Lorentz force [6]. The change in the drift direction is generally described by the opening angle, θ_L, measured with respect to the direction of the electric field. Typical values of the so-called Lorentz angle are normally in the range of 0° to 20° and can be calculated for electrons $\theta_{(L,n)}$ and holes $\theta_{(L,p)}$ as:

$$tan\theta_{(L,n)} = \mu_{(H,n)} B_{\perp} \qquad tan\theta_{(L,p)} = \mu_{(H,p)} B_{\perp} \tag{2.26}$$

where μ_H is the Hall mobility and $B_{\perp}$ is the magnetic field component that is perpendicular to the direction of the charge carrier drift. In turn, the Hall mobility is proportional to the drift mobility: $\mu_H = r\mu$. The Lorentz factor r is slightly temperature dependent, and its value is 1.15 for electrons and 0.72 for holes at 0°C [6].

2.5.2 Induced Signals

As already mentioned, ideally a silicon sensor should be fully depleted, so that the electrons and holes created by radiation will drift under the effect of an applied electric field. In this condition, it should be stressed that the current signals induced on the electrodes will appear immediately as the charges start moving, and not when they reach the respective electrode. With reference to a system with an arbitrary electrode configuration, the current signal induced by a single elementary charge on the electrode of interest can be expressed by Shockley–Ramo's theorem as [21, 22]:

$$i(t) = -q\vec{v}(t) \cdot \overrightarrow{E_W} \tag{2.27}$$

where two vectors appear, representing the velocity $\vec{v}(t)$ and the weighting field $\overrightarrow{E_W}$. Despite the similarity of their name, the weighting field is very different from the electric

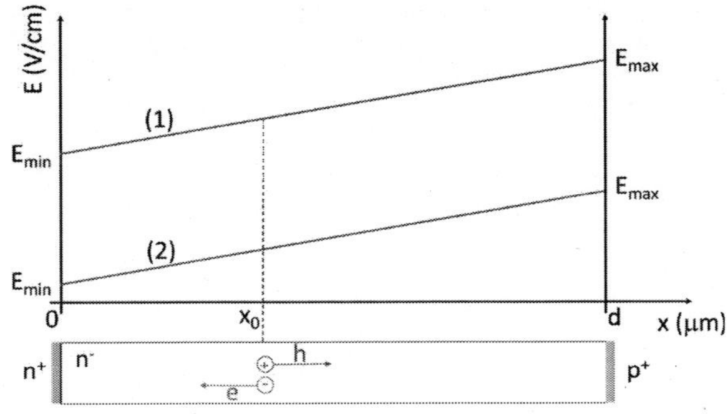

FIGURE 2.12 Schematic diagram of the electric field in a P-I-N diode slightly above full depletion.

field. The electric field depends on the bias, and it determines the charge carrier trajectory from its point of generation to the point of collection at the electrode and the carrier velocity (through the mobility). The weighting field depends only on the electrode geometry and determines the fraction of charge induced on each electrode. The weighting field depends on the weighting potential V_W. In particular, $\overrightarrow{E_W} = -\nabla V_W$. The spatial distributions of these quantities for a given electrode are obtained by solving Laplace's equation $\nabla^2 V_W = 0$ (i.e., the Poisson's equation in the absence of charge distribution), setting the electrode of interest at 1 V and keeping all the other electrodes grounded.

As an example, let's start with a simple structure, that is, a diode on an n-type substrate, with only two electrodes (one for readout and one for bias) at a distance d equal to the substrate thickness (see Fig. 2.12). Let's assume the bias voltage (V_{bias}) to be larger than the full depletion voltage (V_{fd}). In this condition, the electric field shows a linear dependence on depth and can be written as:

$$E(x) = \frac{2V_{fd}}{d^2}(2x - d) + \frac{V_{bias}}{d} \quad \text{or} \quad E(x) = \frac{E_{max} - E_{min}}{d} x + E_{min} \tag{2.28}$$

where:

$$E_{min} = E(0) = \frac{V_{bias} - V_{fd}}{d} \quad \text{and} \quad E_{max} = E(d) = \frac{V_{bias} + V_{fd}}{d} \tag{2.29}$$

In case $V_{bias} \gg V_{fd}$ (case 1 in Fig. 2.12), Eqn. 2.28 can be simplified, yielding an approximately constant electric field across the whole diode's volume $\left(E \cong \frac{V_{bias}}{d}\right)$, so that the electron ($v_e$) and hole ($v_h$) drift velocities are also constant and can be written as:

$$v_e = \mu_e E = \mu_e \frac{V_{bias}}{d} \qquad v_h = \mu_h E = \mu_h \frac{V_{bias}}{d} \tag{2.30}$$

where μ_e and μ_h are the electron and hole mobility, respectively. The weighting field is also a constant, $E_W = 1/d$ for both electrodes. As a result, the induced current signals will also be constant during the motion of carriers to the electrodes:

$$i = qvE_W = q\mu EE_W = q\mu\frac{V_{bias}}{d^2} \tag{2.31}$$

Let's now assume that an electron-hole pair is generated at a certain position x_0 between the n-type electrode ($x = 0$) and the p-type electrode ($x =$ d). In this case, the collection times of electron and hole will be:

$$t_e = \frac{x_0}{v_e} = \frac{x_0}{\mu_e E} = \frac{x_0 d}{\mu_e V_{bias}} \qquad t_h = \frac{d - x_0}{v_h} = \frac{d - x_0}{\mu_h E} = \frac{(d - x_0)d}{\mu_h V_{bias}} \tag{2.32}$$

Because of their opposite charge sign and drift direction, the electron and the hole induce a signal of the same polarity on one electrode. From Eqns. 2.31 and 2.32, the integrated charge signals for electron and hole can be written as in Eqn. 2.33:

$$Q_e = \int_0^{t_e} i_e\,dt = i_e t_e = q\frac{x_0}{d} \qquad Q_h = \int_0^{t_h} i_h\,dt = i_h t_h = q\frac{(d - x_0)}{d} \tag{2.33}$$

As expected, the sum of the two charge signals is the elementary charge q; more importantly, the contribution to the total induced charge by the electron and the hole only depends on the generation point. As a result, if the charges were generated very close to the p-type (n-type) electrode, the hole (electron) would be collected instantaneously, inducing a negligible signal, and the entire signal would be induced by the electron (hole) motions.

Note that if one is only interested in the total charge signals induced on the electrodes rather than on their time evolution, integration of Eqn. 2.27 offers a direct solution, as shown in Eqn. 2.34:

$$Q_e = -q\int_0^{t_e} v_e E_W\,dt = -q\int_{x_0}^{0} E_W\,dx = -q\left[V_W(0) - V_W(x_0)\right] \tag{2.34}$$

$$Q_h = q\int_0^{t_h} v_h E_W\,dt = q\int_{x_0}^{d} E_W\,dx = q\left[V_W(d) - V_W(x_0)\right]$$

More generally, an elementary charge $e_0 = \pm\, q$, moving from an initial point x_i to a final point x_f, will induce a mirror charge on the electrode:

$$Q_{i,f} = e_0\left[V_W(x_f) - V_W(x_i)\right] \tag{2.35}$$

In case the considered moving charges are able to complete their trajectories from the point of origin to the electrodes, the result of Eqn. 2.35 is the same as in Eqn. 2.33. However, if the moving charges are not able to reach the respective electrodes, either because of recombination or trapping, part of the induced charge signal will be lost. In this case, Eqn. 2.27 should include a correction factor for the moving charge:

$$i(t) = -q \cdot exp\left(-\frac{t}{\tau_{eff}}\right) \cdot \vec{v}(t) \cdot \overrightarrow{E_W} \tag{2.36}$$

where τ_{eff} is the mean free drift time. Similarly, if the integration time of the readout circuit is not long enough as compared to the collection time, part of the charge will be lost, causing the so-called ballistic deficit.

Let's now consider a more realistic situation of a diode with the same characteristics as before but reverse biased only slightly above full depletion (case 2 in Fig. 2.12). While the weighting field is still constant, because it only depends on the thickness, the linear dependence of the electric field on the depth can no longer be neglected, as illustrated in Fig. 2.12. As a result, the current signals depend on the position and, in turn, on time:

$$i(x) = qv(x)E_W = q\mu E(x)E_W = q\mu\left[\frac{2V_{fd}}{d^2}(2x-d) + \frac{V_{bias}}{d}\right]\frac{1}{d} \tag{2.37}$$

Following [18], it is possible to calculate the current signals induced by an electron-hole pair generated at a certain depth x_0. In fact, the time required by a charge carrier originated at x_0 to reach a generic point x can be expressed as:

$$t(x) = \int_{x_0}^{x} \frac{1}{v(x)} dx = \int_{x_0}^{x} \frac{1}{\mu E(x)} dx = \frac{1}{\mu}\int_{x_0}^{x} \frac{1}{E(x)} dx \tag{2.38}$$

The linear dependence of the electric field on the position yields a logarithmic dependence of the time on the position, which in turn implies that the position shows an exponential dependence on time. In particular, since the electron drifts toward the n-side electrode (lower electric field), it induces a current signal following an exponential decay with time; on the contrary, the hole drifts toward the p-side electrode (higher electric field), thus inducing a current signal, which exponentially increases with time. The total induced current is therefore the sum of two exponentials. At the end of the charge transit, consistently with Eqn. 2.35, the total induced charge on the p-side electrode still depends only on the generation position x_0 as in the previous simplified case, but the collection time (signal duration) now exhibits a more complex dependence on the generation position. As an example, Fig. 2.13 shows the time evolution of the current signals induced by an electron-hole pair generated at two different depths in a 300-μm thick sensor. Figure 2.13a refers to a generation depth of 250 μm, that is, 50 μm away from the p-side electrode. The hole has to drift only a small distance to the junction, so it is collected very fast and induces only a small fraction of the total charge. On the contrary, the electron has to drift for 250 μm toward the n-side contact, so it induces the major part of the signal. The total collection time is limited

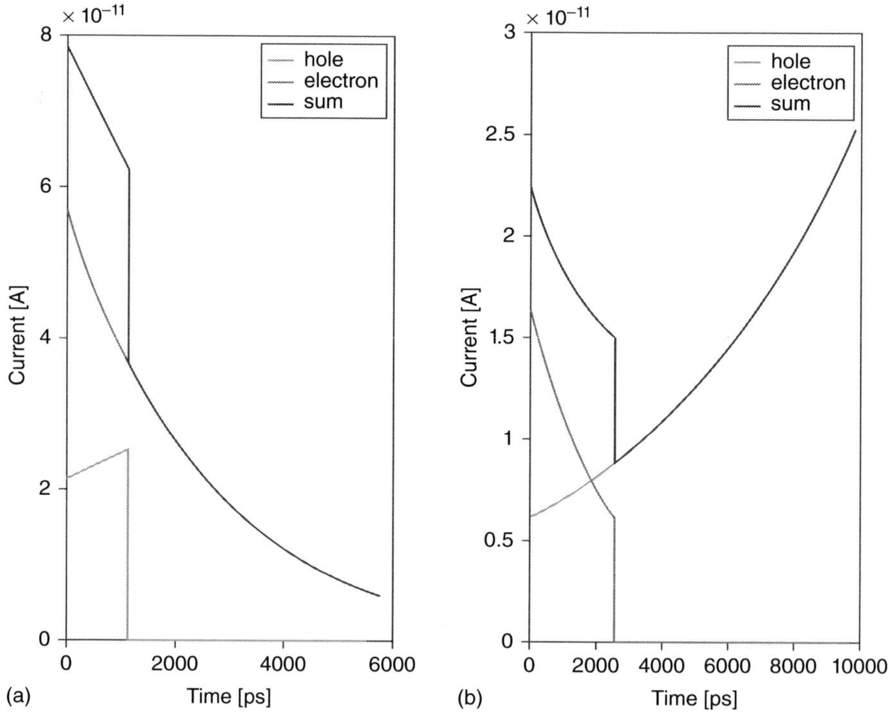

FIGURE 2.13 Time dependence of current signals induced on the p-side electrode of device in Fig. 2.11 by an electron-hole pair generated at: (a) $x_0 = 250$ μm (i.e., 50 μm from the p-side electrode), and (b) $x_0 = 50$ μm (i.e., 250 μm from the p-side electrode).

by the electron, which has to travel a much longer distance than the hole, taking about 6 ns. Figure 2.13b refers to a symmetric case, where the electron-hole pair is generated at a depth of 50 μm, that is, 250 μm from the p-side electrode. In this case, the electron is collected very fast and induces only a small fraction of the total charge. The hole has to travel for most of the depletion region: it induces the largest part of the signal and, since its mobility is lower than that of the electron, it causes the total collection time to be longer (~10 ns).

In conclusion, in a simple sensor with only one readout electrode as a diode, the contribution to the total signal from the hole and the electron depends only on the position of the generation point. Due to the different mobility values, the choice of the readout polarity (either p or n) can affect the collection time, thus also affecting the radiation tolerance, as it will be shown later.

The application of Shockley–Ramo's theorem becomes much more complex in segmented sensors like strips or pixels. Because of the presence of multiple readout electrodes, the weighting potential (field) profile is not linear (constant) across the sensor volume. As an example, let's consider a 300 μm thick p-on-n strip sensor. The strip pitch and width are 100 μm and 50 μm, respectively. Figure 2.14 shows the two-dimensional weighting potential distribution calculated with reference to the strip electrode centered at $x = 0$ (Strip 1). The plots are obtained by solving the Poisson's equation after replacing the silicon substrate with a dielectric layer, so as to null the charge.

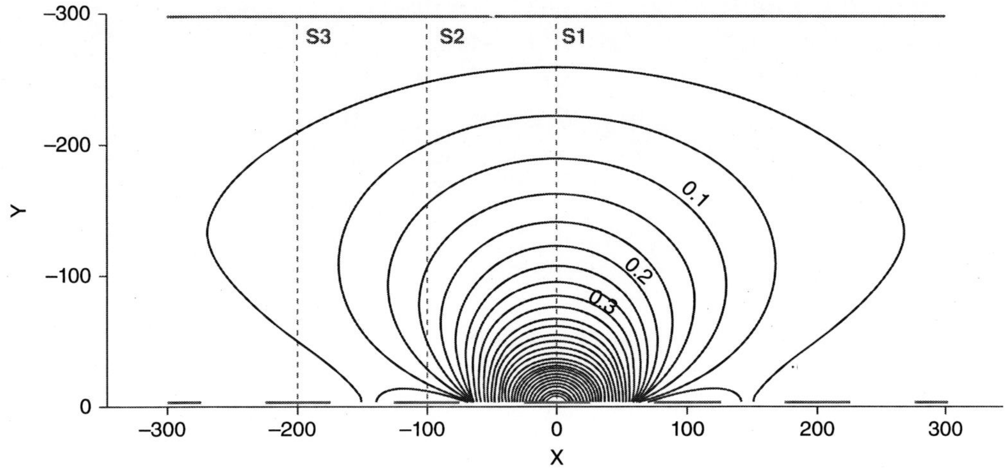

FIGURE 2.14 Simulated weighting potential distribution in a p-on-n strip sensor (upside down). All geometrical dimensions are given in micrometers, whereas the weighting potential is given in volts.

It can be seen that the weighting potential rapidly decreases as the distance from the central strip electrode increases. This trend is more evident when we view the one-dimensional cuts of the weighting potential of Fig. 2.15, taken along lines S_1, S_2, and S_3 of Fig. 2.14, which correspond to the centers of the strip ($x=0$), first neighbor strip (Strip 2, $x=-100$ μm), and second neighbor strip (Strip 3, $x = -200$ μm). The weighting potential in a pad sensor (diode) of the same thickness is also shown as a reference. Looking at cut S_1, it can be seen that the weighting potential is peaked in a narrow region below the electrode and then steeply decreases (at a depth of −100 μm, the value is 0.2 V). Thus, the corresponding weighting field is high only close to the electrode, and it is much lower elsewhere. As a result, most of

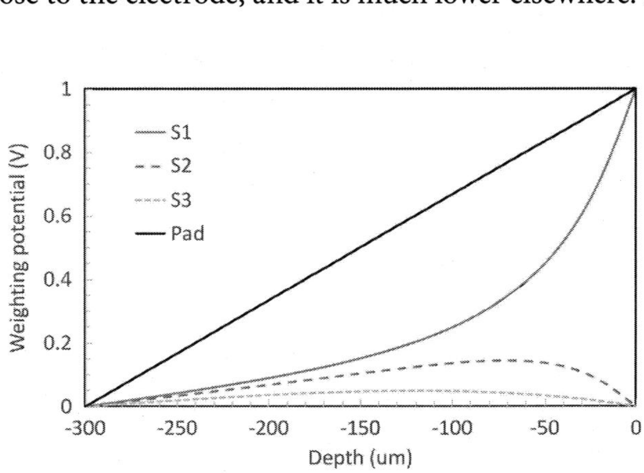

FIGURE 2.15 1-D cuts of the weighting potential along lines S_1, S_2, and S_3 of Fig. 2.13, corresponding to the center of three adjacent strips. The weighting potential in a pad sensor (diode) of the same thickness is also shown.

the signal is induced on Strip 1 by charge carriers that are near enough to it, so only those charge carriers that drift toward that electrode are really significant.

The other carrier type, which moves toward the backside, induces only a small fraction of the signal, unless its origin point is very close to the strip electrode. In case of an MIP crossing the considered strip sensor perpendicularly to the surface and generating an approximately uniform charge distribution along its path, about 80% of the signal is induced by holes drifting toward the high weighting field region.

Figure 2.15 also shows that the weighting potentials calculated along cuts S_2 and S_3 exhibit a nonmonotonic trend, starting and ending at 0V (due to boundary conditions). The corresponding weighting field along those lines will have a bipolar shape (i.e., with both positive and negative values), depending on the depth. Note that charges generated along lines S_2 and S_3 will drift toward and finally be collected by Strip 2 and Strip 3, respectively, following the electric field lines (not shown); these lines are uniform throughout most of the bulk, since all strips are biased at the same potential. Although those charges will be collected by neighbor strips, they will also induce a bipolar current pulse on Strip 1, though with zero net integral (null total induced charge). More generally, at the end of the charge motion, a net total induced charge not equal to zero is only observed at the collecting electrode, but while the charge is moving also other electrodes can experience non-negligible bipolar current pulses.

2.6 READOUT ELECTRONICS AND NOISE

Silicon sensors used for single-particle tracking are mostly readout with charge-sensitive amplifiers followed by pulse-shaping circuits to optimize the noise performance. This is also the classical structure of a spectroscopic chain. Some basics on energy resolution and noise issues are summarized in the following section.

2.6.1 Energy Resolution

A measurement of the incident radiation energy distribution is required in spectroscopy applications. In this respect, a key parameter for performance evaluation is the energy resolution that represents the detector's ability to distinguish between two or more photons having nearly equal energies. In order to obtain these measurements, the detection system, consisting of the sensor plus the readout electronics, must provide an output signal (voltage pulse) of amplitude proportional to the amount of energy deposited in the sensor by each incident photon. In theory, in case the photon is emitted by a monoenergetic source, the detector response should be a single impulse, which can be represented as an intensity versus energy curve. Yet, the detector output spectrum of such a source is a rather broad peak of counts (proportional to intensity) versus pulse heights (proportional to energy) curve. In other words, unequal pulse heights are produced for photons of equal energy. This effect is called line broadening and can be explained by taking into account the statistical fluctuations of the number of generated carriers and the magnitude of the possible noise sources [3]. The creation of electron-hole pairs within a silicon sensor is indeed a statistical process. If we assume that such a process follows Poisson statistics, the standard deviation of the number of electron-hole

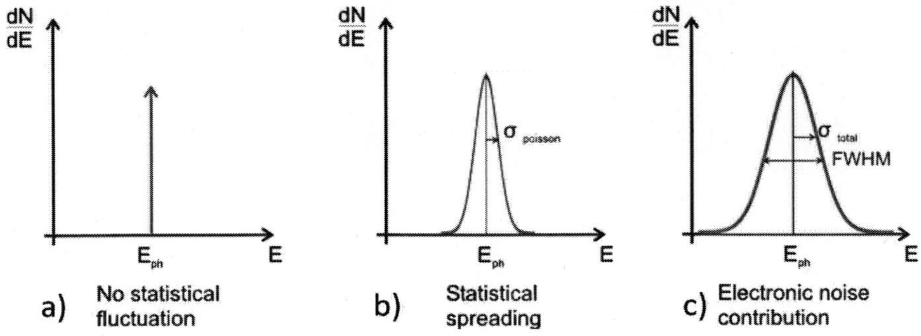

FIGURE 2.16 (a) Ideal energy spectrum measurement without fluctuations, (b) with Poisson fluctuations only, and (c) with both Poisson and electronic noise fluctuations.

pairs, N, would be equal to the square root of this number. The standard deviation, $\sigma_{Poisson}$, for the Gaussian energy pulse is then given by [3]:

$$\sigma_{Poisson} = \sqrt{N_{ave}} \cdot E_{ion} \tag{2.39}$$

where N_{ave} is the average number of charge carriers produced. The intrinsic energy resolution, defined as the full width at half maximum (FWHM) of the Gaussian pulse, as shown in Fig. 2.16, is related to the standard deviation by Eqn. 2.40:

$$FWHM = \frac{dE}{E} N_{ave} \sigma_{Poisson} = 2.35 \sigma_{Poisson} \tag{2.40}$$

Substituting Eqn. 2.39 into Eqn. 2.40, and considering that $N_{ave} = E_{ph} / E_{ion}$, it follows that:

$$\frac{dE}{E} = 2.35 \sqrt{\frac{E_{ion}}{E_{ph}}} \tag{2.41}$$

This clearly shows the advantages associated with a low-ionization energy; moreover, it is evident that the energy resolution deteriorates with decreasing photon energy.

In fact, careful measurements of the energy resolution have resulted in values lower by a factor of 3 or 4 than those given by Eqn. 2.41, indicating that the mechanisms leading to the generation of each individual charge carrier are not independent. Therefore, simple Poisson statistics is not suitable for describing these processes. This discrepancy can be taken into account by inserting into Eqn. 2.41 the so called Fano factor, F, as an adjusting parameter [3]:

$$\frac{dE}{E} = 2.35 \sqrt{F \frac{E_{ion}}{E_{ph}}} \tag{2.42}$$

Although a complete explanation of all the processes that lead to a nonunity value for the Fano factor has not been provided yet, some models have been developed that qualitatively account for experimental observations. Numerical values of the Fano factor close to 0.12 have been reported for silicon [3].

2.6.2 Electronic Noise

The energy resolution of a silicon detector depends on two factors: (1) the fluctuation in the number of electron-hole pairs that are generated by incident radiation, and (2) the fluctuation in the amount of charge that is effectively detected by the readout electronics system [3]. The first factor, arising from the mechanisms involved in the conversion of radiation energy into electric charge, sets an intrinsic limit for energy resolution. The second factor is related to the physics of the electric charge transport and measurement; in other words, it is determined by the "electronic noise" present in the detection system and sets an extrinsic limit to the energy resolution that might also vary with time (drift). If all those sources of fluctuation were independent, the overall energy resolution (FWHM) would be given by the quadrature sum of the FWHM values for each individual source:

$$FWHM_{total}^2 = FWHM_{statistical}^2 + FWHM_{noise}^2 + FWHM_{drift}^2 + \ldots \tag{2.43}$$

where the dots stand for any other independent source of fluctuation. In the field of radiation detection and measurement, the energy resolution due to electronic noise is customarily expressed in terms of equivalent noise charge (ENC) instead of $(FWHM)_{noise}$. ENC is defined as the amount of charge, which, if applied to the input terminal of the detection system, would give rise to a unitary SNR—that is, to an output signal equal to the rms level of the output due to noise only. Recalling Eqns. 2.39 and 2.40, we can write for silicon at room temperature:

$$FWHM_{noise} = 2.35 \cdot 3.6 \cdot ENC = 8.5\ ENC \tag{2.44}$$

where $(FWHM)_{noise}$ is expressed in units of eV, and ENC is expressed in units of electrons rms.

The analysis of the different noise contributions related to the detector and the readout electronics can be carried out with respect to a typical spectroscopy system, whose simplified schematic is depicted in Fig. 2.17. The complete study of the ENC is rather complicated because it depends on a large number of variables. A basic analysis is reported here, whereas additional details can be found in [23, 24].

Since the time scale for the release of radiation energy in silicon is in the order of nanoseconds, the sensor output signal can be represented by a current pulse of intensity $Q\delta(t)$, where Q is the total amount of generated charge and $\delta(t)$ is the delta-Dirac function. The current pulse $Q\delta(t)$ is fed to a feedback amplifier that converts it to a voltage signal. If the preamplifier and the following processing stages are properly designed, their contribution to the total system noise is negligible. In order to preserve the basic information carried by the magnitude of the generated charge, Q, the most common configuration of the input preamplifier is that of a charge-sensitive amplifier (CSA). After preamplification, signal

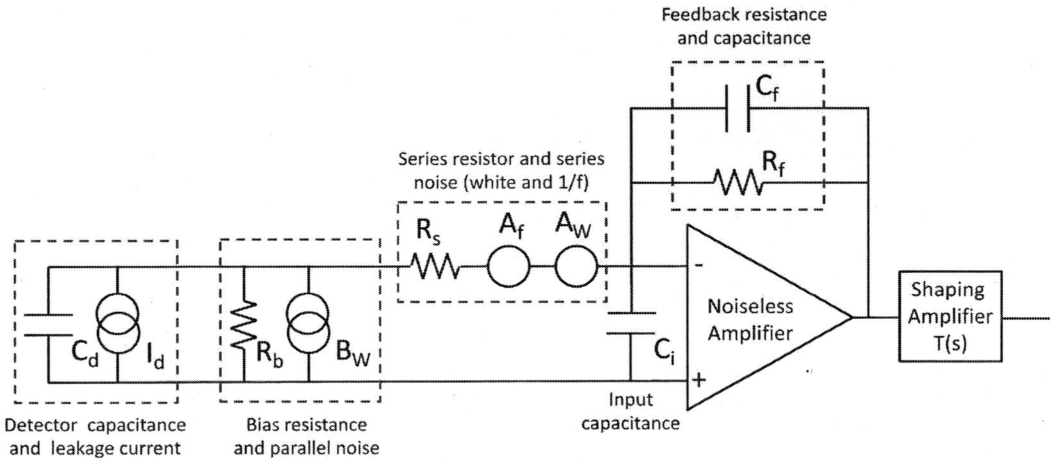

FIGURE 2.17 Sketch of a typical radiation spectroscopy chain, including the sensor, the bias, and readout circuits, emphasizing the different electronic noise sources.

and noise have to be filtered to optimize the SNR. This is accomplished by eliminating the low- and high-frequency components of the signal, while preserving the most important information, that is, the amount of generated charge, in the output voltage pulse height. In principle, the best SNR would be obtained if the pulse shape were an infinite cusp [23], but practical filters implements triangle, Gaussian, semi-Gaussian, and other shaping functions that exhibit S/N ratios that are only slightly worse than the theoretical limit.

As outlined in [24], the noise sources can be categorized into those that are effectively in series and in parallel with the signal source. Both of them can be described by their power densities, $(A_W + A_f)$ and B_W, respectively, as shown in Fig. 2.17, where other important circuit elements are also indicated: the detector capacitance (C_d) and leakage current (I_d), the bias resistance (R_b), the input capacitance (C_i) and feedback network of the CSA (the feedback capacitance C_f and resistance R_f), and the shaping amplifier transfer function $T(s)$.

The white series noise A_W is almost entirely contributed by the thermal noise in the current of the input transistor of the CSA and by the stray resistance ($\dot{R}_S$) in series with the input, if its value is non-negligible:

$$A_W = \frac{4k_B T \Gamma}{g_m} + 4k_B T R_S \tag{2.45}$$

where g_m is the transconductance of the input transistor and the parameter Γ also depends on the type of input transistor ($\Gamma \cong 2/3$ for long channel transistor and tends to 1 for short channel ones). With regard to the parallel noise, its main contributions are the shot noise in the detector leakage current, the shot noise in the input transistor gate current (I_g), and the thermal noise of the CSA feedback resistor and the detector bias resistor:

$$B_W = 2qI_d + 2qI_g + \frac{4k_B T}{R_f} + \frac{4k_B T}{R_b} \tag{2.46}$$

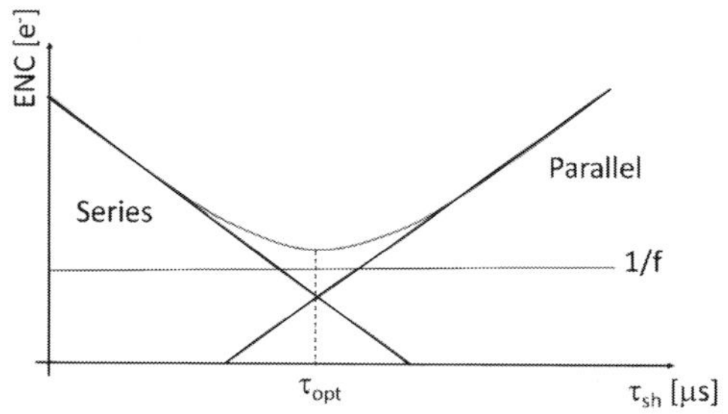

FIGURE 2.18 Summary of the different contributions to the ENC. Parallel, series, and 1/f noise are plotted versus the shaping time.

Following [24], the square of the equivalent noise charge can be expressed as:

$$ENC^2 = A_W C_T^2 \frac{A_1}{\tau_{sh}} + A_f C_T^2 A_2 + B_W A_3 \tau_{sh} \tag{2.47}$$

where $C_T = C_d + C_i + C_f$ is the total capacitance at the input of the amplifier, τ_{sh} is the shaping time, whereas A_1, A_2, and A_3 are the "shaping factors" that depend on the type of shaping and can be found in literature for the most common configurations. The coefficient A_f is related to the 1/f noise, which is only dependent on the type of input transistor used in the CSA. It is generally available in data sheets or extracted from appropriate measurements [5].

Based on Eqn. 2.47 and using the correct set of parameters, it is possible to estimate the noise of almost any readout system. The three terms in Eqn. 2.47 have differing degrees of dependence on the shaping time: (1) The series noise is inversely proportional to τ_{sh} and directly proportional to the total capacitance, (2) the parallel noise is directly proportional to τ_{sh}, whereas (3) the 1/f noise depends only on the total capacitance and not on τ_{sh}.

These trends are illustrated in Fig. 2.18 as a function of the shaping time. The quadrature sum of the three components yields a total ENC, which initially decreases with the shaping time, reaches a minimum, and then starts increasing again. By finding the minimum of this function it is possible to determine the ideal shaping time for each system (τ_{opt}). However, the shaping time also sets a limit to the time resolution of the detection system, which in turn determines the maximum pulse count rate.

In spectroscopic applications, the optimum shaping is chosen mainly on the basis of noise requirements without significant constraints on the speed of the system, since the rate of incoming photons is normally low.

In tracking applications, the shaper output is normally fed to a comparator that fires only if the signal exceeds a certain threshold. In addition, the time stamp of the incoming particle is stored. Since the amplitude of a signal from an MIP follows a Landau distribution, the threshold of the comparator should be set low enough to capture most of the signal. At the same time, it should be high enough to prevent spurious peaks due to noise and threshold dispersion from being recorded as a hit into the system. Having a low ENC or, better, a good signal-to-noise ratio, is therefore very important; using the optimum shaping time would help, but this is not always possible. In particular, in high-energy physics experiments, the shaping time is strongly limited by the bunch crossing, which can be very short (e.g., 25 ns at LHC), causing the noise to be strongly dependent on detector capacitance.

REFERENCES

1. S. M. Sze, K. K. Ng, "Physics of semiconductor devices," 3rd Edition, J. Wiley & Sons, Hoboken, New Jersey, 2007.
2. S. M. Sze, "Semiconductor sensors," J. Wiley & Sons, New York, 1994.
3. G. F. Knoll, "Radiation detection and measurement," 3rd Edition, J. Wiley & Sons, New York, 2000.
4. G. Lutz, "Semiconductor radiation detectors—Device Physics," Springer-Verlag, Berlin, 2007.
5. H. Spieler, "Semiconductor detector systems," Oxford University Press, 2005.
6. L. Rossi, P. Fischer, T. Rohe, N. Wermes, "Pixel detectors from fundamentals to applications," Springer-Verlag, Berlin, 2006.
7. F. Hartmann, "Evolution of silicon sensor technology in particle physics," Springer Tracks in Modern Physics, Springer-Verlag, Berlin, 2009, vol. 231.
8. S. Adachi, "Handbook on physical properties of semiconductors," Springer-Verlag, Berlin, 2004.
9. C. A. Klein, "Bandgap dependence and related features of radiation ionization energies in semiconductors," Journal of Applied Physics, vol. 39(4), pp. 2029–2038, 1968.
10. W. L. Leo, "Techniques for nuclear and particle physics experiments—A how to approach," 2ndEdition, Springer-Verlag, Berlin, 1994.
11. L. Landau, "On the energy loss of fast particles by ionization," Journal of Physics USSR, vol. 8, pp. 201–205, 1944.
12. H. Bichsel, "Straggling in thin silicon detectors," Review of Modern Physics, vol. 60(3), pp. 663–699, 1988.
13. G. R. Lynch, O. I. Dahl, "Approximations to multiple Coulomb scattering," Nuclear Instruments and Methods B, vol. 58, pp. 6–10, 1991.
14. J. F. Ziegler, M. D. Ziegler, J. P. Biersack, "SRIM—The stopping and range of ions in matter," Nuclear Instruments and Methods B, vol. 268, pp. 1818–1823, 2010.
15. A. Chilingarov, "Temperature dependence of the current generated in Si bulk," Journal of Instrumentation, vol. 8, P10003, 2013.
16. E. Gatti, P. Rehak, "Semiconductor drift chamber—An application of a novel charge transport scheme," Nuclear Instruments and Methods, vol. 225, pp. 608–614, 1983.
17. E. Gatti, P. Rehak, J. T. Walton, "Silicon drift chambers—First results and optimum processing of signals," Nuclear Instruments and Methods, vol. 226, pp. 129–141, 1984.
18. E. A. Hijzen, E. M. Schooneveld, C. W. E. van Eijk, R. W. Hollander, P. M. Sarro, A. van den Bogaard, "New ideas for two dimensional position sensitive silicon drift detectors," IEEE Transactions on Nuclear Science, vol. 41(4), pp. 1058–1062, 1994.
19. E. Gatti, P. Rehak, "Review of semiconductor drift detectors," Nuclear Instruments and Methods A, vol. 541, pp. 47–60, 2005.
20. P. Lechner, S. Eckbauer, R. Hartmann, S. Krisch, D. Hauff, R. Richter, H. Soltau, L. Strüder, C. Fiorini, E. Gatti, A. Longoni, M. Sampietro, "Silicon drift detectors for high resolution room temperature X-ray spectroscopy," Nuclear Instruments and Methods A, vol. 377, pp. 346–351, 1996.
21. W. Shockley, "Currents to conductors induced by a moving point charge," Journal of Applied Physics, vol. 9, p. 635, 1938.
22. S. Ramo, "Currents induced by electron motion," Proceedings of the I.R.E., vol. 27. pp. 584–585, 1939.
23. E. Gatti, P. F. Manfredi, "Processing the signal from solid-state detectors in elementary-particle physics," La rivista del Nuovo Cimento, vol. 9(1), pp. 1–146, 1986.
24. V. Radeka, "Low-noise techniques in detectors," Annual Review of Nuclear Particle Science, vol. 38, pp. 217–277, 1988.

CHAPTER 3

Radiation Effects in Silicon Sensors

3.1 INTRODUCTION

Charged particles moving through matter interact with the atomic electrons in the material. If the energy of the interacting particle is high enough, the atoms will ionize, leading to an energy loss of the traversing particle and the release of free electrons and charged ions. The basic objective of all tracking devices is to detect the movement of the free charges resulting from the ionization of a medium from a passing charged particle.

In silicon, the average energy needed to create an electron-hole pair is 3.6 eV, which is about three times larger than its band gap of 1.12 eV, since part of the energy released by the particle is dissipated thermally for the creation of phonons. For an MIP, the most probable number of generated electron-hole pairs is about 76 per micron of material traversed by the charged particle. However the same particles traversing the silicon lattice can damage it, causing its properties to change dramatically. These changes need to be understood to allow proper mitigations to be applied.

3.2 RADIATION DAMAGE IN SILICON

When silicon devices are operated in highly radioactive environments, they undergo severe damage to their structure. Detailed studies performed on this topic by several groups working in many applications have led to a fairly complete understanding of radiation damage in silicon devices, both from the microscopic formation and evolution of defects and the macroscopic effects on signal and noise degradation. For details on the extensive work performed by many groups and collaborations, see the following papers and references therein [1, 2, 3, 4, 5, 6]. The authors would like to apologize in advance if any reference for this important work has been omitted. Because of the limited space available, we decided to concentrate mainly on recent results obtained as a consequence of the need for robust tracking devices for the Large Hadron Collider experiments at European Organization for Nuclear Research (CERN) [7, 8, 9, 10]. In this context, it should be stressed that the radiation level accumulated by tracking devices in colliders' experiments is directly proportional to the beam-integrated luminosity measured in inverse femtobarns. The inverse femtobarn (fb^{-1}) is a measurement of particle-collision events and accounts for both the collision number and the amount of data collected. One inverse femtobarn corresponds to approximately 100 trillion (10^{12}) proton–proton collisions.

Tracking detectors are usually located in the innermost part of the experiments and are exposed to a mixed radiation field following a negative exponential law, depending on their radial distance from the center of the beam and from their distance along the beam axis (Z) with respect to the beam's interaction point. The ultimate radiation tolerance target for tracker detectors is therefore the one corresponding to a beam-integrated luminosity of about 4000 fb^{-1}, which will be reached after about 10 years of LHC operation following luminosity upgrades, the last of which should start in 2026 with instantaneous peak luminosity of $7.5 \times 10^{34} cm^{-2} s^{-1}$. This last parameter depends on many factors but in summary represents the instantaneous number of interactions per second.

From the practical, macroscopic point of view, radiation effects on silicon detectors can be divided in two well-distinguished categories: *surface damage* and *bulk damage*. On the one hand, surface damage causes modification to the device's breakdown behavior, the inter electrode isolation, and the surface carrier recombination, and therefore to the overall device leakage current. Bulk damage, on the other hand, affects the junction effective space charge and consequently the device full depletion voltage, the leakage current, now bulk generated, and the loss of signal due to charge carriers trapped by radiation-induced defects, some of which are the same ones responsible for the radiation-induced leakage current [1]. In the following analysis, we will try to show some of the relevant effects of these radiation-induced modifications on the operation of silicon detectors.

3.2.1 Surface Damage

Surface passivation is a necessary step in any silicon device. It protects the semiconductor surface from external agents and from any possible mechanical damage. Passivation in silicon devices is made mainly by about 1 micron of silicon dioxide (SiO_2), sometimes accompanied by silicon nitride and an overglass. Traditionally, silicon dioxide has been the preferred passivation method used in silicon devices since it is readily available in any silicon-related process fabrication facility, with a relatively simple high-temperature step consisting of exposing the device to an oxidizing atmosphere. More details will be given in Chapter 5.

Silicon dioxide is known to contain defects [11]. These defects, which are generated both inside it and at its interface with the silicon bulk, usually act as very effective charge traps. Such trapped charge can be divided into two different categories:

1. Interface trapped charge, which can be positive or negative and is caused by structural defects or oxidation defects, as grown open bonds or radiation-induced bond breaking. Interface traps, located at the Si–SiO_2 interface, are empty or filled with

electrons, depending on the applied surface potential, and they often generate so-called interface states. The main effect of interface states on the device properties is a possible increase of the surface carriers' generation–recombination, depending on their energy level within the band gap and, therefore, an increase of the surface leakage current.

2. Oxide charge, which in turn can be divided into:
   - Mobile oxide charge, which is mainly caused by the presence of ionic impurities such as sodium (Na), lithium (Li), and possibly hydrogen (H). In state-of-the-art detector technologies, this is no longer an issue.
   - Fixed oxide charge, which is positive and near the Si-SiO_2 interface. This charge normally forms a layer, which attracts negatively charged electrons on the opposite side of it, in the silicon bulk. The presence of the electron layer can lead to current paths and shorts in the device, possibly affecting the functionality and segmented information of the sensor.
   - Oxide trapped charge, which can be either positive or negative, which results from exposure to ionizing radiation, avalanche injection (e.g., during breakdown), or other mechanisms and can be annealed at low temperature.

Ionizing radiation modifies the intrinsic properties of these defects [1]. When a particle crosses a SiO_2 layer with sufficient energy, it generates charge through ionization. Most generated electron-hole pairs will recombine instantly, but a fraction of them will not. Because of the very different mobility of electrons (μ_n~20 $cm^2\,V^{-1}s^{-1}$) and holes (μ_p~2×10^{-5} $cm^2\,V^{-1}s^{-1}$) in silicon dioxide, electrons are able to escape quickly while holes will diffuse to the Si-SiO_2 interface where they get trapped, resulting in an increase of the total positive oxide charge. An additional radiation-induced effect on SiO_2 is the increase in interface states concentration, with a consequential increase in surface recombination velocity and surface current. More detailed and comprehensive information on the evolution of these defects with fluence can be found in [12].

It has been shown that for <111> crystal orientation substrates, the oxide charge concentration starts, even in very good oxides, from a few units of $10^{11}cm^{-2}$, while lower values have been observed for <100> ones [13]. The exposure to ionizing radiation causes this value to increase by at least one order of magnitude, and therefore up to about 10^{12} cm^{-2}. At the same time, a standard value for surface recombination velocity (s_0) is about 10 cm/s or less before irradiation, while it increases several orders of magnitude after irradiation [12].

A very important aspect related to surface damage that is often underestimated is its strong dependence on the device's operational conditions while exposed to ionizing radiation. This is the case for all silicon devices used in collider experiments or those used as beam monitors at beam lines, which require biasing the silicon p-n junction to obtain its full depletion and, therefore, an effective signal collection. Depending on the substrate polarity, when exposed to radiation under bias causing the electric field through the oxide to point to the Si-SiO_2 interface, the oxide charge concentration and surface recombination velocity start increasing considerably already at low radiation doses. Even more remarkably, the oxide charge concentration can reach values in the order of 4×10^{12} cm^{-2} at the doses of interest for tracking detectors and attain even higher values in X-ray free electron laser applications [13].

As already indicated, the main consequences of the exposure to ionizing radiation on the silicon device's oxide surfaces can be summarized as follows:

1. Change in overall charge concentration due to the accumulation of positive charge trapped in the oxide, which attracts negative charge on the other side of the interface leading to:
   - compromised isolation between segmentation defined by n^+ implanted electrodes. This could cause shorts in pixels or strip readout electronics;
   - increased parasitic capacitance between adjacent regions that could degrade the noise performance of the device; and
   - modified electric field distribution at the silicon–silicon oxide interface, which could affect the voltage-handling capabilities of the sensor.
2. Modified interface states and surface recombination velocity leading to an increase of the surface related leakage current.

   Some of these effects can be mitigated with particular counteractions, as it will be described later.

3.2.2 Bulk Damage

In silicon detectors used for tracking in highly radioactive environments, the damage to the bulk is caused by charged hadrons like protons, pions or neutrons or highly energetic leptons like electrons or muons. The damage mechanism is characterized mainly by the loss of kinetic energy of the traversing particle inside the silicon lattice through multiple collisions with the silicon atoms. If the energy of the considered particle is larger than the displacement threshold energy $E_d = 25$ eV, the result is the displacement of a primary knock-on atom (PKA), which forms an interstitial and vacancy (Frenkel) pair [1]. Both interstitials and vacancies can migrate in the silicon lattice with different mobilities, depending on temperature [14].

Finally, they form stable defects when they interact with impurities or doping atoms present in the material. Such defects can act as effective acceptors or donors. If the particle energy is larger than ~5 keV, then highly dislocated defect clusters can be formed. The recoil atom from a highly energetic collision can potentially cause further damage to the silicon lattice [15]. Its energy loss can occur through ionization or further displacements. At the end of a heavy recoil range, displacement prevails and defect clusters are formed, as can be seen in Fig. 3.1, which shows the mean displacement produced by a 1MeV neutron, with a PKA of 50 KeV.

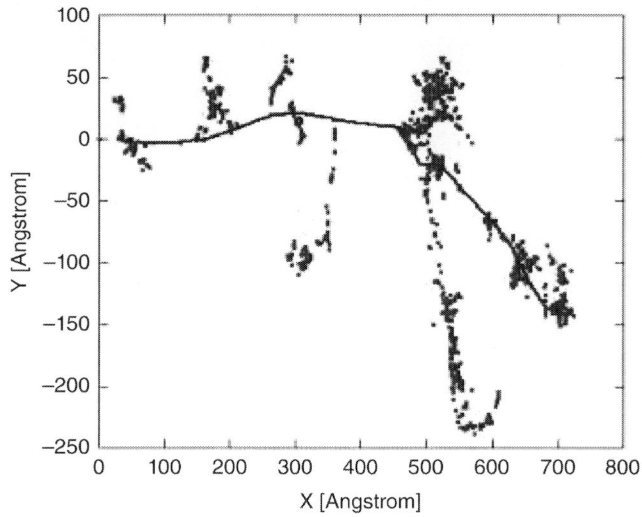

FIGURE 3.1 Displacement produced by 1 MeV neutron in silicon. This event generates 1159 displaced silicon atoms, 1053 become vacancies, 106 replace the displaced atom, and 1159 interstitials are formed. [16]

The simulation is performed using TRIM, an open-source simulation. This event produces 1159 displaced silicon atoms, of which 1053 become vacancies and 106 replace the displaced atom; 1159 interstitials are also formed [16].[1]

Since different particles and energies have shown to generate a variety of defects that impact, in a multitude of ways, the sensor's parameters, it was crucial to find a way to scale radiation damage with respect to the macroscopic effects observed in silicon [17, 18, 19]. The non-ionizing energy loss (NIEL) [20] was introduced to calculate the displacement damage in the material, which have shown to scale linearly with the amount of energy imparted in the displacing collisions. By using the displacement damage cross section D(E), it was possible to define a hardness factor "*k*," which allows comparison of the damage efficiency of different radiation sources with different particles and individual energy spectra Φ(E):

$$k = \frac{1}{D\ (E_n = 1\ MeV)} \cdot \frac{\int D(E)\Phi(E)dE}{\int \Phi(E)dE} \tag{3.1}$$

where E is the energy of the considered particle, $\Phi(E)$ is the energy spectrum of the radiation field, and $D(E)$ is the displacement damage cross section. Figure 3.2 shows how the

[1] SRIM results from the original work by J. P. Biersack on range algorithms (see J. P. Biersack and L. Haggmark, Nucl. Instr. and Meth., vol. 174, 257, 1980) and the work by J. F. Ziegler on stopping theory (see "The Stopping and Range of Ions in Matter," volumes 2–6, Pergamon Press, 1977–1985). The various versions of SRIM are described briefly in the file VERSION on the SRIM package. The SRIM program originated in 1983 as a DOS-based program and was converted to Windows in 1989. If you use SRIM programs in a scientific publication, please mail a copy to the authors. This will help continued support of SRIM in the future.

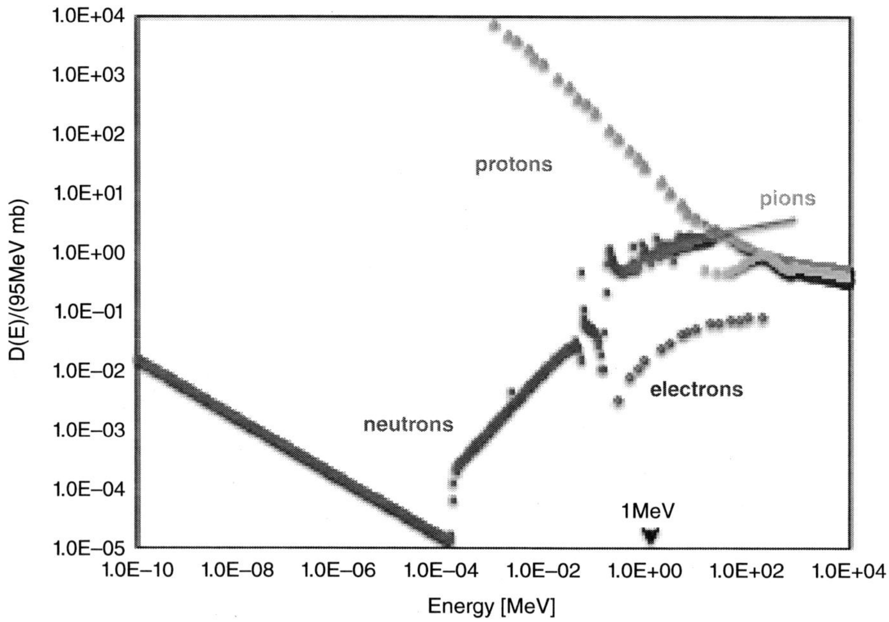

FIGURE 3.2 Displacement damage function $D(E)$ normalized to 95 MeV mb, for different types of particles and energies. Plotted from data in [20].

displacement damage cross section of different particles relates to one another at different energies. The hardness factor is commonly used to compare the damage produced by a specific type of irradiation to the damage that would be caused by the same fluence of 1 MeV monoenergetic neutrons. The 1 MeV neutron equivalent fluence Φ_{eq} corresponding to a specific irradiation can be calculated using the following equation:

$$\Phi_{eq} = k\Phi = k\int \Phi(E)dE \tag{3.2}$$

The displacement-induced defects in the silicon lattice translate into the creation of energy levels in the forbidden band gap, behaving like acceptors or donors, that can capture or emit electrons. In thermal equilibrium, the charge state of the defect is generally related to the Fermi level. An acceptor is negatively charged if it is occupied by an electron, while a donor in the same condition is neutral. If the Fermi level is higher than the defect level, the acceptor will have negative charge and the donor will be neutral. In contrast, if the Fermi level is lower than the defect level, the acceptor will be neutral and the donor will be positively charged. Deviations to this standard can happen for very deep levels.

Figure 3.3 is a representation of the measured defects distribution within the band gap after irradiation. As will be described shortly, the introduction of donor-like and acceptor-like defects, in particular in the middle of the band gap, can influence the effective doping concentration of the silicon bulk and the leakage current because of the enhancement of the generation recombination rate.

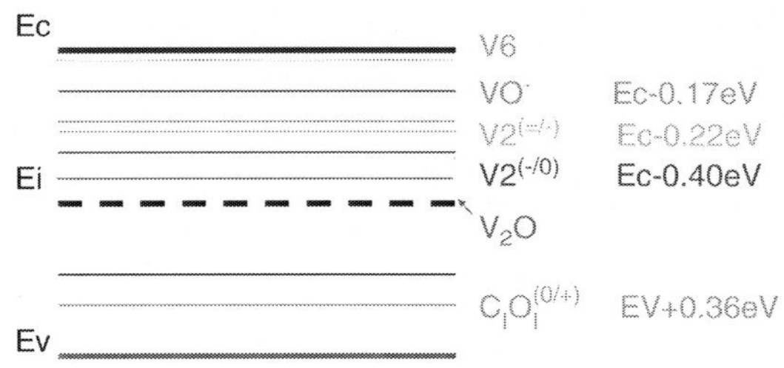

FIGURE 3.3 Representation of energy levels of stable defects observed after neutron and proton irradiation within the silicon band gap. Vacancies and interstitials can move in the lattice until they meet impurities and dopants present in the material forming stable centers, which act as effective donors and acceptors.

In particular, considering a defect energy level E_t, the total energy required to excite a captured electron to the conduction band will correspond to $\Delta E = Ec - Et$.

- *Leakage current*

The creation of radiation-induced energy levels in the forbidden band gap enhances the generation–recombination processes in silicon. In reverse-biased p-n junctions, this translates to an increase in the reverse leakage current, which will ultimately affect the final signal-to-noise detector performance and the system power budget [1]. The increase in leakage current after exposure to radiation was measured on different devices, with n- and p-type substrates irradiated with different particles to be linearly proportional to the irradiation fluence following the relationship:

$$\Delta I = \alpha \cdot \Phi_{eq} \cdot V \tag{3.3}$$

where α is the current-related damage parameter, Φ_{eq} the 1 MeV neutron equivalent fluence, and V the total depleted sensor volume, as can be seen in Fig. 3.4. The leakage current is strongly temperature dependent, so all measurements are scaled to the same reference temperature, usually 20°C, where the damage parameter tends to be $\alpha = (3.99 \pm 0.03) \times 10^{-17} A/cm$ [1]. However, it is impossible to explain the measured current intensity based on simple Shockley–Read–Hall generation–recombination processes using measured midgap defects such as the di-vacancy V_2. Corrected current values were obtained by including the so-called intercenter charge transfer process where electrons are "hopping" to physically close defects toward the conduction band [21].

As already stated in chapter 2, leakage currents measured at different temperatures T can be correlated using the relation [22]:

$$I(T_R) = I(T) \cdot \left(\frac{T_R}{T}\right)^2 exp\left[-\frac{E_{eff}}{2k_B}\left(\frac{1}{T_R} - \frac{1}{T}\right)\right] \tag{3.4}$$

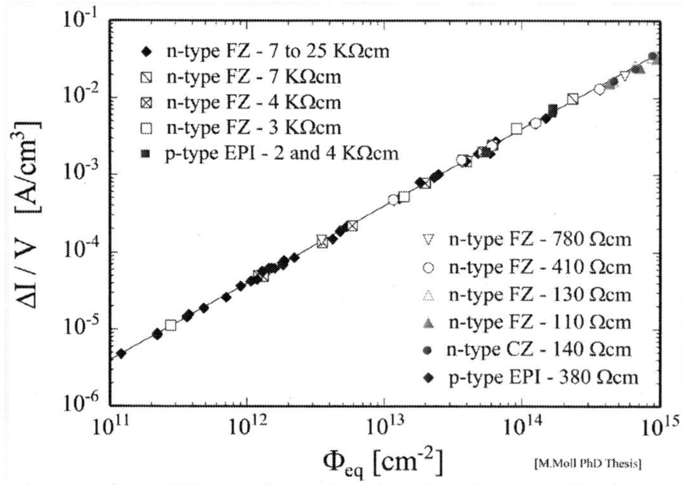

FIGURE 3.4 (a) Leakage currents of devices produced in different materials and exposed to different types of radiation versus fluence after a thermal treatment at 60°C for 80 minutes [1].

where T_R is the reference temperature (usually 20°C) and k_B is the Boltzmann constant. The damage-induced leakage current can be annealed at high temperatures where the damage constant α was shown to decrease with time at different annealing temperatures [1].

These findings led to the conclusion that silicon devices exposed to high radiation fluences should be operated below 0°C to mitigate the increase of leakage current and consequent signal-to-noise ratio and possibly thermal runaway, but they should be left at room temperature for a few days during downtimes to allow leakage current beneficial annealing.

The material modification after irradiation results in a substantial change in the current-voltage (I-V) characteristics of silicon devices. The device does not show the characteristic forward and reverse current behavior any more but, rather, an increasing resistive dependence of the substrate exposure to radiation, which becomes symmetric at high fluences. The material is said to become “relaxation like” as a result of the Fermi level being “pinned” near midgap, the location for minimum conductivity [23, 24, 25]. The point of breakdown in the forward bias I-V curve has a direct relationship with the fluence as well as the charge collection characteristics [25].

- *Effective bulk doping concentration*

Silicon detectors are usually operated with the junction slightly overdepleted (i.e., at a bias voltage slightly larger than the full depletion voltage), such that the active region is as much as possible free of charge carriers. In Chapter 2, it was shown that in planar sensors the depletion voltage is related to the substrate thickness and the substrate doping concentration. If the junction is not fully depleted, the overall signal produced by a particle traversing the device will be reduced.

Since displacement defects can behave either like acceptors or donors, depending on their position in the band gap, they can affect the full depletion bias voltage and in certain

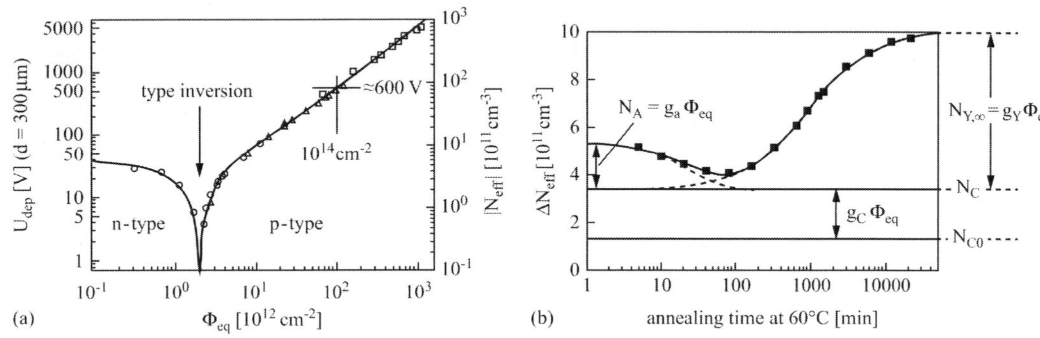

FIGURE 3.5 (a) Variation of the effective doping concentration of an n-type silicon sensor as a function of the equivalent particle fluence (plot from [1] based on data from R. Wunstorf, PhD thesis, University of Hamburg, 1992), and (b) its annealing behavior as a function of time at 60°C after irradiation at a fluence of 1×10^{13} n_{eq}/cm^2 [1].

cases the space charge's "effective" polarity. Several tests have shown that the radiation-induced doping concentration can vary depending on the original material characteristics. A different behavior was also observed following irradiation with different particle types (more specifically with neutron vs. proton irradiation) since they induce an imbalance of point defects versus defect clusters in the latter due to Coulomb interaction. An example of the variation of effective doping concentration (N_{eff}) for an n-type starting substrate is shown in Fig. 3.5a. The effective doping concentration is initially reduced by a possible donor removal, accompanied by simultaneous effective acceptor insertion at low fluences that finally lead to a p-type effective space charge-type inversion and a further increase of N_{eff} proportional to radiation fluence.

Unlike what happens to the current-related damage parameter α, N_{eff} undergoes both a beneficial and a reverse annealing. The cause of reverse annealing is still disputed but could be related to the formation of a multi-interstitial defect observed with photoluminescence [26]. Both of these effects are shown in Fig. 3.5b. Oxygenation has been shown to control the variation of the space charge and reverse annealing after proton irradiation, acting as a very efficient getter for vacancies while carbon has been shown to enhance negative effects [27].

The effective doping concentration in p-type materials was also observed to increase with fluence, as can be seen in Fig. 3.6 [28].

At the relatively large irradiation fluences of interest for collider experiments, for n-type substrates the stable damage term can be written considering only the acceptor introduction that occurs after type inversion and ignoring the initial donor removal.

The recent availability of pure p-type substrates allows their use for radiation hard sensors since the junction electrode in p-type devices remains on the same side also after heavy irradiation, avoiding the significant signal degradation and the loss of spatial resolution due to underdepletion that affects p-on-n sensors [29, 30].

Before irradiation, p-on-n sensors deplete from the p^+ electrode toward the backplane. After irradiation (left sketch in Fig. 3.7), the bulk undergoes type inversion, becoming p-type. The depleted region grows from the n^+ electrode, and due to irradiation, the substrate has a

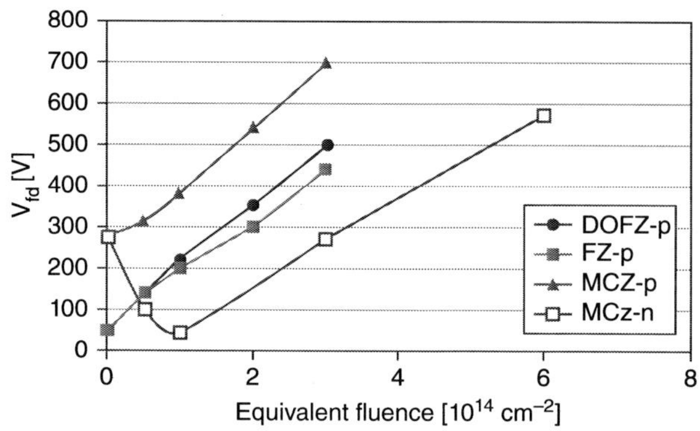

FIGURE 3.6 Dependence of full depletion voltage (V_{fd}) on neutron fluence for MCz and FZ p-type materials (From Kramberger, G. et al., Nucl. Instrum. Methods A, 2007. With permission of ELSEVIER.)[28].

higher N_{eff}, which increases the full depletion voltage. Incidentally, it should be stressed that after irradiation the weighting field, fundamental for signal calculation as shown in Chapter 2 and as will be seen later in this chapter, will not change its shape since it depends only on the electrodes' geometry and it peaks near the collecting electrode for a segmented device. Holes and electrons generated in the depleted region now close to the back region are inducing a very low signal on the collecting electrode since the weighting field peaks near the surface and goes to zero rapidly. At high-radiation levels, segmented planar p-on-n sensors will have difficulty depleting up to the collecting electrode with a substrate thickness useful to deliver enough charge (200 microns), so the signal will be very small.

At the same time, n-on-p detectors can overcome this problem since no type inversion occurs after irradiation [28] as well as n-on-n detectors after type inversion, since the junction will always be close to the readout electrode where both electric and weighting field are strong, as can be seen in Fig. 3.7 (right sketch). Since the major part of the signal (≈ 80%) is induced by carriers moving to the readout electrode that collects electrons, these detectors are intrinsically faster than p-on-n, also counteracting trapping probability. The subject of trapping probabilities and their effect will become clearer in the next section.

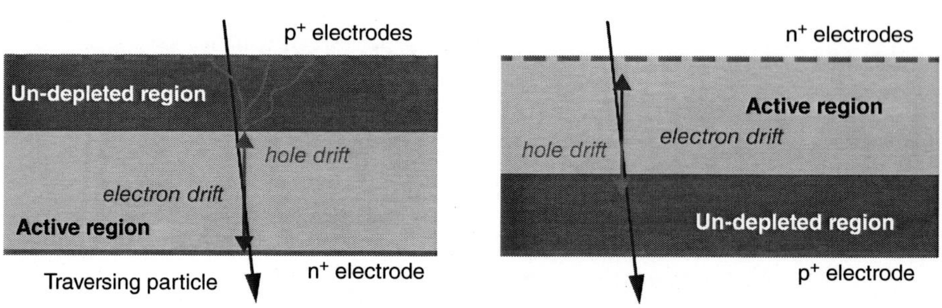

FIGURE 3.7 Schematic cross sections of p-on-n (left) and n-on-p (right) segmented sensors, both with effective p-type bulk after irradiation and operated in an underdepleted condition.

It has been observed that n-type materials after type inversion have shown a persistence of the main junction. This double electric field peak, known as a "double junction," has been shown to slightly increase the total collected charge [31, 32].

- *Signal degradation due to charge trapping*

The predominant cause of signal loss in detectors irradiated to above a fluence of 1×10^{15} $n_{eq.}cm^{-2}$ has been seen to be trapping, as most defects act as very efficient trapping centers for free carriers. If free carriers generated from an impinging particle are captured, they are usually not released before hours or longer at room or lower temperature [33]. Such carriers therefore cease contributing to the final sensor's signal within the current electronics readout times (~25 ns for the Large Hadron Collider experiments).

The total trapping probability is inversely proportional to the irradiation fluence following the equation:

$$\frac{1}{\tau_{eff,(n,p)}} = \beta_{n,p}\ (t, T)\Phi_{eq} \tag{3.5}$$

where τ_{eff}, is the effective trapping time and β is the trapping time constant, which is time and temperature dependent and is also influenced by the type of particles used for irradiation. The effective drift length (or path before trapping) for electrons and holes depends on their effective trapping times (or lifetimes), so the higher the fluence the silicon bulk receives, the shorter the carrier's effective drift length (λ) according to the formula:

$$\lambda = \tau_{eff} v_D \tag{3.6}$$

where v_D is the drift velocity, meaning that shorter trapping times will result in a larger amount of carriers being trapped. The total amount of drifting charge that induces current on the readout electrodes follows Shockley's and Ramo's theorems and takes into account the trapping probability in a negative exponential that can be expressed as:

$$N_{e,h} = N\,(0)\exp\left(-\frac{t}{\tau_{eff,(n,p)}}\right) \tag{3.7}$$

As shown in Chapter 2, defining a signal induced by generated charge in a detector was a serious business before Ramo and Shockley concluded at the same time in 1939 that a signal generated by moving charges in a conductor could be simply derived using their "weighting potentials" and "weighting fields." As a reminder, the weighting potential describes the coupling of a charge at any position with respect to the collecting electrode and is obtained by setting the potential of that electrode to 1 while setting all other electrodes to potential 0.

In a parallel-plate detector of thickness L with an applied bias voltage V_b, the electric field, which is different from the weighting field, across it would be $E = V_b/L$.

A charge generated within the detector volume would therefore have a velocity of $v_D = \mu E = \mu \frac{V_b}{L}$. If we assume a potential equal to 1 on the collecting electrode and zero on the bias one, then we have seen that a weighting field would be constant and equal to $E_W = 1/L$. The induced charge Q_0 on the collecting electrode by a charge q was calculated using electrostatic laws to be $Q_0 = -qV_W$ where V_W is the *weighting potential.* By definition, the relationship between the weighting potential and the *weighting field* is $E_W = -\nabla V_W$. If we are only considering a one-dimensional case along the x coordinate, then $E_W = -dV_w/dx = 1/L$.

As demonstrated in Chapter 2, the induced current on the collecting electrode (the one at potential = 1) is given by the equation:

$$i(t) = qv_D E_W = qv_D / L \tag{3.8}$$

Now, if we assume that an electron-hole pair was generated at position x within the sensor's volume, the induced charge will depend on the time the electrons and holes will take to reach their respective electrodes. As demonstrated in Chapter 2, the induced charge for electrons and holes would therefore be:

$$Q_e = it_e = \frac{q_e}{L}x \quad and \quad Q_h = it_h = \frac{q_h}{L}(L-x) \tag{3.9}$$

Respectively, with q_e and q_h the electron and hole charge.

In the presence of trapping, the charge induced to the collecting electrode will now be reduced by the lifetime of the carriers. The fraction of charge Q_{tr} remaining after trapping can then be expressed as $Q_{tr}(t) = Q_0 \exp(-t/\tau)$, with Q_0 the charge without trapping and where the time to traverse a distance x is defined as $t = x/v_D = x/\mu E$. By substituting the expression of t in the previous exponential equation, we obtain the induced charge as a function of distance x as:

$$Q(x) = Q_0 e^{-\frac{x}{\mu\tau E}} = Q_0 e^{-\frac{x}{\lambda}} \tag{3.10}$$

The mobility lifetime product $\mu\tau$ is an indicator of the quality of the sensor, while $\mu\tau E$ is the effective drift length λ that was defined earlier. The signal development with trapping, assuming a generated unit charge $q = 1$ in position x inside the active volume, can therefore be defined as follows:

$$\frac{dS}{dt} = \frac{dQ(x)}{dt} = \frac{d}{dt}\left(Q_0 e^{-\frac{x}{\lambda}}\right) = \frac{d}{dt}\left(-V_W e^{-\frac{x}{\lambda}}\right) = \frac{d}{dt}\left(-V_W e^{-\frac{x}{\lambda}}\right)\frac{dx}{dx} = \left(-\frac{dV_W}{dx}\right)e^{-\frac{x}{\lambda}}\frac{dx}{dt} = E_W e^{-\frac{x}{\lambda}}\frac{dx}{dt} \tag{3.11}$$

where the derivative over distance of the weighting potential became the weighting field and the minus sign disappeared. It is now possible to extract the resulting signal by integrating the previous equation from position x to the collecting electrode at position 0:

$$\begin{aligned} S &= \int dS = \int_x^0 qE_W \frac{dx}{dt} \, exp\left(-\frac{x}{\lambda}\right) dt = \frac{1}{L}\int_x^0 exp\left(-\frac{x}{\lambda}\right) dx \\ S &= \frac{1}{L}\left[\left(\lambda \exp\frac{-0}{\lambda}\right) - \left(\lambda \exp\frac{-x}{\lambda}\right)\right] \\ S &= \frac{\lambda}{L}\left[1-\exp\left(-\frac{x}{\lambda}\right)\right] \end{aligned} \tag{3.12}$$

Note that for a charge that is trapped before reaching the collecting electrode (x>>λ), $S=\lambda/L$. In other words, even the charge that gets trapped generates some signal.

It should be stressed that the drift velocity v_D after irradiation is usually saturated to a value ~10^7 cm s^{-1} and therefore is constant in position due to the need to operate irradiated detectors with a high electric field. Thus, the effective drift length is also constant and independent on position since a high bias voltage is applied. The effective trapping probability for electrons and holes has been measured [34] and can be seen in Fig. 3.8.

From the data in Fig. 3.8, it is possible to extract the trapping time constant τ_{eff}, for electrons and holes to be for electrons 3.7×10^{-16}cm^{-2} s^{-1} and 5.4×10^{-16}cm^{-2} s^{-1} and for holes 5.7×10^{-16}cm^{-2} s^{-1} and 6.6×10^{-16}cm^{-2} s^{-1} after neutron and proton irradiation, respectively. Thus, since a lower trapping time constant means longer drift, electrons have a longer lifetime before being trapped and therefore can contribute to a larger signal. Recent studies of fluences up to 10^{17}n$_{eq}$cm^{-2} seem to point to a saturation of the trapping probability. This result would explain some observation of saturated signal efficiency at high fluences [35].

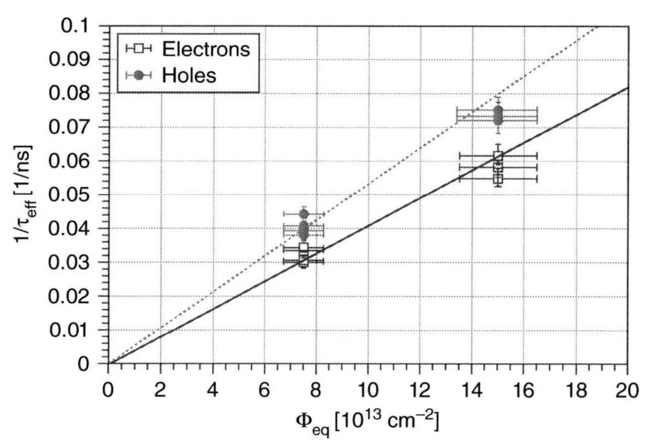

FIGURE 3.8 Dependence of the effective trapping probability of electrons and holes versus fluence, showing that electrons have a longer lifetime and therefore a higher probability of contributing to a larger signal than holes. A slight difference in trapping probability has been observed with irradiation performed with neutrons and protons (From Kramberger, G. et al., Nucl. Instrum. Methods A, 2002. With permission of ELSEVIER.)[34].

Effective trapping probability, as well as doping concentration and leakage current, depends on time and on annealing temperature. The simplest model that describes the evolution of the trapping probability is the decay from one electron or hole trap state to another stable defect. When annealing is performed at 60°C [34], it is observed that the long-term behavior is beneficial for electrons and not beneficial for holes. In fact, with time, the trapping probability decreases for electrons (~15%) and increases for holes (~30%). The temperature of the annealing step does not change the value of the trapping probability at the end of the transitory state, but only the time constant to the transition from one state into the other. The time constant of the transition is ≈ 600 minutes at 60°C. In case of annealing at room temperature, the constant time increases up to 50 days. It can be concluded that again, electrons readout has to be preferred to holes readout since electrons are 3 times faster than holes and have a lower trapping time constant τ_{eff} with respect to holes. In addition, the annealing tends to lower the trapping probability for electrons while it increases it for holes. A full description of these important measurements can be found in [34].

For the situation in which charge is generated uniformly between the p^+ and n^+ electrodes, one can show that the signal efficiency (SE), defined as the ratio of the maximum signal with trapping to the maximum signal with no trapping, is:

$$SE = \frac{\lambda}{L} - \left(\frac{\lambda}{L}\right)^2 + \left(\frac{\lambda}{L}\right)^2 exp\left(-\frac{\lambda}{L}\right) \tag{3.13}$$

This is for one charge carrier and has a maximum value of 0.5. One must sum for electrons and holes that have different effective drift lengths, λ_e and λ_h, respectively. Equation. 3.13 is derived by integrating Eqn. 3.12 over the inter electrode length L for a uniform charge density for both irradiated and not irradiated diodes. The ratio of these two integrations gives the SE result of Eqn. 3.13.

On the basis of calculated effective drift lengths [36] one can relate the hole (λ_h) and electron (λ_e) effective drift lengths as follows:

$$\lambda_h = 0.4\lambda_e \tag{3.14}$$

The factor 0.4 is largely due to the difference in hole and electron mobilities and trapping times.

A fit to SEs can be made using the simplified equation:

$$SE = \frac{1}{1 + K_C\Phi} \tag{3.15}$$

where K_C is the damage constant for the signal efficiency. K_C depends on the trapping damage constant (K_τ), the inter electrode distance (L) and the drift velocity (v_D):

$$K_C = 0.6LK_L = \frac{0.6\ LK_\tau}{v_D} \tag{3.16}$$

keeping in mind that $\frac{1}{\lambda_e} = \frac{1}{\lambda_0} + K_L\Phi \approx K_L\Phi$ and $K_L = \frac{K_\tau}{v_D}$ and that λ_0 is large, v_D is the drift velocity saturated and K_τ is the damage constant for the electron effective trapping time.

Short inter electrode spacing explains the superior radiation tolerance performance of 3D sensors whose geometry benefits from this, together with the possibility of tailoring the thickness of the substrate, which provides the original amount of charge released by an ionizing particle.

A compilation of signal efficiences versus fluence for planar and 3D sensors processed on float zone (FZ) and epitaxial substrates is shown in Fig. 3.9. The SEs measured after irradiation at various fluences, are reported in the literature [37, 38, 39, 40, 41, 42, 43, 44, 45, 46, 47]. Signal Efficiencies were measured using either low-intensity infra red lasers (3D) or minimum ionising particles (3D and planar). Since both generate charge uniformly across the substrate thickness the formalism described above applies. As can be seen in Fig. 3.9, the SE degrades less after heavy irradiation for devices with a shorter inter electrode distance L.

Thin silicon substrates can therefore be a good solution to radiation damage in harsh environments if the noise of the readout electronics is low enough to guarantee a good signal-to-noise ratio [48, 49, 50]. A lot of impressive progress has been made recently in this direction with the development of new readout chips [52].

Another interesting mechanism, which has been observed mainly in irradiated thin silicon sensors, is the presence of excess charge due to charge multiplication at high bias voltages, making the signal amplitudes even exceed the pre-irradiation values. The multiplication

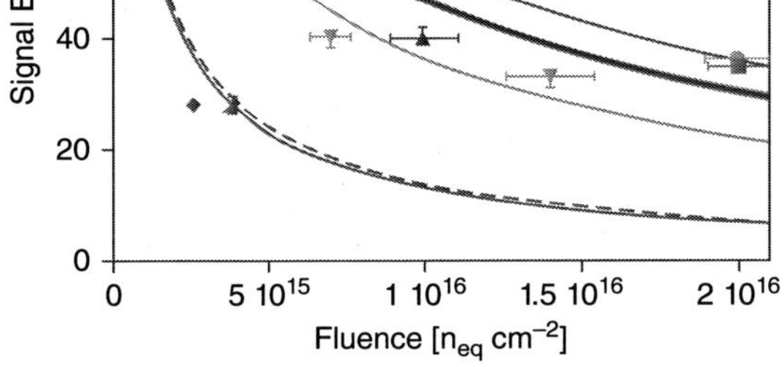

FIGURE 3.9 Compilation of signal efficiencies versus fluence of silicon sensors planar and 3D processed on either float zone or epitaxial substrates with inter electrode distances from 35 microns (top curve) corresponding to 3D sensors to bottom curve corresponding to 285 microns planar sensors. Data are available in the literature [37–47] The data follow the predicted trend of inverse proportionality with the fluence. The increased SE with reduced inter electrode spacing L is clearly visible in the plot.

is caused by impact ionization due to hot electrons moving in the high electric field that develops near the junction in irradiated sensors. Charge multiplication has been independently observed by several research groups in planar pad and strip detectors after exposure to large radiation fluences [52, 53, 54, 55, 56, 57].[2] Very high bias voltages, in the order of 1000 V, are normally applied in order to reach the critical electric field values (of the order of 20 V/μm), causing the onset of avalanche effects. In parallel with measurements of collected charge with LHC-like readout chips, further studies (e.g., by the edge transient current technique [57]) aim at a deep understanding of the charge multiplication phenomenon and its modeling [58, 59]. Moreover, modified sensor designs easing the electric field control are being investigated in order to allow reliable exploitation of charge multiplication in future experiments [60, 61].

3.3 CAN RADIATION DAMAGE BE CONTROLLED?

To prevent some of the outcomes of radiation damage in silicon, several techniques can be implemented. This section briefly describes the most important ones of such techniques.

3.3.1 Surface Damage

The increase of positive oxide charge up to very large values can cause several problems. Among them the most relevant are the following:

i. In p-in-n devices, a strong modification of the electric field distribution at the surface, possibly leading to anticipated breakdown, as well as an increased inter electrode capacitance after irradiation.

ii. In n-in-p and n-in-n devices, the induction of a surface electrons layer that will eventually short all the segmented n^+ readout electrodes, catastrophically compromising position resolution.

To cope with the second problem, well-known solutions such as (a) p-spray, (b) p-stop, and (c) moderated p-spray [61] are available to prevent the surface inversion layer and to avoid the shorting of n^+ electrodes. Thanks to numerical simulations the behavior of these techniques is very well understood [62]. The perfect choice of isolation technique is not possible, but by properly weighing the different options, it is possible to find the best one for a specific application. The main parameters that should be kept in mind are:

i. breakdown voltage both before and after irradiation, in order to operate the device at optimal bias;

ii. inter electrode capacitance to control the noise budget;

iii. inter electrode resistance, which is necessary for proper isolation.

[2] Charge multiplication has also been observed in 3D sensors, as will be detailed in Chapter 6.

Each of these solutions brings advantages and disadvantages:

1. ***p-spray:*** consisting of a medium-dose implantation of p dopant on the n^+(which is usually also the readout) side of the wafer. This solution has the main advantage of being very simple. The desired result is a charge compensation of the surface electron layer that is formed as a consequence of the positive charge trapped in the oxide. The main disadvantages of this technique are related to a rather low breakdown voltage before irradiation, caused by the relatively high doping concentration of the p-spray region in contact with the n^+ electrodes, whose potential tends to follow the substrate potential and by a larger inter electrode capacitance. It should be noted, however, that the reduced breakdown voltage pre-irradiation is not necessarily a big issue since all devices can be operated at rather low bias voltages. After irradiation, the total oxide charge concentration increases, which results in lower electric fields, larger breakdown voltages, and lower inter electrode capacitances.
2. ***p-stop:*** a solution intended to interrupt the continuity of the surface electron layer with localized high-dose p^+ implants surrounding the readout electrodes. The best geometrical implementation of p-stops is with narrow implants to limit the coupling with the substrate bias and therefore increase the sensors' breakdown voltage. In this particular case, the surface electron layer is in direct contact on one side with the n^+ junctions and will therefore be grounded, and on the other side with the high-concentration p^+ implant. The breakdown voltage of the p-stop is typically high before irradiation, but, as the oxide charge increases, it starts decreasing. As far as the inter electrode capacitance is concerned, it increases with the charge oxide concentration. The main advantage of the p-stop over the p-spray is the strong isolation between electrodes.
3. ***Moderated p-spray:*** a technique that consists in combining both p-stop and p-spray with a medium/low dose of uniform implant and a high dose of localized implant to fuse both properties. As far as breakdown voltage is concerned, it starts low, and it increases with irradiation (typical of p-spray). If the p-spray layer is processed with a low-dose implant, to maximize breakdown before irradiation, it might happen that the increased oxide charge will result in complete compensation and surface inversion. At this point, the structure behaves like a p-stop, with breakdown voltage decreasing as the total radiation dose increases. Comparably, the inter electrode capacitance will initially decrease (p-spray) and then increase (p-stop) [63].

3.3.2 Bulk Damage

Bulk damage can be prevented in three main ways:

i. material engineering;

ii. device engineering;

iii. device operational conditions.

Material engineering consists mainly of the growth of substrates with the type and amount of impurities capable of mitigating the defect formation responsible for leakage current generation and trapping. Several studies were performed in the 1990s leading to important conclusions on the subject like the use of oxygen. Oxygen is believed to capture vacancies in stable and neutral point defects after irradiation, leading to a more homogeneous effective doping change in different silicon substrates [3, 4, 64, 65, 66]. The beneficial effect of oxygen is limited to the stable damage, and it was shown that reverse annealing could be better controlled after proton but not after neutron irradiation.

Other materials like GaAs [67] and recently diamond [68] have attracted attention owing to their very low leakage current due to larger band gaps. Despite the high cost, in particular that of monocrystalline diamond substrates, several impressive advances have been made using the less expensive polycrystalline diamond with laser-ablated 3D geometries [69].

Device engineering consists in the design of layouts capable of overcoming the structural drawbacks of irradiation. The first groundbreaking design modification for harsh radiation environments of silicon devices consisted in moving from the traditional p-on-n to n-on-n structures [70]. This choice, together with the use of oxygenated substrates, allowed the current pixel devices at LHC experiments to survive up to an unprecedented fluence of 1×10^{15} n_{eq} cm^{-2}. The next breakthrough was then to move to n-in-p structures, thanks to the production of high-resistivity p-type substrates [71]. Both new layouts gain from the enhanced electrons collection because of their higher mobility, their larger signal due to the longer electrons' effective drift length, and preserved spatial resolution since the junction always depletes from the n^+ side. The next important milestone in device engineering was the use of 3D layouts, which allows important modifications to the traditional silicon detector concept. Chapter 6 is fully dedicated to exploring the benefits of these devices in terms of radiation hardness.

Device operational conditions can help mitigate some of the effects of irradiation. The operation at high bias voltages allows a longer carrier drift before trapping and therefore a larger signal and possible charge multiplication with enhanced signal; the operation at temperatures just below 0°C can control the amplitude of radiation-induced leakage current and breakdown due to thermal runaway; and forward bias operation after high fluences, making use of the material, with resistive I-V characteristics at both forward and reverse bias can be used to guarantee particle detection in forward bias with moderate voltages [25]. Finally, the operation at cryogenic temperatures, precisely at 130 K, has shown promising results with enhanced speed and recovery of the charge at both reverse and forward bias after heavy irradiation. The so called *Lazarus effect* was extensively studied in the 1990s, but it was dropped from the main LHC arena due to the difficulty of operating a large system at liquid nitrogen temperatures [72, 73, 74].

REFERENCES

1. M. Moll, "Radiation damage in silicon particle detectors—microscopic defects and macroscopic properties," PhD Thesis, University of Hamburg, Germany, 1999.
2. G. Lindström, M. Moll, E. Fretwurst, "Radiation hardness of silicon detectors—A challenge from high energy physics," Nuclear Instruments and Methods A, vol. 426, pp. 1–15, 1999.

3. G. Lindström, et al. (the CERN RD48 Collaboration), "Developments for radiation hard silicon detectors by defect engineering—results by the CERN RD48 (ROSE) Collaboration," Nuclear Instruments and Methods A, vol. 465, pp. 60–69, 2001.
4. G. Lindström, et al. (the CERN RD48 Collaboration), "Radiation hard silicon detectors—developments by the RD48 (ROSE) collaboration," Nuclear Instruments and Methods A, vol. 466, pp. 308–326, 2001.
5. H. Dijkstra, J. Libby, "Overview of silicon detectors," Nuclear Instruments and Methods A, vol. 494, pp. 86–93, 2002.
6. M. Moll, "Development of radiation hard sensors for very high luminosity colliders—CERN-RD50 project," Nuclear Instruments and Methods A, vol. 511, pp. 97–105, 2003.
7. The LHC study group, The Large Hadron Collider, conceptual design, CERN/AC/95-05 (LHC), October 20, 1995.
8. The Compact Muon Solenoid (CMS), Technical Proposal, CERN/LHCC 94-38 LHCC/P1, 1994.
9. A Toroidal LHC Apparatus (ATLAS), Technical Proposal, CERN/LHCC/94-43, LHCC/P2, 1994.
10. P. A. Aarnio, M. Huhtinen, "Hadron fluxes in inner parts of LHC detectors," Nuclear Instruments and Methods A, vol. 336, pp. 98–105, 1993.
11. D. K. Schroder, "Semiconductor material and device characterization," 3rd Edition, John Wiley & Sons, Hoboken, NJ, 2006.
12. R. Wunstorf, H. Feick, E. Fretwurst, G. Lindström, G. Lutz, C. Osius, R.H. Richter, T. Rohe, A. Rolf, P. Schlichthärle, "Damage-induced surface effects in silicon detectors," Nuclear Instruments and Methods A, vol. 377, pp. 290–297, 1996.
13. J. Zhang, "X-ray radiation damage studies and design of a silicon pixel sensor for Science at the XFEL," PhD Thesis, University of Hamburg, Germany, 2013.
14. G. D. Watkins, "Intrinsic defects in silicon," Materials Science in Semiconductor Processing, vol. 3, pp. 227–235, 2000.
15. M. Huhtinen, "Simulation of non-ionising energy loss and defect formation in silicon," Nuclear Instruments and Methods A, vol. 491, pp. 194–215, 2002.
16. http://www.srim.org
17. J. R. Bilinski, E. H. Brooks, U. Cocca, R. J. Maier, "Proton-neutron damage correlation in semiconductors," IEEE Transactions on Nuclear Science, vol. 10, pp. 20–26, 1963.
18. T. Angelescu, A. Vasilescu, "Comparative radiation hardness results obtained from various neutron sources and the NIEL problem," Nuclear Instruments and Methods A, vol. 374, pp. 85–90, 1996.
19. A. Chilingarov, D. Lipka, J. S. Meyer, T. Sloan, "Displacement energy for various ions in particle detector materials," Nuclear Instruments and Methods A, vol. 449, pp. 277–287, 2000.
20. A. Vasilescu, G. Lindström, "Displacement damage in silicon," online compilation, https://rd50.web.cern.ch/rd50/NIEL/default.html
21. S. Watts, J. Matheson, I. H. Hopkins-Bond, A. Holmes-Siedle, A. Mohammadzadeh, R. Pace, "A new model for generation-recombination in silicon depletion regions after neutron irradiation," IEEE Transactions on Nuclear Science, vol. 43(6), pp. 2587–2594, 1996.
22. A. Chilingarov, "Temperature dependence of the current generated in Si bulk," Journal of Instrumentation, JINST 8 P10003, 2013.
23. V. N. Brudnyi, S. N. Grinayaev, V. E. Stepanov, "Local neutrality conception: Fermi level pinning in defective semiconductors," Physica B, pp. 429–435, 1995.
24. A. Chilingarov, T. Sloan, "Operation of heavily irradiated silicon detectors under forward bias," Nuclear Instruments and Methods A, vol. 399, pp. 35–37, 1997.
25. L. J. Beattie, A. Chilingarov, T. Sloan, "Forward-bias operation of Si detectors: A way to work in high-radiation environment," Nuclear Instruments and Methods A, vol. 439, pp. 293–301, 2000.
26. G. Davies, S. Hayama, L Murin, R. Krause-Rehberg, V. Bondarenko, A. Sengupta, C. Da Vià, A. Karpenko, "Radiation damage in silicon exposed to high-energy protons," Physical Review B 73, 165–202, 2006.

27. G. Lindström, "Radiation damage in silicon detectors," Physical Review A, vol. 512, pp. 30–43, 2006.
28. G. Kramberger, on behalf of CERN RD50 collaboration, "Recent results from CERN RD50 collaboration," Physical Review A, vol. 583, pp. 49–57, 2007.
29. K. Borer, S. Janos, V. G. Palmieri, J. Buytaert, V. Chabaud, P. Chochula, P. Collins, H. Dijkstra, T. O. Niinikoski, C. Lourenço, C. Parkes, S. Saladino, T. Ruf, V. Granata, S. Pagano, F. Vitobello, W. Bell, P. Bartalini, O. Dormond, R. Frei, L. Casagrande, T. Bowcock, B. M.Barnett, C. Da Vià, I. Konorov, S. Paul, L. Schmitt, G. Ruggiero, I. Stavitski, A. Esposito, "Charge collection efficiency and resolution of an irradiated double-sided silicon microstrip detector operated at cryogenic temperatures," Physical Review A, vol. 440, pp. 17–37, 2000.
30. P. P. Allport, L. Andricek, C. M. Buttar, J. R. Carter, M. J Costa, L. M. Drage, T. Dubbs, M. J. Goodrick, A. Greenall, J. C. Hill, T. Jones, G. Moorhead, D. Morgan, V. O'Shea, P. W. Phillips, C. Rain, P. Riedler, D. Robinson, A. Saavedra, H. F.-W. Sadrozinski, J. Sánchez, N. A. Smith, S. Stapnes, S. Terada, Y. Unno, "A comparison of the performance of irradiated p-in-n and n-in-n silicon microstrip detectors read out with fast binary electronics," Physical Review A, vol. 450, pp. 297–306, 2000.
31. E. Verbitskaya, V. Eremin, Z. Li, J. Härkönen, M. Bruzzi, "Concept of double peak electric field distribution in the development of radiation hard silicon detector," Nuclear Instruments and Methods A, vol. 583, pp. 77–86, 2007.
32. Z. Li, E. Verbitskaya, G. Carini, W. Chen, V. Eremin, R. Gul, J. Härkönen, M. Li, "Space charge sign inversion and electric field reconstruction in 24GeV/c proton-irradiated MCZ Si p+-n(TD)-n+ detectors processed via thermal donor introduction," Nuclear Instruments and Methods A, vol. 598, pp. 416–421, 2009.
33. G. Kramberger, V. Cindro, I. Mandic, M. Mikuž, M. Zavrtanik, "Effective trapping time of electrons and holes in different silicon materials irradiated with neutrons, protons and pions," Nuclear Instruments and Methods A, vol. 481, pp. 297–305, 2002.
34. G. Kramberger, M. Batic, V. Cindro, I. Mandic, M. Mikuž, M. Zavrtanik, "Annealing studies of effective trapping times in silicon detectors," Nuclear Instruments and Methods A, vol. 571, pp. 608–611, 2007.
35. M. Mikuž, G. Kramberger, V. Cindro, I. Mandić, M. Zavrtanik, "Extreme radiation tolerant sensor technologies," to appear in Proceedings of Science (Vertex 2017), 2018.
36. C. Da Vià, S. J. Watts, "The geometrical dependence of radiation hardness in planar and 3D silicon detectors," Nuclear Instruments and Methods A, vol. 603, pp. 319–324, 2009.
37. C. Da Vià, E. Bolle, K. Einsweiler, M. Garcia Sciveres, J. Hasi, C. Kenney, V. Linhart, S. Parker, S. Pospisil, O. Rohne, T. Slavicek, S. Watts, N.Wermes, "3D active edge silicon sensors with different electrode configurations: Radiation hardness and noise performance," Nuclear Instruments and Methods A, vol. 604, pp. 505–511, 2009.
38. G. Kramberger, V. Cindro, I. Dolenc, E. Fretwurst, G. Lindström, I. Mandic, M. Mikuž, M. Zavrtanik, "Charge collection properties of heavily irradiated epitaxial silicon detectors," Nuclear Instruments and Methods A, vol. 554, pp. 212–219, 2005.
39. G. Kramberger, "Trapping in silicon detectors," Workshop on Defect Analysis in Silicon Detectors, Hamburg, Germany, August 2006. http://wwwiexp.desy.de/seminare/defect.analysis.workshop.august.2006.html
40. G. Casse, P. P. Allport, S. Marti, I Garcia, M. Lozano, P. R. Turner, "Performances of miniature microstrip detectors made on oxygen enriched p-type substrates after very high proton irradiation," Nuclear Instruments and Methods A, vol. 535, pp. 362–365, 2004.
41. P. Allport, G. Casse, M. Lozano, P. Sutcliffe, J. J. Velthuis, J. Vossebeld, "Performance of P-type micro-strip detectors after irradiation to 7.5×10^{15} p/cm^2," IEEE Transactions on Nuclear Science, vol. 52(5), pp. 1903–1906, 2005.
42. T. Rohe, D. Bortoletto, V. Chiochia, L. M. Cremaldi, S. Cucciarelli, A. Dorokhov, C. Hörmann, D. Kim, M. Konecki, D. Kotlinski, K. Prokofiev, C. Regenfus, D. A. Sanders, S. Son, T. Speer,

M. Swartz, "Fluence dependence of charge collection of irradiated pixel sensors," Nuclear Instruments and Methods A, vol. 552, pp. 232–238, 2005.
43. S. Marti i Garcia, P. P. Allport, G. Casse, A. Greenall, "A model of charge collection for irradiated p^+n detectors," Nuclear Instruments and Methods A, vol. 473, pp. 128–135, 2001.
44. F. Lemeilleur, S. J. Bates, A. Chilingarov, C. Furetta, M. Glaser, E. H. M. Heijne, P. Jarron, C. Leroy, C. Soave, I. Trigger (the CERN Detector R&D Collaboration RD2), "Study of characteristics of silicon detectors irradiated with 24 GeV/c protons between –20°C and +20°C," Nuclear Instruments and Methods A, vol. 360, pp. 438–444, 1995.
45. G. Lindström, E. Fretwurst, F. Hönniger, G. Kramberger, M. Möller Ivens, I. Pintilie, A. Schramm, "Radiation tolerance of epitaxial silicon detectors at very large proton fluencies, "Nuclear Instruments and Methods A, vol. 556, pp. 451–458, 2006.
46. A. Macchiolo, R. Nisius, N. Savic, S. Terzo, "Development of n-in-p pixel modules for the ATLAS upgrade at HL-LHC," Nuclear Instruments and Methods A, vol. 831, pp. 111–115, 2016.
47. J. Albert, et al. (The ATLAS IBL Collaboration), "Prototype ATLAS IBL modules using the FE-I4A front-end readout chip," Journal of Instrumentation, JINST 7, P11010, 2012.
48. Y. Unno, C. Gallrapp, R. Hori, J. Idarraga, S. Mitsui, R. Nagai, T. Kishida, A. Ishida, M. Ishihara, S. Kamada, T. Inuzuka, K. Yamamura, K. Hara, Y. Ikegami, O. Jinnouchi, A. Lounis, Y. Takahashi, Y. Takubo, S. Terada, K. Hanag, N. Kimura, K. Nagaij, I. Nakano, R. Takashima, J. Tojo, K. Yorita, "Development of novel n^+-in-p Silicon Planar Pixel Sensors for HL-LHC," Nuclear Instruments and Methods A, vol. 699, pp.72–77, 2013.
49. S. Terzo, L. Andricek, A. Macchiolo, H. G. Moser, R. Nisius, R. H. Richter, P. Weigell, "Heavily irradiated N-in-p thin planar pixel sensors with and without active edges," Journal of Instrumentation, JINST 9, C05023, 2014.
50. K. Kimura, D. Yamaguchi, K. Motohashi, K. Nakamura, Y. Unno, O. Jinnouchi, S. Altenheiner, A. Blue, M. Bomben, A. Butter, A. Cervelli, S. Crawley, A. Ducourthial, A. Gisen, M. Hagihara, K. Hanagaki, K. Hara, M. Hirose, Y. Homma, Y. Ikegami, S. Kamada T. Kono, A. Macchiolo, G. Marchiori, F. Meloni, M. Milovanovic, A. Morton, G. Mullier, F. J. Munoz, C. Nellist, B. Paschen, A. Quadt, T. Rashid, J. Rieger, A. Rummler, K. Sato, K. Sato, N. Savic, H. Sawai, K. Sexton, M. E. Stramaglia, M. Swiatlowski, R. Takashima, Y. Takubo, S. Terzor, K. Todome, J. Tojo, K. Van Houten, J. Weingarten, S. Wonsak, K. Wraight, K. Yamamura, "Test beam evaluation of newly developed n-in-p planar pixel sensors for use in a high radiation environment," Nuclear Instruments and Methods A, vol. 831, pp. 140–146, 2016.
51. M. Garcia Sciveres, RD53A Integrated Circuit Specifications, CERN-RD53-PUB-15-001 https://cds.cern.ch/record/2113263?ln=en
52. I. Mandić, V. Cindro, G. Kramberger, M. Mikuž, "Measurement of anomalously high charge collection efficiency in n^+p strip detectors irradiated by up to 10^{16} n_{eq}/cm^2," Nuclear Instruments and Methods A, vol. 603, pp. 263–267, 2009.
53. M. Mikuž, V. Cindro, G. Kramberger, I. Mandić, M. Zavrtanik, "Study of anomalous charge collection efficiency in heavily irradiated silicon strip detectors," Nucl. Instrum. Methods A, vol. 636, pp. S50–S55, 2011.
54. J. Lange, J. Becker, E. Fretwurst, R. Klanner, G. Lindström, "Properties of a radiation-induced charge multiplication region in epitaxial silicon diodes," Nuclear Instruments and Methods A, vol. 622, pp. 49–58, 2010.
55. G. Casse, A. Affolder, P. P. Allport, H. Brown, M. Wormald, "Enhanced efficiency of segmented silicon detectors of different thicknesses after proton irradiations up to 1×10^{16} n_{eq} cm^{-2}," Nuclear Instruments and Methods A, vol. 624, pp. 401–404, 2010.
56. G. Casse, A. Affolder, P. P. Allport, H. Brown, I. McLeod, M. Wormald, "Evidence of enhanced signal response at high bias voltages in planar silicon detectors irradiated up to 2.2×10^{16} n_{eq} cm^{-2}," Nuclear Instruments and Methods A, vol. 636, pp. S56–S61, 2011.

57. G. Kramberger, V. Cindro, I. Mandic, M. Mikuž, M. Milovanovic, M. Zavrtanik, K. Zagar, "Investigation of Irradiated Silicon Detectors by Edge-TCT," IEEE Transactions on Nuclear Science, vol. 57(4), pp. 2294–2302, 2010.
58. V. Eremin, E. Verbitskaya, A. Zabrodskii, Z. Li, J. Härkönen, "Avalanche effect in Si heavily irradiated detectors: Physical model and perspectives for application," Nuclear Instruments and Methods A, vol. 658, pp. 145–151, 2011.
59. E. Verbitskaya, V. Eremin, A. Zabrodskii, Z. Li, P. Luukka, "Restriction on the gain in collected charge due to carrier avalanche multiplication in heavily irradiated Si strip detectors," Nuclear Instruments and Methods A, vol. 730, pp. 66–72, 2013.
60. G. Casse, D. Forshaw, T. Huse, I. Tsurin, M. Wormald, M. Lozano, G. Pellegrini, "Charge multiplication in irradiated segmented silicon detectors with special strip processing," Nuclear Instruments and Methods A, vol. 699, pp. 9–13, 2013.
61. G. Kramberger, V. Cindro, I. Mandić, M. Mikuž, M. Zavrtanik, "Charge collection studies on custom silicon detectors irradiated up to 1.6×10^{17} n_{eq}/cm^2," Journal of Instrumentation, JINST 8, P08004, 2013.
62. R. Richter, L. Andricek, T. Gebhart, D. Hauff, J. Kemmer, G. Lutz, R. Weiβ, A. Rolf, "Strip detector design for ATLAS and HERA-B using two-dimensional device simulation," Nuclear Instruments and Methods A, vol. 377, pp. 412–421, 1996.
63. C. Piemonte, "Device simulations of isolation techniques for silicon microstrip detectors made on p-type substrates," IEEE Transactions on Nuclear Science, vol. 53(3), pp. 1694–1705, 2006.
64. Z. Li, E. Verbitskaya, V. Eremin, B. Dezillie, W. Chen, M. Bruzzi, "Radiation hard detectors from silicon enriched with both oxygen and thermal donors: improvements in donor removal and long-term stability with regard to neutron irradiation," Nuclear Instruments and Methods A, vol. 476, pp. 628–638, 2002.
65. G. Kramberger, D. Contarato, E. Fretwurst, F. Hönniger, G. Lindström, I. Pintilie, R. Röder, A. Schramm, J. Stahl, "Superior radiation tolerance of thin epitaxial silicon detectors," Nuclear Instruments and Methods A, vol. 515, pp. 665–670, 2003.
66. J. Härkönen, E. Tuovinen, P. Luukka, E. Tuominen, K. Lassila-Perini, P. Mehtälä, S. Nummela, J. Nysten, A. Zibellini, Z. Li, E. Fretwurst, G. Lindström, J. Stahl, F. Hönniger, V. Eremin, A. Ivanov, E. Verbitskaya, P. Heikkilä, V. Ovchinnikov, M. Yli-Koski, P. Laitinen, A. Pirojenko, I. Riihimäki A. Virtanen, "Radiation hardness of Czochralski silicon, float zone silicon and oxygenated float zone silicon studied by low energy protons," Nuclear Instruments and Methods A, vol. 518, pp. 346–348, 2004.
67. P. J. Sellin, J. Vaitkus, "New materials for radiation hard semiconductor detectors," Nuclear Instruments and Methods A, vol. 557, pp. 479–489, 2005.
68. F. Bachmair, "Diamond sensors for future high energy experiments," Nuclear Instruments and Methods A, vol. 831, pp. 370–377, 2016.
69. F. Bachmair, L. Bäni, P. Bergonzo, B. Caylar, G. Forcolin, I. Haughton, D. Hits, H. Kagan, R. Kass, L. Li, A. Oh, S. Phan, M. Pomorski, D. S. Smith, V. Tyzhnevyi, R. Wallny, D. Whiteheade, "A 3D diamond detector for particle tracking," Nuclear Instruments and Methods A, vol. 786, pp. 97–104, 2015.
70. F. Hügging, C. Gößling, J. Klaiber-Lodewigs, J. Wüstenfeld, R. Wunstorf, "Prototype performance and design of the ATLAS pixel sensors," Nuclear Instruments and Methods A, vol. 465, pp. 77–82, 2001.
71. A. Affolder, "Latest developments in planar n-on-p sensors," Proceedings of Science (Vertex 2011) 031, 2012.
72. V. G. Palmieri, K. Borer, S. Janos, C. Da Vià, L. Casagrande, "Evidence for charge collection efficiency recovery in heavily irradiated silicon detectors operated at cryogenic temperatures," Nuclear Instruments and Methods A, vol. 413, pp. 475–478, 1998.

73. K. Borer, S., Janos, V. G. Palmieri, B. Dezillie, Z. Li, P. Collins, T. O. Niinikoski, C. Lourenço, P. Sonderegger, E. Borchi, M. Bruzzi, S. Pirollo, V. Granata, S. Pagano, S. Chapuy, Z. Dimcovski, E. Grigoriev, W. Bell, S. R. H. Devine, V. O'Shea, K. Smith, P. Berglund, W. de Boer, F. Hauler, S. Heising, L. Jungermann, L. Casagrande, V. Cindro, M. Mikuž, M. Zavartanik, C. Da Vià, A. Esposito, I. Konorov, S. Paul, L. Schmitt, S. Buontempo, N. D'Ambrosio, S. Pagano, G. Ruggiero, V. Eremin, E. Verbitskaya, "Charge collection efficiency of irradiated silicon detector operated at cryogenic temperatures," Nuclear Instruments and Methods A, vol. 440, pp. 5–16, 2000.
74. T. O. Niinikoski et al. (the CERN RD39 Collaboration), "Low-temperature tracking detectors," Nuclear Instruments and Methods A, vol. 520, pp. 87–92, 2004.

CHAPTER 4

3D Sensors

4.1 BASIC CONCEPT

3D sensors were proposed by Sherwood Parker and collaborators in 1997 [1]. Unlike planar sensors, where electrodes are implanted on the wafer's top and bottom surfaces (see Fig. 4.1a), in 3D an array of columnar (or trench-shaped) electrodes of both doping types are arranged in adjacent cells, penetrating partially or entirely through a high-resistivity silicon substrate, perpendicularly to the wafer surface (see Fig. 4.1b). Electric field lines, which are parallel to the wafer's surface, begin at one electrode type (e.g., n) and end at the closest electrode of the opposite type (p). Similarly to standard planar sensors, the electric field strength is controlled by an applied reverse-bias voltage. Various connection layers at the surface processed by diffusion, polysilicon or metal, can be used to join electrodes together to form either single-column or multicolumn arrangements, as the ones shown in Figs. 4.1c. In this way, several types of detectors can be obtained using the same electrode layout, such as pixels, strips, and pads.

The 3D configuration offers many advantages over the planar configuration:

1. Full depletion voltage (V_{depl}). On the one hand, in planar sensors, the depletion region grows vertically, as we said previously, perpendicularly to the wafer's surfaces, and the full depletion voltage depends on the substrate thickness (Δ), as can be seen in Fig. 4.1a. On the other hand, in 3D sensors, the depletion region grows laterally between the columnar electrodes, whose distance (L) can be decoupled from Δ to a large extent[1]: in particular, L can be made much smaller than Δ by design, so that the full depletion voltage can be dramatically reduced compared to the planar one of the same thickness. This property is extremely important in applications in which full depletion of planar sensors can be difficult to achieve due to junction breakdown and/or thermal runaway. This is particularly the case for sensors with thick substrates, which are interesting for X- and γ-ray detection, and detectors exposed to large radiation fluences, as the ones used in the innermost tracking layers

[1] This assumption neglects the constraints due to the maximum aspect ratio (i.e., depth/width) achievable for columnar electrodes, which are discussed in Chapter 5.

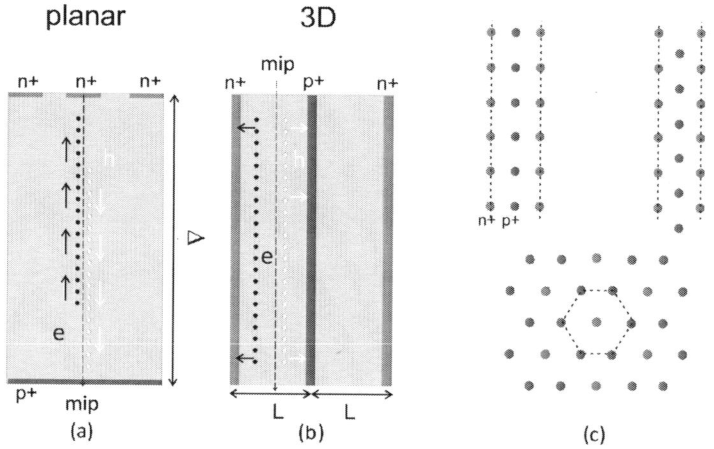

FIGURE 4.1 Schematic cross sections of (a) planar and (b) 3D detectors, emphasizing the decoupling of active thickness (Δ) and collection distance (L) in 3D detectors; (c) sketch of different column arrangements for strips and pixels.

of particle physics experiments at accelerators. In the latter case, since the radiation-induced leakage currents can be very high, the related power dissipation reduction can be crucial to ease the realization of complex systems, in terms of power supply as well as cabling and cooling needs. More information on this subject can be found in Chapters 3 and 6.

2. Charge collection efficiency (CCE). A minimum ionizing particle (MIP), traversing a detector, produces a uniform electron-hole (e-h) pair distribution along its track, as shown in Figs. 4.1a and 4.1b. The total amount of generated charge only depends on the substrate thickness, since silicon generates about 76e− per micron traversed by a MIP, so it is the same for both detector types. However, the charge collection distance is much shorter in 3D, and high electric fields, as well as carrier velocity saturation, which depend on the voltage applied per unit distance, can be achieved already at low bias voltages, so that the charge collection times can be much shorter. These are usually of the order of a few nanoseconds in 3D, while they can be as much as 10 ns for planar with a standard 300 micron thickness. Using Ramo's theorem, one can estimate the amount of generated current from the carrier velocity and the sensor's weighting field [2]. In planar detectors, each charge carrier is generated at different distances from the collecting electrode, where the weighting field peaks; thus, the signal waveform is broader in time. This effect is strongly attenuated in 3D detectors, as can be seen in Fig. 4.1b, where all charges along the ionization track are generated within a much shorter distance from the electrodes throughout the bulk and therefore induce very fast signals [3]. Besides being very useful for applications requiring good time resolutions, this property can effectively counteract radiation-induced charge loss from trapping [4].

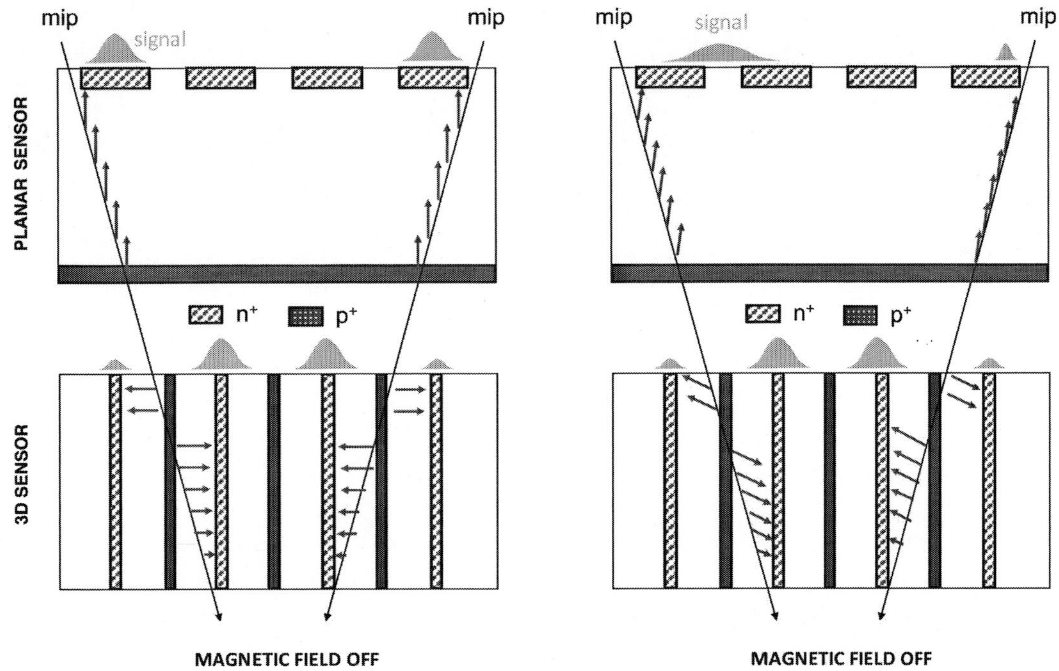

FIGURE 4.2 Effect of magnetic field on planar (top) and 3D (bottom) pixel sensors stressing the higher signal focal property for 3Ds.

3. Electric field shape. The unique structure of 3D detectors, where electric field lines extend radially from each cylindrical electrode throughout its length (see Fig. 4.4c), induces a self-shielding effect in each unitary cell, which is beneficial in several respects. As an example, charge sharing between adjacent cells is considerably reduced compared to planar sensors with small pixel sizes. Moreover, the sensitivity of 3D detectors to magnetic fields, which are always used in High Energy Physics (HEP) experiments to identify the particles charge, is also lower than that in planar sensors, as sketched in Fig. 4.2. In planar sensors, due to the Lorentz force, the charge carrier trajectories bending induces signals in adjacent pixels (Fig. 4.2, top right), while in 3Ds signals are mostly confined to one pixel (Fig. 4.2, bottom right) [5]. This effect can be important when the signal is reduced due to radiation since sharing might imply a charge height below the electronic threshold and consequent information loss. With enough charge, however, charge sharing could be used to improve the spatial resolution.

Beyond the technological complications, which will be described in detail in Chapter 5, 3D sensors also present some functional disadvantages when compared to planar sensors:

1. Signal responses to particles are not spatially uniform due to the presence of very low field regions within the active volume, which are due to (i) the null points in between electrodes of the same doping type, and (ii) the electrodes themselves. As a result,

charge carriers generated in these regions have to diffuse until they reach a region with a sufficiently high electric field, therefore delaying the overall signal response and possibly lowering the charge collection efficiency, especially after irradiation, if the delay is longer than the electronics shaping time;

2. Small interelectrode spacing and the electrodes extension throughout the silicon substrate cause the sensor capacitance to be quite high and a possible cause of degraded noise performance at short shaping times. A higher capacitance would also degrade the signal rise time and preclude the use of 3D sensors in modern, very-fast detectors, unless the electronics is properly designed.

One important characteristic of the 3D design is the possibility of processing so-called active edges. In planar sensors, the active volume, in which a particle impinging it could actually be seen, is normally kept far away from the scribe line, which defines the cutting marks that will be used to separate the sensor from the wafer. In this way, the bulge of the electric field does not reach defects like cracks, chipped regions, and the like, generated at the edge by the saw cut, as can be seen in Fig. 4.3. If the electric field's bulge would reach this region, which is highly conductive due to the mechanically generated atomic broken (dangling) bonds, a current path would form and generate high leakage currents at the readout input. To prevent this from happening, most planar sensors' designs allow for additional space at the surface, in the region close to the edge, to host one or more current protective "guard rings."

Guard rings can be designed in different configurations:

a. A main guard ring, which is to be kept at the same potential of the charge-collecting electrodes, so that a uniform electric field configuration at the outermost electrodes can be ensured and guarantee sinking of leakage currents coming from the periphery.

b. Several floating guard rings, aimed at evenly distributing the lateral voltage drop and at enhancing the breakdown bias voltage and, therefore, the long-term stability of the device (for more details on floating guard rings see [6]).

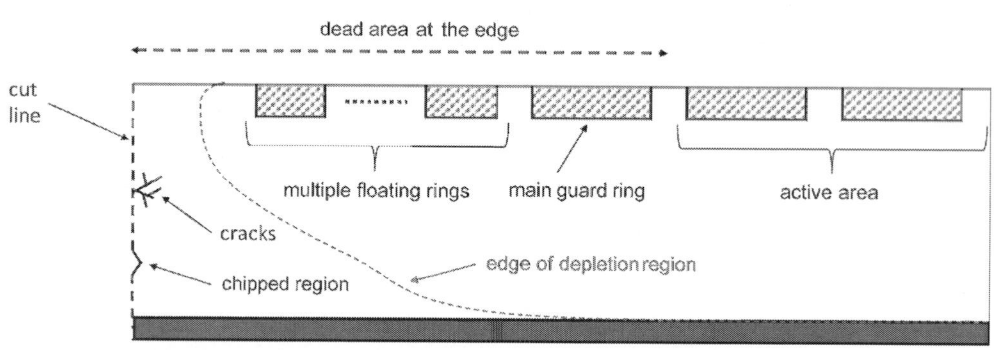

FIGURE 4.3 Schematic cross section of the edge region in a standard planar detector showing different guard rings and a wide dead area at the edge.

As a result of the presence of guard rings in planar devices, a dead region exists at the sensor's edges, which can extend for a few hundred micrometers and sometimes can reach about 1 mm in some designs.

In 3D sensors, heavily doped trenches, having the same characteristics of the ohmic columnar electrodes and surrounding all the physical perimeter of the device, can be used to terminate both physically and electrically the active volume without dicing. At the same time, this solution eliminates the dead region at the edges and any unwanted leakage current generated by the saw cut. This idea, which was included in the original 3D paper, was further developed in the study of the so-called wall electrodes [7] and has shown, as expected, that the insensitive edge region width can be reduced to the width of the trench, which is commonly just a few micrometers. The same approach can also be used to produce "planar active-edge" sensors, which would have collection electrodes processed on the surface using a standard planar design, but with scribe lines replaced by doped trenches, which are effectively an extension of the ohmic electrode on the sides [8]. In this way, the active volume is effectively "encapsulated" inside a doped implantation on five sides (back and four lateral edges) and has the opposite doping segmented implantations on the remaining "top" side for readout. Because these fully sensitive planar devices are gaining increasing importance in many imaging applications, planar sensors with active (or at least slim) edges will be extensively discussed in Chapter 8.

4.2 DEVICE SIMULATIONS

Device simulation was key in gaining a deep understanding of the behavior of 3D sensors before fabrication and tests of the first prototypes. Several commercial software packages are now available; the two most commonly used packages in the sensors' design community are SILVACO and SYNOPSYS [9].

The first simulations of the static and dynamic characteristics of 3D sensors were reported in [1] in which the unit cell was designed with 25 μm spacing between same-type and opposite-type columns, as can be seen in Fig. 4.4a. Since the cell is fully symmetric, it is possible to reduce the simulation domain to a quarter of the unit cell and therefore limit the overall computational time. Moreover, symmetry within a cell is also present in depth, throughout the silicon volume, owing to the three-dimensional structure of the cylindrical electrodes. Therefore, it is possible to accurately extract the detector properties of the internal region, which is normally not affected by surface effects, by two-dimensional simulations along a plane of unit thickness parallel to the wafer's surface. In the case of the sensors discussed above, with p-type substrates and doping concentration $N_A=10^{12}$ cm^{-3}, simulations predict a full depletion voltage of just 1.6 V (including the contribution from the built-in voltage, which is normally about 0.3 V). It is also possible to predict the change in bias voltage after the sensor is used in a defined environment. For example, the full depletion voltage increases to 8.8 V when the effective doping concentration reaches $N_A=10^{13}$ cm^{-3}, which represents the value expected after 10 years of operation of pixel sensors in LHC experiments due to radiation damage. Figure 4.4b also shows the equipotential lines in a quarter-cell with $N_A=10^{12}$ cm^{-3} and a reverse bias (V_{rev}) of 5V, while Fig 4.4c shows the corresponding carrier drift lines.

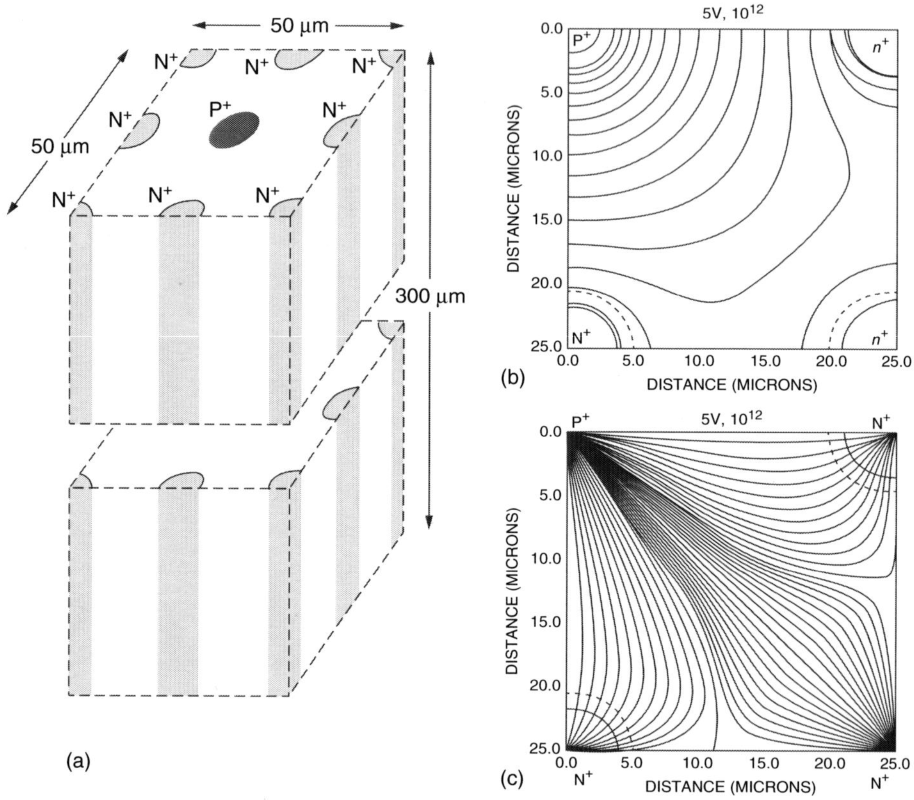

FIGURE 4.4 (a) 3D view of a unit cell in a 3D detector; (b) equipotential lines for one-quarter of the unit cell with 10^{12} cm^{-3} p-type substrate doping concentration and 5-V reverse-bias voltage; (c) drift lines in the same condition as in (b). (From Parker et al., Nucl. Instrum. Methods A, 395, 1997. With permission of ELSEVIER.) [1]

Low-field regions can be clearly seen in both figures in between two electrodes of the same type. The same regions are also visible in the simulation of Fig. 4.5, which shows electric field profiles at different V_{rev} along two lines connecting the electrodes. In more detail, looking at Fig. 4.5a, which simulates the field along a line from the p^+ to the adjacent n^+ electrode, we see that the electric field magnitude is high enough to induce fast carrier drifts, while the field peaks at the electrode edges are still safely below those critical for breakdown in all bias conditions. On the other hand, in Fig. 4.5b, which shows the field along a line connecting two adjacent n^+ electrodes, it is possible to observe an almost-zero field region, about 2–3 μm wide, at the midpoint between the two electrodes. These electric field nonuniformities are the main cause of both carrier drift time and signal shape being strongly dependent on the particle impact position. Looking again at the cell represented in Fig. 4.4a, we observe that the predicted collection time of electrons and holes generated at the cell center is ~1 ns, while a longer time, which is up to ~6 ns for holes, is predicted if a particle impinges at the "null" point between electrodes [1].

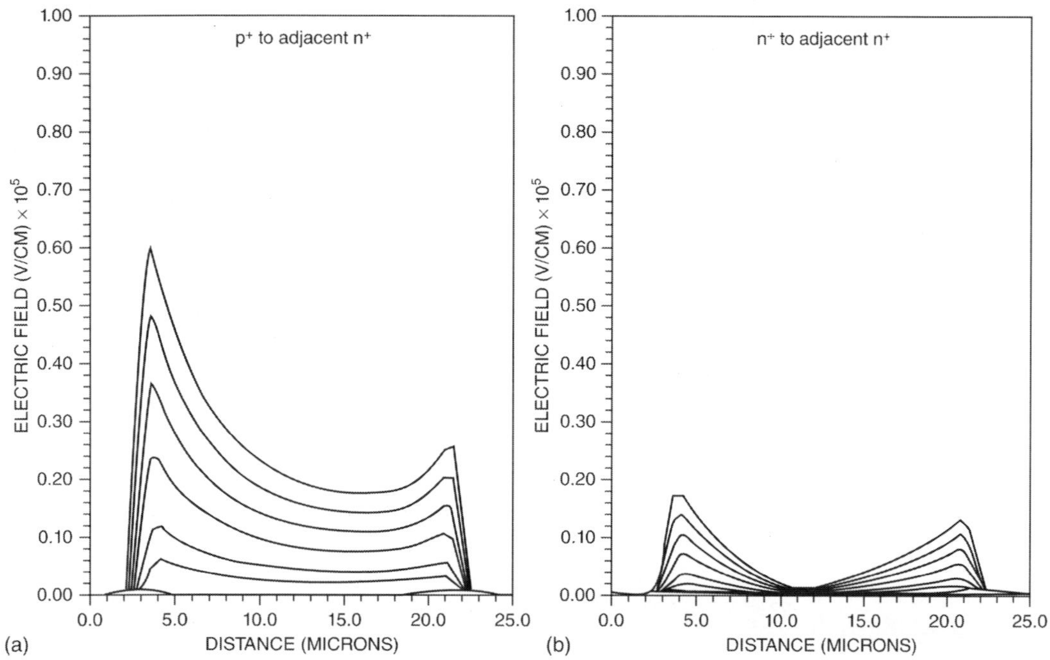

FIGURE 4.5 Electric field profiles for the quarter cell of Fig. 4.4a with 10^{12} cm^{-3} p-type substrate doping concentration and reverse bias voltages of 50, 40, 30, 20, 10, 5, and 0 V (curves from top to bottom), along lines from (a) the p+ electrode to the adjacent n+ electrode and (b) the n+ electrode to the adjacent n+ electrode. (From Parker et al., Nucl. Instrum. Methods A, 395, 1997. With permission of ELSEVIER.) [1]

These position-dependent collection times reflect into markedly different signal shapes, as highlighted in Fig. 4.6, which compares simulations of induced current pulses on collecting electrodes from particles hitting perpendicularly to the surface in two different positions. However, these different signal shapes do not cause too many problems since, even in the worst-case, they are much faster than in planar sensors, even with much lower peak fields.

Device simulations have also proven very effective in predicting the resistance–capacitance (RC) product time constants of the electrodes to be in the order of 100–200 ps for 300 μm thick sensors [1]. This value is usually small enough for most applications. Finally, it should be mentioned that close to the top and bottom surfaces of 3D sensors, both static and dynamic signal characteristics deviate from those observed in the bulk. This is due to the presence of oxide fixed charge, as already mentioned in chapter 3, results in an electron inversion layer at the silicon–oxide interface, which would short all n^+ electrodes. The surface isolation structures, usually "p-spray" or "p-stops" or a combination of the two [10]), that would need to be used to prevent the n^+ columnar electrodes from being resistively connected, would affect the electrical characteristics (namely, breakdown voltage, leakage current, and capacitance), and would also cause distortions of the equipotential lines and electric field profiles, therefore affecting the charge collection properties. To accurately predict these effects, very time-consuming 3d simulations of the entire volume are generally necessary.

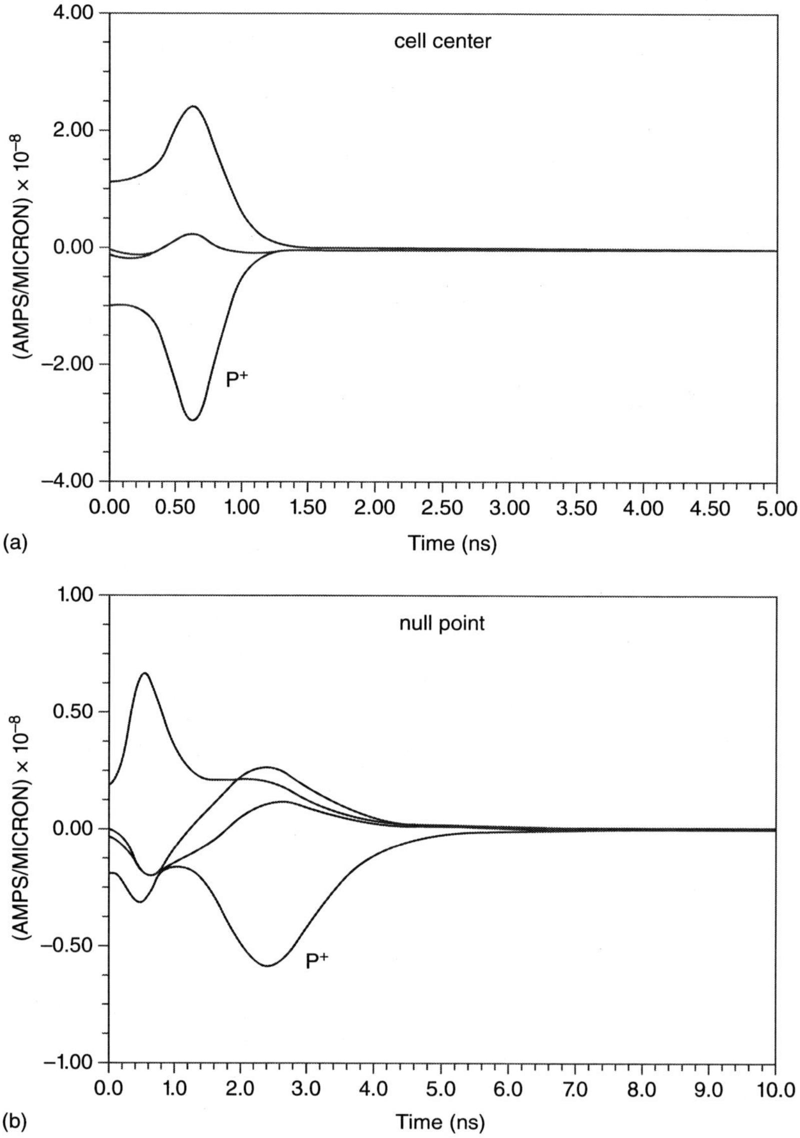

FIGURE 4.6 Current pulses on the electrodes in response to an MIP with a track parallel to the electrodes and passing (a) through the cell center and (b) through the null point in between two n+ electrodes ($N_A = 10^{12}$ cm^{-3}, $V_{rev} = 10$ V). (From Parker et al., Nucl. Instrum. Methods A, 395, 1997. With permission of ELSEVIER.) [1]

4.3 EXPERIMENTAL RESULTS

Test results of the electrical characteristics and response to infrared (IR) light pulses of the first 3D detector prototypes performed on 121 μm thick p-type wafers were reported in [11]. These results included leakage current densities of the order of 1nA per mm^3 and breakdown voltages larger than 60 V. These results should be compared to the full depletion voltages of 5 and 8 V measured for the 100 μm and 200 μm electrodes pitch, respectively. The first charge collection signals in response to X-rays and β electrons are reported in [12] where the full width at half maximum (FWHM) energy resolution at the Mn K line of a ^{55}Fe source was measured to be 652 eV. This value is well explained by the combined effect of the detector capacitance and leakage current on the electronic noise at the considered shaping time of 1 μs. Further measurements confirmed the low charge sharing between adjacent cells predicted by simulations. The effective detection of beta particles using a ^{106}Ru source and fast electronics was also demonstrated at the same time.

Later on, 3D pixel sensors with different electrode configurations were bump bonded to the ATLAS FEI3 front-end chip [13] and were measured in several high-energy pion beams at the CERN Super Proton Synchrotron (SPS) [5, 14, 15, 16, 17]. The spatial resolution with binary readout corresponded to the theoretical predictions of pixel pitch divided by $\sqrt{12}$ in the case of a cell size of $50 \times 400\ \mu m^2$. In the case of multi-hit clusters, the use of charge interpolation algorithms allowed an improvement of the space resolutions up to ~11 μm [17].

On the one hand, the hit efficiency of minimum ionizing particles was measured to be ~95% for orthogonal incidence. The 5% lack of efficiency is due the electrodes, which only respond partially in their interior and had a relatively wide diameter of ~17 μm. On the other hand, hit efficiency reaches almost 100% in the case of inclined tracks. When particles impinge at an angle, the cluster sizes increase due to the larger charge sharing between adjacent pixels.

The inefficiency of electrodes was studied in detail in various test beams, and an example of such studies can be seen in Fig 4.7, which compares measured distributions of the cluster charge in two cases: (1) with tracks orthogonal to the detector surface (Fig. 4.7a) and (2) with tracks impinging with an inclination of 10° (Fig. 4.7b). The two figures show histograms of tracks passing inside and outside the region occupied by electrodes. While the inclination angle has a minor effect for tracks hitting far from the electrodes, for which Landau-like distributions can be observed, it strongly impacts the results in case of tracks

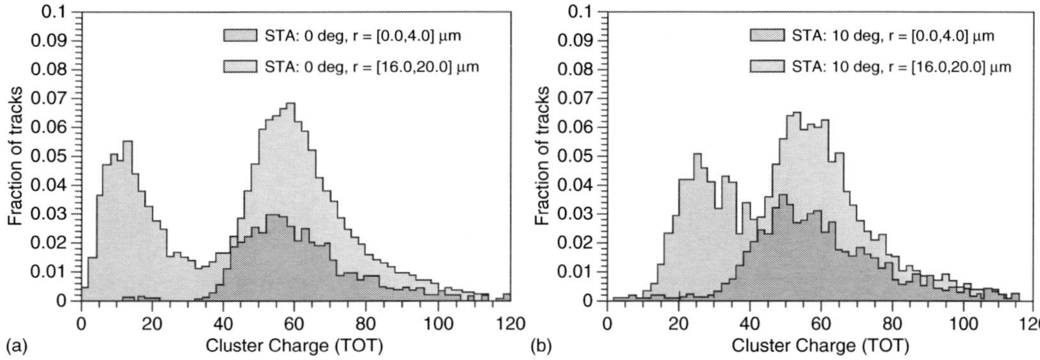

FIGURE 4.7 Measured distributions of cluster charge (expressed in units of time over-threshold—ToT) for pion tracks going inside (yellow histograms) and outside (gray histograms) the electrodes under different inclination angles: (a) 0° and (b) 10°. (From Grenier et al., Nucl. Instrum. Methods A, 638, 2011. With permission of Elsevier.) [5]

going inside the electrodes, due to the limited energy resolution of the FEI3 chip. As can be seen in Figure 4.7a, when particles impinge the surface orthogonally, the distribution is not ideal; therefore, a wide fraction of signals due to the electrode inefficiency populates the lower energy side of the distribution on the left. A smaller number of events following a Landau-like distribution are also present with a higher energy and with the same most probable value as the one observed for tracks hitting far from the electrodes. From Figure 4.7b, however, it can be seen that the high-charge part of the histogram is not significantly changed when particles impinge with a 10° incidence angle, while the low-charge part of the distribution is shifted to larger values as proof of the increased number of particles releasing charge outside the electrodes.

The full 3D with active-edge sensor technology was transferred to the Norwegian Foundation for Industrial and Technical Research (SINTEF) from the Stanford Nanofabrication Facility (SNF) within the 3D Consortium (which was formed in 2006 by Manchester, Stanford with SINTEF and Oslo, and afterward Prague and Purdue universities) and was afterward successfully fabricated, reaching a staggering 75% yield on 6-inch wafers with several large devices in 2018. The first batch of prototypes, however, was made much earlier on 250 μm thick n-type substrates with resistivity of 2 kΩ cm, and they included pixel sensors compatible with the ATLAS FEI3 readout chip with 2E, 3E, and 4E electrode configurations and several test structures [18]. The second batch included 200 μm and 285 μm thick p-type substrates with resistivity of 10 kΩ cm, and additional pixel layouts, among which were 2E and 4E CMS devices, compatible with the PSI46 readout chip [19]. The measured devices showed leakage current of about 200 pA/pixel and breakdown voltages in the range of 80–120 V. Selected CMS detectors were studied in a 120 GeV beam test at the Fermi National Accelerator Laboratory (FNAL). The measured collected charge at full depletion was ~24 ke, a value that was very close to expectations [20].

4.4 ALTERNATIVE 3D DESIGNS

In the early 2000s, a few research institutes started developing modified 3D architectures with simplified fabrication technologies. Some of these alternative structures are schematically summarized in Fig. 4.8 and are reviewed in more detail in the following subsections.

4.4.1 Single-Type-Column 3D Detectors

The first attempt to produce a simplified 3D design was the one sketched in Fig. 4.8a with just one type of columnar electrodes partially passing the wafer's substrate. This design included a uniform ohmic contact on the backside and was independently proposed by Fondazione Bruno Kessler (FBK, Trento, Italy), with the name "single-type column" (3D-STC) [21], and by the Technical Research Centre of Finland (VTT, Helsinki, Finland), with the name of "Semi 3D" [22]. While the 3D-STC/semi-3D design is relatively simple from the technological point of view, its performance is not as good as that of the full 3D, a fact that was accurately predicted by Technology Computer Aided Design (TCAD) simulations [21].

An example is shown in Fig. 4.9, which schematically describes the lateral depletion mechanism between columns (a) and the vertical depletion toward the backside electrode of 3D-STC sensors in a planar-like fashion (b) [23]. This trend is well reflected in the C-V

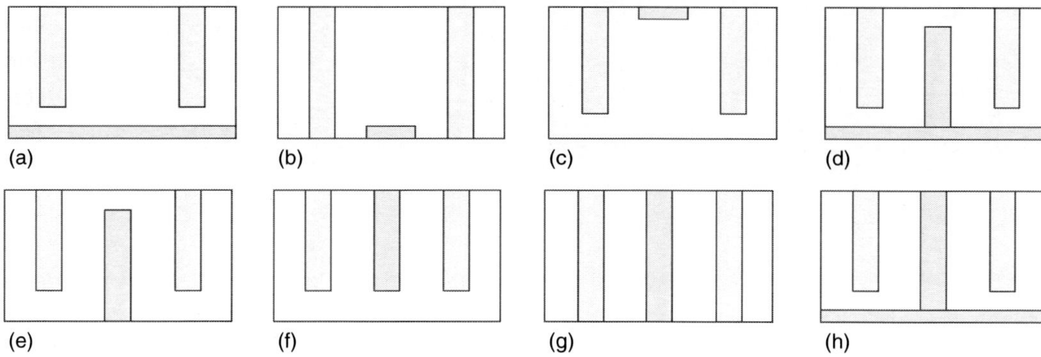

FIGURE 4.8 Schematic cross sections describing the modified 3D detector architectures so far reported: (a) single-type column 3D detectors, also called semi-3D detectors, with backside ohmic contact, independently proposed by FBK and VTT; (b) an alternate version of (a) with passing-through column, proposed by FBK; (c) single-type column 3D detectors with front-side ohmic contact, proposed by BNL/CNM; (d) and (e) double-sided, double-type column detectors with slightly different back-side configuration, independently proposed by FBK and CNM, respectively; (f) single-sided, double-type column detector proposed by BNL; (g) and (h) alternate versions of double-sided, double-type column detectors proposed by FBK featuring passing-through columns.

measurements shown in Fig. 4.9d, where the two depletion contributions are clearly visible in the two flat regions at low bias voltages (lateral depletion) and after about 15 V, which defines the full depletion of the p-n junction.

Since all columns have the same doping type, once the lateral depletion is reached, the electric field between them cannot be further increased as it depends only on the substrate doping concentration. As a result, low-field regions exist midway between two columns [21]. Signals induced by an MIP hitting a 3D-STC structure at different impact points from near a column, toward the middle of the cell between columns (indicated with coordinates and red, blue and

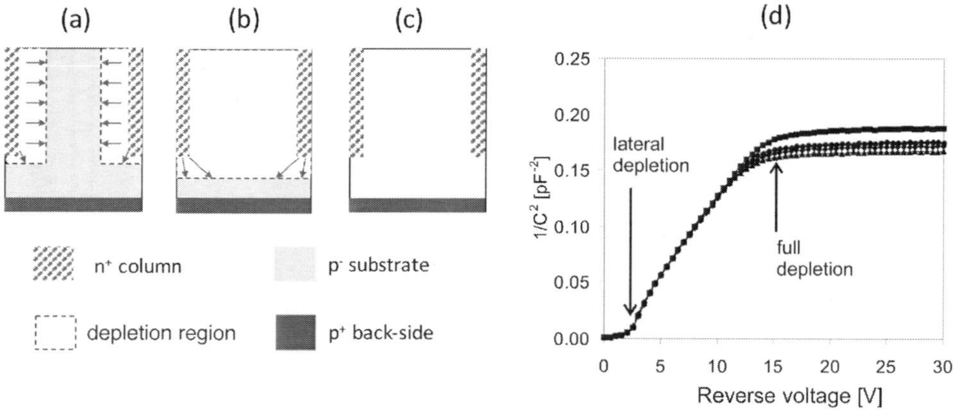

FIGURE 4.9 Cross section of a region between two columns with different depletion conditions in a 3D-STC detector: (a) initial lateral depletion, (b) depletion toward the backplane, (c) full depletion. (d) Extraction of the lateral depletion voltage and full depletion voltage from $1/C^2$ plots.

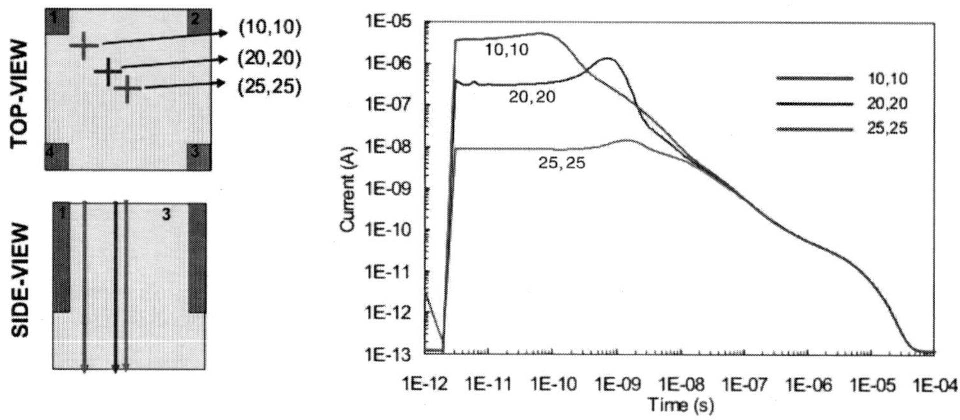

FIGURE 4.10 Simulation of the current signal induced on Electrode-1 by a uniform (MIP-like) charge deposition along three different tracks. A sketch of the simulated cell showing the track positions is also shown. (From Dalla Betta et al., Vertex 2007, Paper 23, 2008.45. Copyright owned by the authors under the terms of the Creative Commons Attribution-NonCommercial-ShareAlike Licence. http://creativecommons.org/licenses/by-nc-sa/3.0/.) [25]

green crosses in the sketch), were simulated, and the position-dependent induced currents can be seen in Fig. 4.10 [21, 24, 25]. All induced currents (indicated with corresponding coordinates and red, blue, and green colors) show a fast rise of the signal lasting a few ns due to the lateral drift of electrons and holes toward the nearest electrode and toward the center of the 3D cell, respectively. Also, the collection times of electrons strongly depend on the particle impact position, as can be seen from the different currents' peak times in the three considered cases. Then, the induced signals have long tails lasting several microseconds due to the slow diffusion of holes toward the back plane. This part of the signals does not depend on the impact position. Holes, which are attracted to the backside, start drifting only in the region below the column tips, where the vertical electric field is greater than zero. Because of this slow signal component, ballistic deficit effects, where signal is lost when its formation is slower than the electronic readout time, can occur when fast electronics are used for the readout. This also limits the radiation hardness of 3D-STC detectors.

Further details about the layout and the electrical characterization of 3D-STC detectors can be found in [23, 26], while details of the fabrication process are addressed in Chapter 5. Here we summarize the main results of the functional characterizations, which were performed using different measurement techniques. For example, transient current technique (TCT) measurements were performed at JSI (Ljubljana, Slovenia) [27] with a system based on a 1060 nm pulsed laser, focused to a few micrometers on 3D microstrip sensors, connected to fast amplifiers and visualized by an oscilloscope. This technique is very effective in the study of the signals induced by the infrared photons on readout and adjacent channels (in this case, aluminum strips connecting a row of adjacent columns).

Figure 4.11a shows the induced signal when the laser beam is focused near the readout electrode (black dot in the central electrode in the insert). As can be seen, the current pulse has a fast component of a few ns and a long tail, which lasts a few microseconds, as

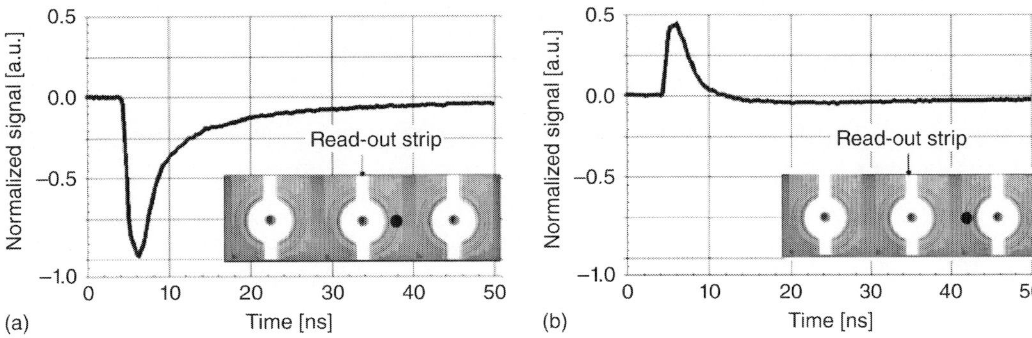

FIGURE 4.11 TCT signal induced on the central strip by a 1060-nm pulsed laser (black spot) in the case of (a) a beam focused near the readout strip and (b) a beam focused near the adjacent strip. (From Dalla Betta et al., Vertex 2007, Paper 23, 2008.45 Copyright owned by the authors under the terms of the Creative Commons Attribution-NonCommercial-ShareAlike Licence. http://creativecommons.org/licenses/by-nc-sa/3.0/.) [25]

predicted by the simulations. When the laser spot is focused on the noncollecting electrode (see Fig. 4.11b, the black dot on the right electrode in the insert), a fast signal peak of the opposite polarity is induced and a long tail of the same polarity of the readout channel is still observed for few microseconds. This behavior is also known in planar strip sensors and can be explained with the aid of Ramo's theorem and the directions in which each electrode sees electrons and holes moving. Since the induced signal on the adjacent strip is not negligible, an increased spatial resolution could be obtained in 3D-STC sensors by using the statistical distribution of all signal heights (without sign). If, however, charge is generated underneath a strip rather than in the interstrip region, the fast signal component will be attenuated since a positive and a negative fast signal would be induced on two columns connected by the same strip and the final signal would be the (reduced) sum of the two.

As already mentioned, sensors similar to 3D-STC were developed at VTT and called "Semi 3D" [22]. Results reported for these detectors included (i) leakage currents in the order of a few pA per column at 100 V (well beyond full depletion); (ii) capacitance at full depletion ranging from 40 to 90 fF/column, which is rather high because of a large contribution from the metal-oxide-semiconductor (MOS) capacitance of the metal interconnection between columns at the surface; (iii) charge collection tests with an X-ray Americium source, with a FWHM energy resolution of 7.7% for the 59.5 keV peak. Additional results were reported for "Semi 3D" pixel detectors coupled to the Medipix2 readout chip where sensors are shown to effectively detect X-rays from a ^{109}Cd source, but the energy resolution of ~1.63 keV does not allow resolving the K lines at 22 and 25 keV. A pulsed IR laser system was used to examine the uniformity in the pixel response and allowed the reduced sensitivity from the pillars to be observed. Finally, the good imaging properties of the device were demonstrated using a W-target X-ray tube operated at 35 kV [28].

FBK proposed another 3D-STC version with passing-through columns and backside, grid-shaped ohmic contact as shown in Fig. 4.8b [21]), but this was never constructed. A 3D with single-type columns was also proposed by the Brookhaven National Laboratory (BNL) (Upton, New York) in collaboration with Centro Nacional de Microelectronica (CNM; Barcelona, Spain) [29], where the design differs from the FBK 3D-STC for the fact that, instead of a blank p^+ implant on the back-side, the p^+ regions are patterned and implanted on the front-side as can be seen in Fig. 4.8c. This one-sided technological solution is useful if one wishes to arrange electrodes to obtain 2d sensitive strip-like sensors, in the so-called stripixel configuration [30]. Simulations also suggest that the electric field configuration could be better than the one of the FBK 3D-STC and could possibly improve the charge collection properties [30]. Unfortunately, only preliminary leakage current measurements were published for these devices. Common to all these versions of 3D with single type columnar electrodes is a major simplification of the fabrication technology with respect to the original 3D, but at the expense of slower signal response times and limited charge collection efficiency with consequent poor radiation tolerance.

4.4.2 Double-Sided Double-Type-Column 3D Detectors

The 3D double-sided double-type column (3D-DDTC) was the next important evolution of 3D design, which was independently proposed by FBK [31] and CNM [32] in slightly different versions, as can be seen in Fig. 4.8d and 4.8e. As will be detailed in Chapter 5, the differences between these two device variants are mainly technological. In the interest of completeness, it should also be noted that BNL proposed a one-sided alternative [33]; see Fig. 4.8f. Several years later, the first prototypes of stripixel detectors made with this approach were fabricated at CNM using a double-metal technology whereby the electrodes were connected by two sets of perpendicular strips providing an X-Y projective readout [34]. Results from the electrical characterization and TCT tests demonstrated a low lateral depletion voltage (~4 V) and confirmed the 2D position sensitivity.

The 3D-DDTCs were designed as a comparable alternative to full 3D sensors with a reduced process complexity. As will be shown in Chapter 7, this proved to be an important advantage for the first medium-volume production of these devices.

TCAD simulation was again the first study of the characteristics of these sensors [35]. Figure 4.12 shows a sketch of a 3D-DDTC with 80 μm pitch between readout columns fabricated on a p-type substrate. A quarter-volume cell of $40 \times 40 \times 250\ \mu m^3$ was simulated again to save mesh points and computational time. The substrate doping concentration was $2 \times 10^{12}\ cm^{-3}$ and included are also oxide charge and p-spray isolation, with values typical of the technology used at FBK. Different geometries were compared to investigate what implication the distance between the column tips and the opposite surface (d) had for performance. For this purpose, two 3D-DDTC designs were simulated with either a small gap value of $d = 25$ μm or a large gap value of $d = 75$ μm, and they were compared with full 3Ds passing through columns ($d = 0$). For the latest structure, p-spray isolation was also included on the back surface.

Figure 4.13 shows the electric field configuration along the diagonal of the simulated cell of Fig. 4.12 at 16 V, which is slightly beyond full depletion. The electric field lines in full 3D are horizontal and very homogeneous between columns, with the exception of the regions close to the surfaces, due to the effect of the p-spray. Taking the full 3D electric

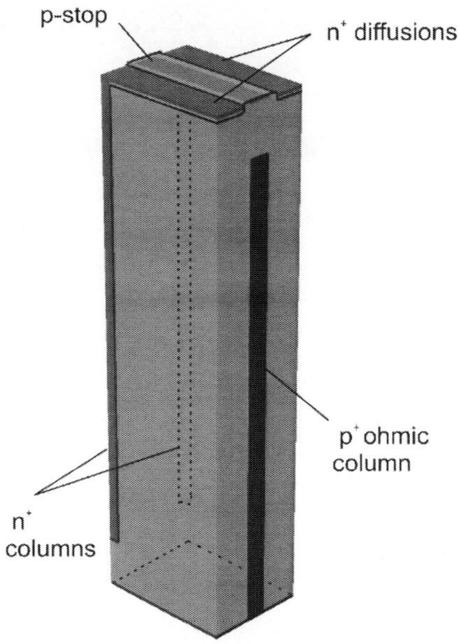

FIGURE 4.12 Sketch of a 3D-DDTC detector on p-type substrate. The cell represents a quarter of a typical pattern present in the layout. The pitch between columns of the same type is 80 μm.

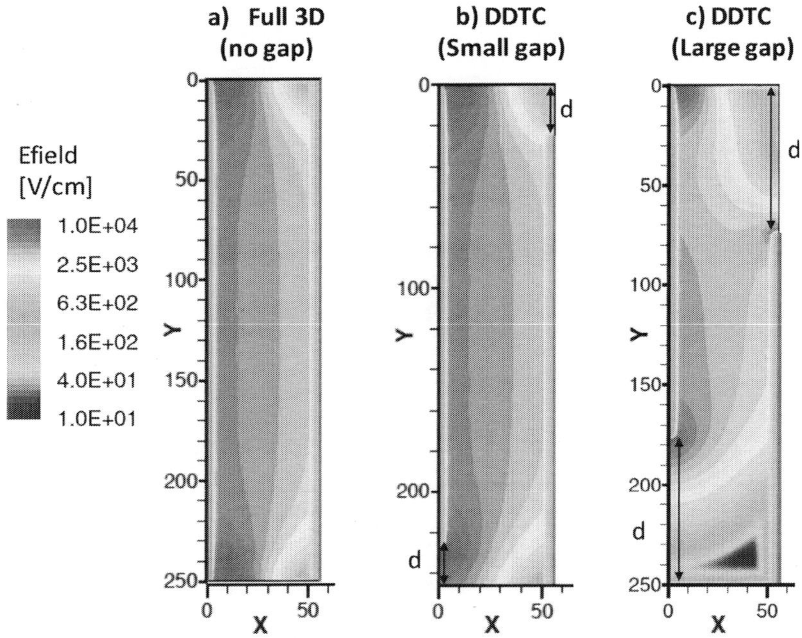

FIGURE 4.13 Electric field distribution of a simulated 2D cross section taken across the diagonal of the quarter cell. All geometrical dimensions are in micrometers. The column junction is placed at x = 0 and the ohmic column is at $x = 40\sqrt{2}$. The three plots are referred to as (a) full 3D, (b) 3D-DDTC with small gap ($d = 25$ μm), and (c) 3D-DDTC with large gap ($d = 75$ μm).

field as a reference, among the simulated 3D-DDTC geometries, the one with a small gap of $d = 25$ μm shows comparable results. When d is large, however, the electric field lines are more distorted and only in a small region at the center, the field lines are similar to those of full 3D, whereas low-field regions can be observed near the top and bottom surfaces of 3D-DDTCs, a fact that degrades the charge collection efficiency. Differently from 3D-STC, in 3D-DDTC increasing the bias voltage beyond full depletion increases the electric field between columns and the carrier drift velocity, therefore reducing the collection time.

Transient simulations of 3D-DDTCs were performed of an MIP impinging the cell at few microns from the ohmic p^+ column. Also, 3D-STCs were simulated in order to have a direct performance comparison among technologies. Figure 4.14a shows the induced currents on the collecting electrode for all the investigated structures at bias voltage of 16 V. As expected from the static electric field simulations, the shorter the distance d, the

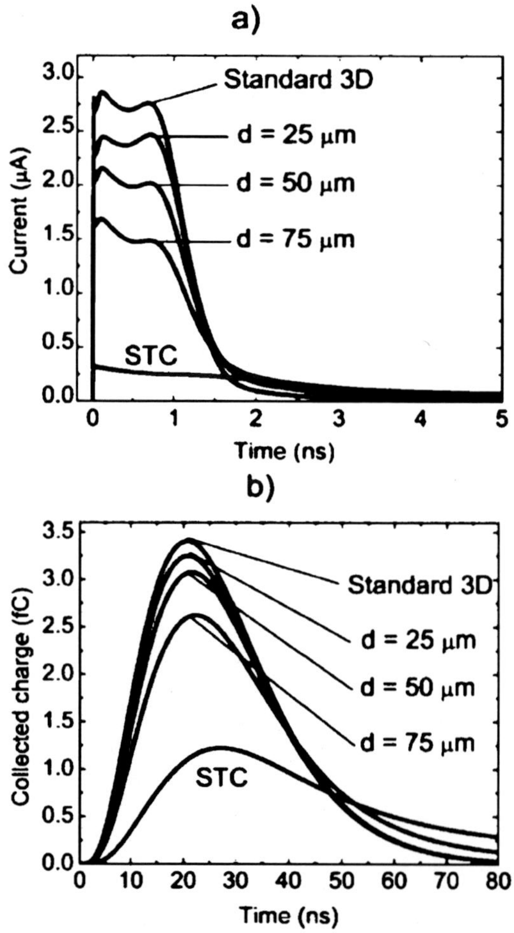

FIGURE 4.14 Transient signals in 3D detectors with different geometries taken from simulation at a bias of 16 V in response to an MIP particle: (a) induced current signal; (b) equivalent charge signal at the output of a semi-Gaussian CR-(RC)^3 shaper amplifier with 20-ns peaking time. (From Zoboli et al., IEEE Trans. Nucl. Sci., 55, 5, 2008. With permission.) [35]

higher the current peaks and the shorter the collection times. The current signals were also postprocessed with an algorithm that reproduces a CR-(RC)^3 filter with a shaping time of 20 ns to emulate the fast readout electronics of the ATLAS semiconductor cracker (SCT) detector [36]. The resulting shaped pulses were extracted and plotted to emphasize the collected charge differences depending on *d*, as can be seen in Fig. 4.14b. The collected charge for the geometry with $d = 25$ μm is only ~10% lower than that for full 3D, confirming the idea that comparable performance can be achieved if *d* is kept small enough. On the other hand, 3D-STC, due to the above-mentioned inefficiencies in the charge collection mechanism, collects only a small fraction of the generated charge within 20-ns. Simulations of high radiation damage of about a 10^{16} cm^{-2} 1-MeV n_{eq} fluence were also performed and confirmed that full 3D and 3D-DDTC with $d = 25$ μm had a comparable charge collection efficiency after irradiation [35]. Similar conclusions were obtained from TCAD simulations at the University of Glasgow, where researchers looked at 3D-DDTC detectors built at CNM [37]: The electrical and charge collection properties were studied, as well as breakdown voltages, pointing out that column tips can indeed play a critical role in 3D-DDTC with a partially through column. Furthermore, in [38], irradiated 3Ds are simulated, with emphasis on the performance of pixel sensors compatible with the current ATLAS pixel design. In this work, a special discussion is dedicated to the impact of electrode configurations from 2E to 8E on depletion voltage, collected charge, capacitance, and related noise.

The results of the first 3D-DDTC prototypes fabricated at FBK were very promising, even if the structures were not optimized from the column overlap point on view. Since the deep reactive ion etching (DRIE) equipment was not available in house at that time, it was not possible to adequately develop the column-etching steps, and so the first batch of FBK 3D-DDTC was processed on 300 μm thick n-type substrates with a column overlap of just 50 μm.

The wafer floor plan contained mainly 3D diode test structures, where arrays of columns are shorted together to form a two-terminal device and 3D strip detectors suitable to the ATLAS SCT front-end electronic electronics with p-readout. Tests with these structures were very good and showed low leakage currents, with values as low as ~0.1 pA/column at full depletion with very low bias voltages [35].

Current and capacitance tests were carried out on strip sensors connected to the ATLAS SCT ABCD3T readout module, with 20 ns shaping time. All tested sensors had an area of 1 cm^2 with 102 strips and 102 columns per strip [39]. As a reference, 3D-STC samples available from the same batch, having a uniform backside contact without the ohmic columns, were tested. From position-resolved laser tests carried out before irradiation, a relatively uniform spatial response was measured, although some regions showed a lower efficiency of about 20% due to the nonoptimized column depth. From tests with a Sr^{90} beta source it was possible to confirm once more that 3D-DDTC sensors had higher charge collection efficiency than 3D-STC with 2.4 fC versus 1.8 fC. However, the nonoptimized column depths prevented the collected charge to reach the theoretically expected values from 300 μm thick substrate, which corresponds to 3.5 fC. The tracking performance obtained during test beams, however, showed very good efficiency, charge sharing, and position resolution [40].

The following 3D-DDTC batches at FBK were processed on 200 and 220 μm thick p-type substrates, respectively, with column depths from 100 to 150 μm and an improved performance, as confirmed by tests on diodes and strip sensors. The electrical characteristics, however, were similar to those measured from the first batch [41, 42, 43]. The wafer layout of these batches contained mostly 3D pixel structures designed to fit the ATLAS FEI3 pixel readout chip [13] with different electrode configurations. Several sensors from both fabrication batches were connected to the ATLAS FEI3 readout chip at SELEX SI [44] and thoroughly characterized, in both lab and beam tests. Noise figures from 200 to 240 electrons rms at full depletion were measured, depending on the electrode configuration. The different values were due to capacitance values resulting from the different number of columns per pixel and the increasing distance between columns. Functional electrical tests with γ- and β-sources showed full charge collection efficiency [45]. The sensor behavior before irradiation was also assessed at high-energy pion beam tests at CERN, with and without a 1.6T magnetic field [5]. 3D-DDTC were found to be comparable to full 3D sensors, with no negligible effects from magnetic field and little charge sharing between cells. Some loss of tracking efficiency was observed for normal incident tracks owing to the well-known inefficiency of the electrodes, but full efficiency was again recovered with tilted tracks.

Finally, the device sketched in Fig. 4.8g, a variant of the FBK DDCT with passing-through columns, was called 3D-DDCT+. This layout was used for the production of the pixel detectors for the ATLAS Insertable B-Layer [46]. The behavior of this layout is the same that of the one of a full 3D sensor, with few exceptions: (i) the columns are not filled by polysilicon, and so there is a dead volume in the electrode's center; and (ii) they are processed without a support wafer and without active edges.

These sensors were thoroughly tested in lab and beam tests with very encouraging results [47, 48, 49, 50, 51, 52, 53, 54]; a notable exception was a relatively low breakdown voltage of about 50 V before irradiation [46]. This weakness was thoroughly studied, both experimentally and with TCAD simulations, and lead to the understanding that the most critical cause of breakdown was the n^+ column to p-spray junction at the backside of the wafer [55]. To mitigate the early breakdown effect, a further design variant was implemented, as illustrated in Fig. 4.8h, with partially through collecting electrodes and full through ohmic electrodes, which significantly improved the breakdown voltage [56].

The first results reported for double-sided 3D devices fabricated at CNM with 250 μm deep electrodes on 300 μm thick n-type Si wafers show few volts for full depletion voltages, leakage currents of ~1pA/column, and breakdown voltages higher than 60 V [57, 58]. Functional electrical characteristics of pixel detectors bump bonded to Medipix2 readouts are reported in [59], where tests with X-rays confirm the low depletion voltage of ~2V lateral depletion, ~9V full depletion. Spectroscopic tests using monochromatic 15-keV X-rays from a synchrotron light source confirmed a substantially reduced charge sharing compared to planar detectors, as can be seen in Fig. 4.15, where the energy distribution tail below the peak at 15 keV is clearly more pronounced for the latter.

3D-DDTC pixels from CNM bump bonded to the TimePix and Medipix2 readout ASICs were also studied in a 120-GeV pion beam test at the SPS at CERN [60]. In particular, the

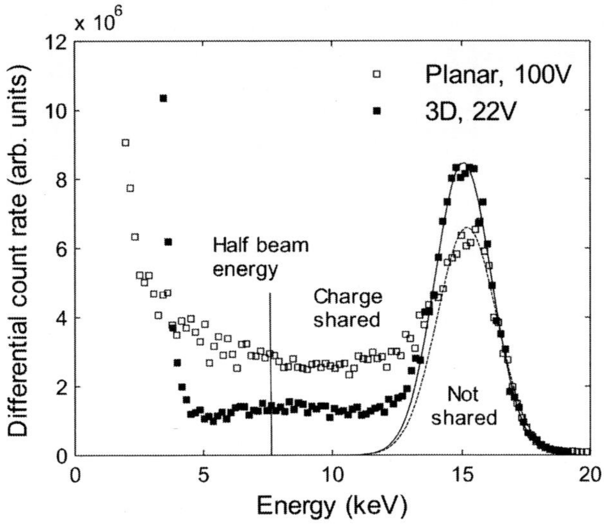

FIGURE 4.15 Comparison of spectra from a 15-keV monochromatic X-ray beam measured with planar and 3D detectors. (From Pennicard et al., Nucl. Instrum. Methods A, 604, 2009. Copyright Elsevier 2009. With permission.) [59]

inefficiencies due to the columnar electrodes were confirmed as a function of the pion beam direction. For orthogonal incidence, the overall efficiency was ~93.0 %, and it increased to almost 100% at a 10° incidence angle when the inefficiencies due to the very electrodes were no longer observed.

4.4.3 Trenched Electrodes

Almost all 3D sensors studied so far have electrodes etched with cylindrical shape. Alternative shapes are also possible, at the expense of additional technological complication, and some of them were already studied in the past with some expected advantages. As already mentioned, wall electrodes were originally proposed at Stanford as precursors to active edges, with very promising results [7]. Later, trench electrodes were also applied to precisely define the active volume in planar detectors for dosimetry applications [61]. Furthermore, as will be shown in more detail in Chapter 9, trench electrodes were proposed to optimize the sensor's timing performance [62].

The advantage of trenches when compared to traditional columnar electrodes is manifold: (1) lower depletion voltage, (2) more uniform electric field distribution, and (3) reduced charge sharing. Later in time, the BNL group, in collaboration with Stony Brook University, proposed a particular design dedicated to tracking detectors upgrades at LHC and X-ray applications. Different design solutions with analytical calculations of relevant parameters and TCAD simulations were reported in [63]. A more systematic simulation study focused on detectors with a columnar electrode at the center of a cell surrounded by hexagonal-shaped trenches of different size is reported in [64], along with preliminary results from the characterization of the first prototypes fabricated at CNM.

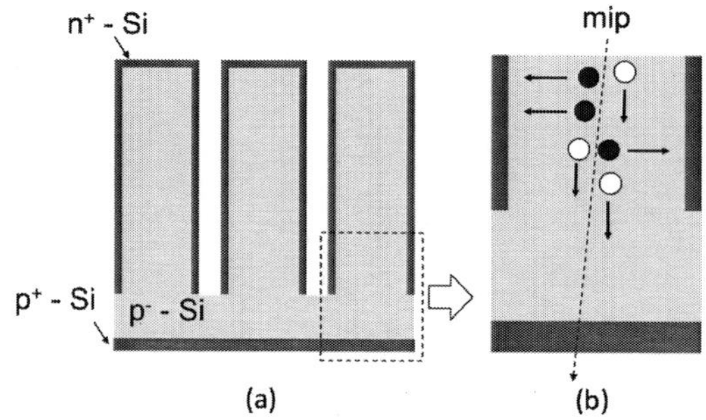

FIGURE 4.16 Sketch of (a) a core–shell detector array, and (b) a magnified basis core–shell structure and carrier collection mechanism.

Recently proposed devices are the so-called core-shell [65]. This very innovative and promising design consists of arrays of high aspect ratio silicon pores (see Fig. 4.16), with small lateral dimensions and consequently, potentially very low full depletion voltages, high-radiation hardness, low power consumption, fast signal response, and high spatial resolution. The fabrication process for these sensors is still being optimized [66, 67].

4.4.4 The Pixelated Vertical Drift Detector

A different way to employ vertical electrodes was proposed in [68] for the so-called pixelated vertical drift detector (PVDD), a new device optimized for X- and gamma-ray detection and imaging. As can be seen in Fig. 4.17, a mixed planar-3D pixel structure is used for the sensor where the signal-collecting electrode is planar to ensure low-output capacitance,

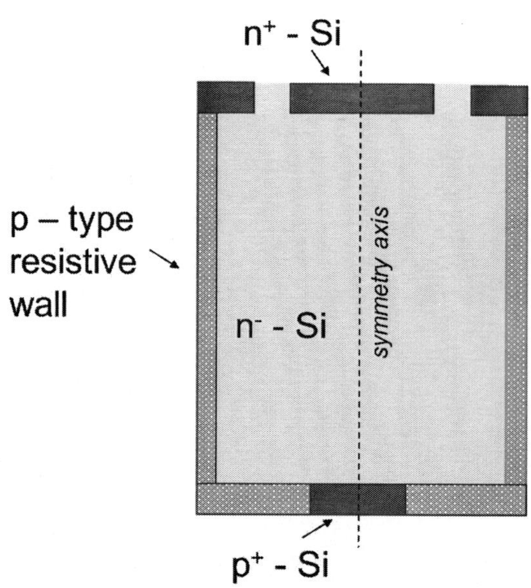

FIGURE 4.17 Schematic cross section of a pixelated vertical drift detector.

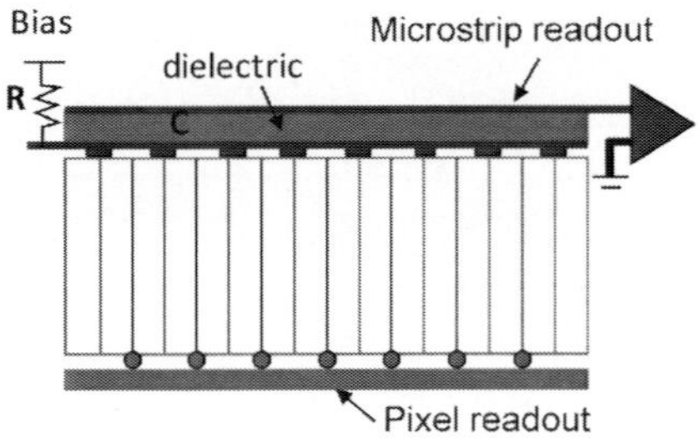

FIGURE 4.18 Sketch of one possible dual electrode readout configuration showing a pixel readout via bump bonding at the bottom and bias electrode microstrip readout (AC coupled) on the top.

whereas vertical columnar electrodes at the pixel boundaries ensure low charge sharing and low cross talk. A potential difference is applied at the end of vertical electrodes at the top and bottom surfaces to induce a vertical electric field along the central symmetry axis of the pixel, so that the charge can drift to the collecting electrode at the top surface. The device concept together with some possible variants is reported in [68], along with TCAD simulations to validate the operation principle and estimated performance. However, the technological feasibility has not been demonstrated yet.

4.4.5 Dual Readout in 3D Sensors

In 3D sensors, signals are normally read out only from one electrode type, which is typically the junction electrode, with the other type being used for biasing purposes. However, a possible alternative exploits reading out signals from both electrode types at the same time using a combination of pixels and strip readouts [69]. The most important benefit of this configuration is an increased spatial resolution while maintaining a reduced material budget. The ohmic column readout could be arranged in a simple strip configuration providing a fast trigger. Such a modified sensor design can suit both single-sided and double-sided 3D technologies, since both n- and p-type columns emerge on either the same or both wafer sides, although its implementation is more straightforward with the latter. As an example, Fig. 4.18 shows the schematic cross section of a detector with pixels bump bonded to a readout chip on one side and strips AC-coupled to another readout on the opposite side [69]. Another possible implementation of the dual readout concept is based on the addition of resistors between the bias supply line and each bias electrode. The latter solution can provide improved spatial resolution in both X and Y directions and lower capacitance [70].

4.5 ACTIVE AND SLIM EDGES IN 3D SENSORS

As mentioned in Section 4.1, active edges were derived as an extension of full 3D sensor technology and are becoming increasingly important for several applications, ranging from high-energy physics to X-ray imaging. In fact, the minimization of the dead area at the sensor's edge allows for large-area seamlessly tiled detector matrices, with major advantages in the detector assembly.

Beam tests of full 3D sensors with active edges confirmed the effectiveness of ensuring high sensitivity in the totality of the sensor's volume. As an example, Fig. 4.19 shows the hit efficiency as a function of the hit position in a 3D pixel terminated with an active

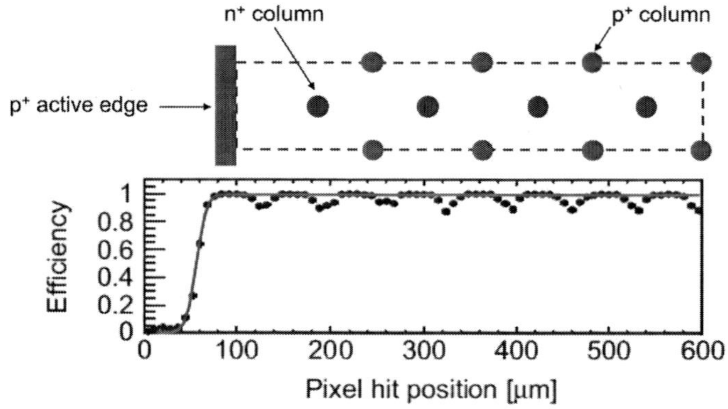

FIGURE 4.19 Mask detail (top) of the edge pixel region compared to the efficiency as a function of the corresponding hit position in the 400 μm direction.

edge, and it confirms that full efficiency can be maintained up to a few micrometers from the sensor's physical edge. In [15], an edge sensitivity of 10–12 μm is reported, probably dominated by tracking resolution and residual misalignment. Similar results were obtained in [17] and [71].

A modified active edge design could also prevent a particular form of failure from occurring in experiments where largely nonuniform radiation fields are involved. This is the case of forward physics experiments, such as ATLAS Forward Physics (AFP) [72] and CT-PPS [73], which measure forward diffracted protons with detector stations placed inside the beam pipes at about 200 m, symmetrically, from the interaction points at the ATLAS and CMS experiments, respectively.

In silicon sensors, the bias voltage necessary for full depletion, maximum signal and breakdown voltage increases after irradiation. If the irradiation pattern is very nonhomogenous, it might be necessary to apply bias voltages well above the breakdown voltage in regions of the same sensor, which have not yet received a very high irradiation. In the case of large radiation fluences, the depletion voltage and the bias voltage for maximum signal could exceed the breakdown voltage of low irradiated regions. A solution to this problem could consist in dividing 3D sensors with active edges into separate sections with a "triple-wall" sandwich of two trench electrodes separated by an oxide insulating layer, as shown in Fig. 4.20. By doing so, each section of the sensor would be completely isolated from the others (note that the oxide layer used for wafer bonding would also ensure a complete sealing at the bottom) and could therefore be biased at different voltages [74].

Alternate slim-edge solutions can be implemented with double-sided processing. In double-sided 3D sensors, active edge trenches are not feasible because of the absence of a support wafer enabling all dice to be held together after the trenches are etched all around each sensor. However, the third dimension inside the silicon substrate still provides some advantages by allowing "slim"-edge terminations. As an example, in CNM sensors, the slim edge includes a guard ring, aimed as usual at sinking the leakage current from the cut line, and it is surrounded by a double fence of ohmic columns [75]. A different slim-edge concept was

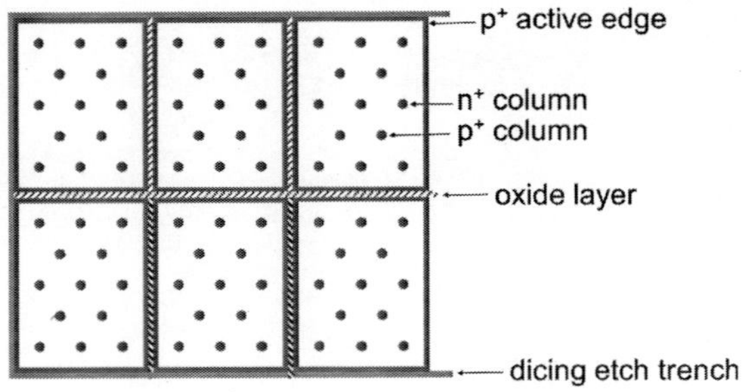

FIGURE 4.20 Simplified schematic view of six bias voltage regions separated by triple-wall voltage isolators and terminated with a dicing etch trench in a multiple-bias-voltage 3D sensor.

developed at FBK without the guard ring [76] but with a multiple ohmic (p^+) column fence, biased at the same substrate potential, surrounding the active area. As a result, the depletion region, spreading from the outermost junction columns near the edge (n^+), is blocked within a short distance from reaching the cut-line, so any edge leakage current is drawn by the detector. This solution was implemented for the first time for the IBL sensors and proved to be very effective [47, 4.75]. In fact, the ~200 μm wide ohmic fence was found to be excessive, and the slim edge properly worked even at a width of 100 μm or less.

Figure 4.21a shows the layout of a double-sided 3D diode with a 200 μm wide slim edge between the active area and the nominal scribe line. I-V curves were measured after dicing along it, and then they were remeasured after multiple cuts (labeled from 1 to 6 in Fig. 4.21a) close to the active area. As can be seen in Fig. 4.21b, the I-V curves start to deviate significantly from their initial behavior from the fourth cut on.

Further insight into the slim-edge operation of double-sided 3Ds was gained from position-resolved laser tests performed on diodes. To monitor the sensors' performance, the

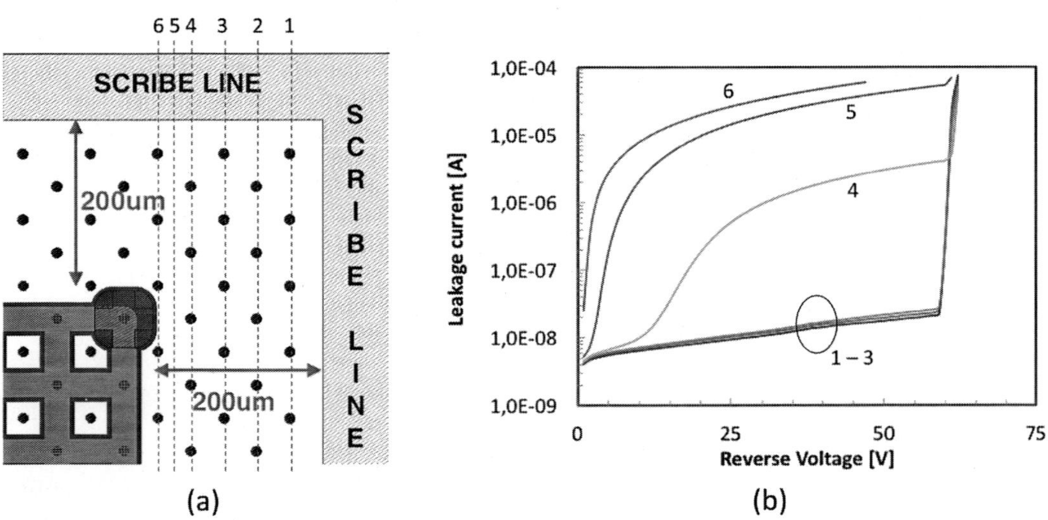

FIGURE 4.21 Results from the electrical characterization performed on 80 μm pitch 3D diodes: (a) layout detail showing the different cuts, and (b) corresponding I-V curves measured after every cut.

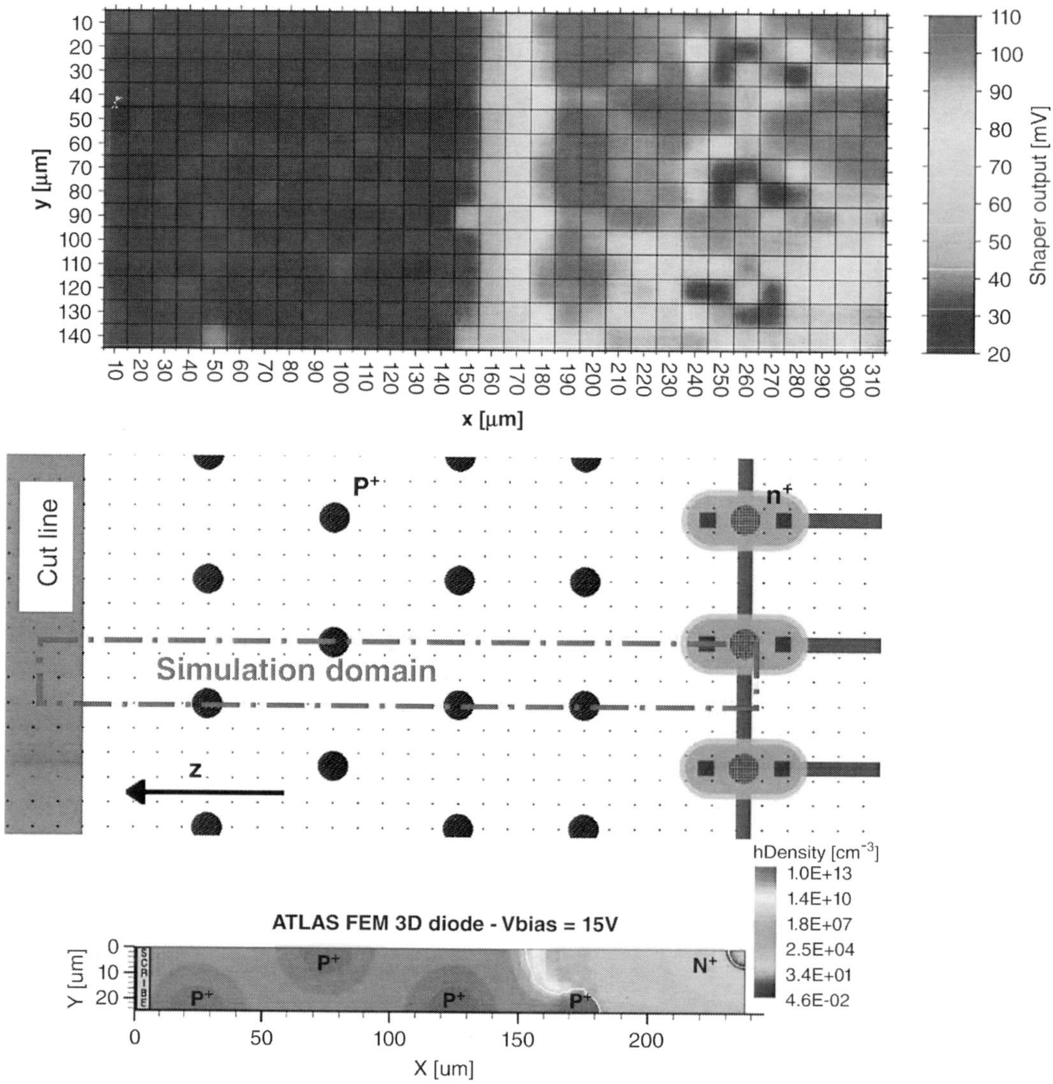

FIGURE 4.22 Results from a position-resolved laser scan on the slim-edge region of a 3D diode biased at 15 V, compared to device simulations: (top) 2d signal map; (center) diode layout and simulation domain; (bottom) simulated hole density along a horizontal plane showing the lateral depletion region extension. (From Dalla Betta et al., Vertex 2013, Paper 42, 2014. Copyright owned by the authors under the terms of the Creative Commons Attribution-NonCommercial-ShareAlike Licence. http://creativecommons.org/licenses/by-nc-sa/3.0/.) [77]

output currents were read out with a charge amplifier and a 20-ns fast shaper. By doing so, only the fast drift-related component of the signal was monitored, thus obtaining information on the lateral spread of the depletion region.

An example of such tests can be seen in Fig. 4.22, which shows a map of the signal induced by a 1060 nm laser beam having a FWHM of ~11 μm at the sensor's surface. The diode was biased at 15 V, which is beyond full depletion [77]. A detail of the slim edge and the simulated hole-density distribution at the same voltage are also shown in Fig. 4.22b. All readout (n^+) columns are covered by metal and can therefore be recognized by the low-signal regions (in blue) on Fig. 4.22 top. The highsignal regions spreading from the n^+ columns extend into the slim edge region beyond the first row of ohmic columns. This is in good agreement with the simulated depletion region of Fig. 4.22 bottom. Simulations also confirm that even at reverse biases as high as 300 V, which are much higher than any realistic bias value for 3D detectors, the extension of the depletion region within the slim edge would be only slightly larger and would remain far enough from the scribe line preventing current injection and breakdowns [76].

The depletion region confinement within the slim edges could be made even more effective by using ohmic-fence electrodes of different shapes. As an example, Fig. 4.23a shows the layout of a 3D diode where columnar electrodes were replaced by dashed trenches [78]. Also in this case, position-resolved laser tests with fast signal readout allowed the lateral extension of the depletion region to be monitored. As shown in Fig. 4.23b, where the sensor was biased at 60 V, the high signal profile within the slim-edge region corresponding to the depletion region is completely blocked by the dashed trenches. So, when compared to columnar electrodes, the latter can be placed closer to the cut line, thus reducing the dead area at the edge to as much as 50 μm.

Alternative ideas were also proposed to transform dashed trenches into active edges without the need for support wafers [79]. This concept requires dashed trenches to be etched with gaps of 10 μm or less. Active edges are then created by thermal diffusion of the p-type dopant until it fills the lateral gaps between adjacent trenches. In this way, a continuous ohmic wall is formed around the active area, and a standard dicing saw can be used to separate sensors by cutting through this wall (see Fig. 4.24).

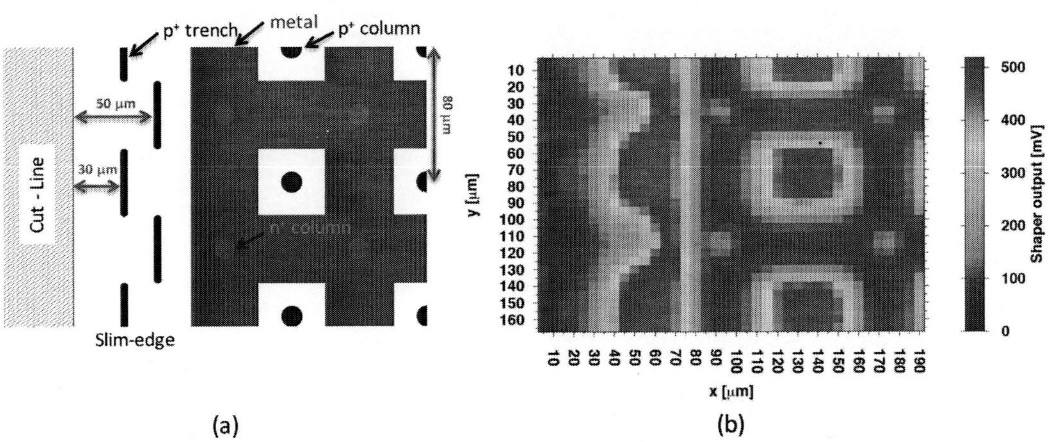

FIGURE 4.23 (a) Layout detail, and (b) position-resolved laser scan for a 3D diode with an improved slim edge based on dashed trenches, measured at 60 V bias voltage. (From Dalla Betta et al., Vertex 2013, Paper 42, 2014. Copyright owned by the authors under the terms of the Creative Commons Attribution-NonCommercial-ShareAlike Licence. http://creativecommons.org/licenses/by-nc-sa/3.0/.) [77]

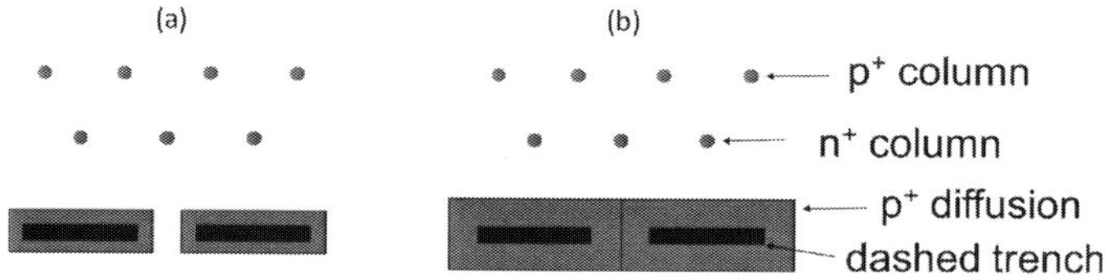

FIGURE 4.24 An active-edge concept for double-sided 3D sensors: (a) initial doping diffusion from dashed trenches, and (b) building of an ohmic wall by lateral dopant diffusion in between dashed trenches.

This concept was successfully tested with TCAD simulations [79]. The simulation domain included the outermost columns, both p^+ and n^+, and the dashed trenches with a gap of 10 μm. To model the damage from the saw cut, a region with a 1 ns carrier lifetime was introduced at the edge. For small lateral diffusion values (see Fig. 4.24a), the simulated I-V curves showed a steep increase at low bias voltages, but for lateral diffusion values of at least 5 μm, which is the minimum value required to build a continuous ohmic wall in case of the 10 μm gap between the trenches, the I-V curve remained almost flat up to very large bias voltages. This demonstrates that the ohmic wall effectively screens the active area from the damage of the saw cut.

REFERENCES

1. S. I. Parker, C. J. Kenney, J. Segal, "3D—A proposed new architecture for solid-state silicon detectors," Nuclear Instruments and Methods A, vol. 395, pp. 328–343, 1997.
2. S. Ramo, "Currents induced by electron motion," Proceedings of the I.R.E., vol. 27. pp. 584–585, 1939
3. A. Kok, G. Anelli, C. Da Vià, J. Hasi, P. Jarron, C. Kenney, J. Morse, S. Parker, J. Segal, S. Watts, E. Westbrook, "3D detectors—state of the art," Nuclear Instruments and Methods A, vol. 560, pp. 127–130, 2006.
4. G. Lindström, M. Moll, E. Fretwurst (The RD48 Collaboration), "Radiation hardness of silicon detectors—A challenge from high energy physics," Nuclear Instruments and Methods A, vol. 426, pp. 1–15, 2001.
5. P. Grenier, G. Alimonti, M. Barbero, R. Bates, E. Bolle, M. Borri, M. Boscardin, C. Buttar, M. Capua, M. Cavalli-Sforza, M. Cobal, A. Cristofoli, G. F. Dalla Betta, G. Darbo, C. Da Vià, E. Devetak, B. DeWilde, B. Di Girolamo, D. Dobos, K. Einsweiler, D. Esseni, S. Fazio, et al., "Test beam results of 3D silicon pixel sensors for the ATLAS upgrade," Nuclear Instruments and Methods A, vol. 638, pp. 33–40, 2011.
6. M. Da Rold, N. Bacchetta, D. Bisello, A. Paccagnella, G. F. Dalla Betta, G. Verzellesi, O. Militaru, R. Wheadon, P. G. Fuochi, C. Bozzi, R. Dell'Orso, A. Messineo, G. Tonelli, P. G. Verdini, "Study of breakdown effects in silicon multiguard structures," IEEE Transactions on Nuclear Science, vol. 46(4), pp. 1215–1223, 1999.
7. C. J. Kenney, S. I. Parker, E. Walckiers, "Results from 3-D silicon sensors with wall electrodes: Near-cell-edge sensitivity measurements as a preview of active-edge sensors," IEEE Transactions on Nuclear Science, vol. 48(6), pp. 2405–2410, 2001.
8. C. J. Kenney, J. D. Segal, E. Weestbrook, S. Parker, J. Hasi, C. Da Vià, S. Watts, J. Morse, "Active-edge planar radiation sensors," Nuclear Instruments and Methods A, vol. 565, pp. 272–277, 2006.
9. SILVACO (http://www.silvaco.com); Synopsys Advanced TCAD Tools (https://www.synopsys.com/silicon/tcad.html)

10. C. Piemonte, "Device simulations of isolation techniques for silicon microstrip detectors made on p-type substrates," IEEE Transactions on Nuclear Science. vol. 53(3), pp. 1694–1705, 2006.
11. C. J. Kenney, S. I. Parker, J. Segal, C. Storment, "Silicon detectors with 3-D electrode arrays: fabrication and initial test results," IEEE Transactions on Nuclear Science, vol. 46(4), pp. 1224–1236, 1999.
12. C. J. Kenney, S. I. Parker, B. Krieger, B. Ludewigt, T. P. Dubbs, H. Sadrozinski, "Observation of Beta and X rays with 3-D architecture silicon microstrip detectors," IEEE Transactions on Nuclear Science, vol. 48(2), pp. 189–193, 2001.
13. I. Perić, L. Blanquart, G. Comes, P. Denes, K. Einsweiler, P. Fischer, E. Mandelli, G. Meddeler, "The FEI3 readout chip for the ATLAS pixel detector," Nuclear Instruments and Methods A, vol. 565, pp. 178–187, 2006.
14. M. Mathes, M. Cristinziani, C. Da Vià, M. Garcia-Sciveres, K. Einsweiler, J. Hasi, C. Kenney, S. I. Parker, L. Reuen, J. Velthuis, S. Watts, N. Wermes, "Test beam characterization of 3-D silicon pixel detectors," IEEE Transactions on Nuclear Science, vol. 55(6), pp. 3731–3735, 2008.
15. O. M. Røhne, "Edge characterization of 3D silicon sensors after bump-bonding with the ATLAS pixel readout chip," 2008 IEEE Nuclear Science Symposium, Dresden (Germany), October 19–25, 2008, Conference Record, paper N27-8.
16. C. Da Vià, E. Bolle, K. Einsweiler, M. Garcia-Sciveres, J. Hasi, C. Kenney, V. Linhart, S. Parker, S. Pospisil, O. Rohne, T. Slavicek, S. Watts, N. Wermes, "3D active edge silicon sensors with different electrode configurations: Radiation hardness and noise performance," Nuclear Instruments and Methods A, vol. 604, pp. 505–511, 2009.
17. P. Hansson, J. Balbuena, C. Barrera, E. Bolle, M. Borri, M. Boscardin, M. Chmeissan, G. F. Dalla Betta, G. Darbo, C. Da Vià, E. Devetak, B. DeWilde, D. Su, O. Dorholt, S. Fazio, C. Fleta, C. Gemme, M. Giordani, H. Gjersdal, P. Grenier, S. Grinstein, et al., "3D silicon pixel sensors: Recent test beam results," Nuclear Instruments and Methods A, vol. 628, pp. 216–220, 2011
18. O. Koybasi, D. Bortoletto, T.-E. Hansen, A. Kok, T. A. Hansen, N. Lietaer, G. U. Jensen, A. Summanwar, G. Bolla, S. Kwan, "Design, simulation, fabrication, and preliminary tests of 3D CMS pixel detectors for the Super-LHC," IEEE Transactions on Nuclear Science, vol. 57(5), pp. 2897–2905, 2010.
19. H. Chr. Kästli, M. Barbero, W. Erdmanna, Ch. Hörmann, R. Horisberger, D. Kotlinski, B. Meier, "Design and performance of the CMS pixel detector readout chip," Nuclear Instruments and Methods A, vol. 565, pp. 188–194, 2006.
20. O. Koybasi, E. Alagoz, A. Krzywda, K. Arndt, G. Bolla, D. Bortoletto, T.-E. Hansen, T. A. Hansen, G. U. Jensen, A. Kok, S. Kwan, N. Lietaer, R. Rivera, I. Shipsey, L. Uplegger, C. Da Vià, "Electrical characterization and preliminary beam test results of 3D silicon CMS pixel detectors," IEEE Transactions on Nuclear Science, vol. 58(3), pp. 1315–1323, 2011.
21. C. Piemonte, M. Boscardin, G. F. Dalla Betta, S. Ronchin, N. Zorzi, "Development of 3D detectors featuring columnar electrodes of the same doping type," Nuclear Instruments and Methods A, vol. A541, pp. 441–448, 2005.
22. S. Eränen, T. Virolainen, I. Luusua, J. Kalliopuska, K. Kurvinen, M. Eräluoto, J. Härkönen, K. Leinonen, M. Palvianen, M. Koski, "Silicon semi 3D radiation detectors," 2004 IEEE Nuclear Science Symposium, Rome (Italy), October 16–22, 2004, Conference Record, Paper N28-3.
23. A. Pozza, M. Boscardin, L. Bosisio, G. F. Dalla Betta, C. Piemonte, S. Ronchin, N. Zorzi, "First electrical characterisation of 3D detectors with electrodes of the same doping type," Nuclear Instruments and Methods A, vol. 560, pp. 317–321, 2007.
24. C. Piemonte, M. Boscardin, L. Bosisio, G. F. Dalla Betta, A. Pozza, S. Ronchin, N. Zorzi, "Study of the signal formation in single-type-column 3D silicon detectors," Nuclear Instruments and Methods, vol. 579, pp. 633–637, 2007.
25. G.-F. Dalla Betta, M. Boscardin, L. Bosisio, M. Bruzzi, V. Cindro, S. Eckert, G. Giacomini, G. Kramberger, S. Kühn, U. Parzefall, M. K. Petterson, C. Piemonte, et al., "Development of 3D detectors at FBK-irst," Proceedings of Science–16th Workshop on Vertex Detectors (Vertex 2007), Paper 23, 2008.

26. S. Ronchin, M. Boscardin, C. Piemonte, A. Pozza, N. Zorzi, G. F. Dalla Betta, G. Pellegrini, L. Bosisio, "Fabrication of 3D detectors with columnar electrodes of the same doping type," Nuclear Instruments and Methods A, vol. 573, pp. 224–227, 2007.
27. M. Zavrtanik, V. Cindro, G. Kramberger, J. Langus, I. Mandic, M. Mikuz, M. Boscardin, G.-F. Dalla Betta, C. Piemonte, A. Pozza, S. Ronchin, N. Zorzi, "Position sensitive TCT evaluation of irradiated 3d-stc detectors," 2007 IEEE Nuclear Science Symposium, Honolulu (USA), October28–November 3, 2007, Conference Record, paper N24-150.
28. L. Tlustos, J. Kalliopuska, R. Ballabriga, M. Campbell, S. Eranen, X. Llopart, "Characterisation of a semi 3-D sensor coupled to Medipix2," Nuclear Instruments and Methods A, vol. 580, pp. 897–901, 2007.
29. Z. Li, W. Chen, Y.H. Guo, D. Lissauer, D. Lynn, V. Radeka, G. Pellegrini, "Development, simulation and processing of new 3D Si detectors," Nuclear Instruments and Methods A, vol. 583, pp. 139–148, 2007.
30. T. Grönlund, Z. Li, G. Carini, M. Li, "Full 3D simulations of BNL one-sided silicon 3D detectors and comparisons with other types of 3D detectors," Nuclear Instruments and Methods A, vol. 586, pp. 180–189, 2008.
31. G. F. Dalla Betta, M. Boscardin, L. Bosisio, C. Piemonte, S. Ronchin, A. Zoboli, N. Zorzi, "New developments on 3D detectors at IRST," 2007 IEEE Nuclear Science Symposium, Honolulu (U.S.A.), October 28–November 3, 2007, Conference Record, paper N18-3.
32. C. Fleta, D. Pennicard, R. Bates, C. Parkes, G. Pellegrini, M. Lozano, V. Wright, M. Boscardin, G. F. Dalla Betta, C. Piemonte, A. Pozza, S. Ronchin, N. Zorzi, "Simulation and test of novel 3D silicon radiation detectors," Nuclear Instruments and Methods A, vol. 579, pp. 642–647, 2007.
33. Z. Li, W. Chen, Y. H. Guo, D. Lissauer, D. Lynn, V. Radeka, M. Lozano, G. Pellegrini, "Development of new 3D Si detectors at BNL and CNM," 2006 IEEE Nuclear Science Symposium, San Diego (USA), October 29–November 4, 2006, Conference Record, paper N34-5.
34. D. Bassignana, Z. Li, M. Lozano, G. Pellegrini, D. Quirion, T. Tuuva, "Design, fabrication and characterization of the first dual-column 3D stripixel detectors," Journal of Instrumentation, JINST 8, P08014, 2013.
35. A. Zoboli, M. Boscardin, L. Bosisio, G.-F. Dalla Betta, C. Piemonte, S. Ronchin, N. Zorzi, "Double-sided, double-type-column 3D detectors at FBK: Design, fabrication and technology evaluation," IEEE Transactions on Nuclear Science, Vol. 55(5), pp. 2775–2784, 2008.
36. F. Campabadal, C. Fleta, M. Key, M. Lozano, C. Martinez, G. Pellegrini, et al., "Design and performance of the ABCD3TA ASIC for readout of silicon strip detectors in the ATLAS semiconductor tracker," Nuclear Instruments and Methods A, vol. 552, pp. 292–328, 2005.
37. D. Pennicard, G. Pellegrini, M. Lozano, R. Bates, C. Parkes, V. O'Shea, V. Wright, "Simulation results from double-sided 3-D detectors," IEEE Transactions on Nuclear Science, vol. 54(4), pp. 1435–1443, 2007.
38. D. Pennicard, G. Pellegrini, C. Fleta, R. Bates, V. O'Shea, C. Parkes, N. Tartoni, "Simulation of radiation-damaged 3D detectors for the Super-LHC," Nuclear Instruments and Methods A, vol. 592, pp. 16–25, 2008.
39. A. Zoboli, M. Boscardin, L. Bosisio, G.-F. Dalla Betta, S. Eckert, S. Kühn, et al., "Laser and Beta Source setup characterization of p-on-n 3D-DDTC detectors fabricated at FBK-IRST," Nuclear Instruments and Methods A, vol. 604, pp. 238–241, 2009.
40. M. Koehler, R. Bates, M. Boscardin, G.-F. Dalla Betta, C. Fleta, J. Harkoenen, S. Houston, K. Jacobs, S. Kuehn, M. Lozano, P.-R. Luukka, T. Maenpaa, H. Moilanen, C. Parkes, U. Parzefall, G. Pellegrini, D. Pennicard, S. Ronchin, A. Zoboli, N. Zorzi, "Beam test measurements with double-sided 3D silicon strip detectors on n-type substrate," IEEE Transactions on Nuclear Science, vol. 57(3), pp. 2987–2994, 2010.
41. A. Zoboli, M. Boscardin, L. Bosisio, G. F. Dalla Betta, C. Piemonte, S. Ronchin, N. Zorzi, "Initial results from 3D-DDTC detectors on p-type substrates," Nuclear Instruments and Methods A, vol. 612, pp. 521–524, 2010.

42. G.-F. Dalla Betta M. Boscardin, L. Bosisio, M. Koehler, U. Parzefall, S. Ronchin, L. Wiik, A. Zoboli, N. Zorzi, "Performance evaluation of 3D-DDTC detectors on p-type substrates," Nuclear Instruments and Methods A, vol. 624, pp. 459–464, 2010.
43. A. Zoboli, M. Boscardin, L. Bosisio, G. F. Dalla Betta, P. Gabos, C. Piemonte, S. Ronchin, N. Zorzi, "Characterization and modeling of signal dynamics in 3D-DDTC detectors," Nuclear Instruments and Methods A, vol. 617, pp. 605–607, 2010.
44. Leonardo, Roma, Italy. http://www.leonardocompany.com
45. G.-F. Dalla Betta, M. Boscardin, G. Darbo, C. Gemme, A. La Rosa, H. Pernegger, C. Piemonte, M. Povoli, S. Ronchin, A. Zoboli, N. Zorzi, "Development of 3D-DDTC pixel detectors for the ATLAS upgrade," Nuclear Instruments and Methods A, vol. 638-S1, pp. S15–S23, 2011.
46. G. Giacomini, A. Bagolini, M. Boscardin, G.-F. Dalla Betta, F. Mattedi, M. Povoli, E. Vianello, N. Zorzi, "Development of double-sided full-passing-column 3D Sensors at FBK," IEEE Transactions on Nuclear Science, vol. 60, no. 3, pp. 2357–2366, 2013.
47. J. Albert, et al. (The ATLAS IBL Collaboration), "Prototype ATLAS IBL modules using the FE-I4A front-end readout chip," Journal of Instrumentation, JINST 7, P11010, 2012.
48. E. Alagoz, M. Bubna, A. Krzywda, G. F. Dalla Betta, A. Solano, M. M. Obertino, M. Povoli, K. Arndt, A. Vilela Pereira, G. Bolla, D. Bortoletto, M. Boscardin, S. Kwan, R. Rivera, I. Shipsey, L. Uplegger, "Simulation and laboratory test results of 3D CMS pixel detectors for HL-LHC," Journal of Instrumentation, JINST 7, P08023, 2012.
49. M. M. Obertino, A. Solano, A. Vilela Pereira, E. Alagoz, J. Andresen, K. Arndt, G. Bolla, D. Bortoletto, M. Boscardin, R. Brosius, M. Bubna, G.-F. Dalla Betta, F. Jensen, A. Krzywda, A. Kumar, S. Kwan, C.M. Lei, D. Menasce, L. Moroni, J. Ngadiuba, I. Osipenkov, L. Perera, M. Povoli, A. Prosser, R. Rivera, I. Shipsey, P. Tan, S. Terzo, L. Uplegger, S. Wagner, M. Dinardo, "3D-FBK pixel sensors with CMS readout: First test results," Nuclear Instruments and Methods A, vol. 718, pp. 342–344, 2013.
50. M. M. Obertino, A. Solano, E. Alagoz, J. Andresen, K. Arndt, G. Bolla, D. Bortoletto, M. Boscardin, R. Brosius, M. Bubna, G.-F. Dalla Betta, F. Jensen, A. Krzywda, A. Kumar, S. Kwan, C.M. Lei, D. Menasce, L. Moroni, J. Ngadiuba, I. Osipenkov, L. Perera, M. Povoli, A. Prosser, R. Rivera, I. Shipsey, P. Tan, S. Terzo, L. Uplegger, S. Wagner, A. Vilela Pereira, M. Dinardo, "Performance of CMS 3D pixel detectors before and after irradiation," Nuclear Instruments and Methods A, vol. 730, pp. 33–37, 2013.
51. M. Povoli, C. Betancourt, M. Boscardin, G.-F. Dalla Betta, G. Giacomini, B. Lecini, S. Kuehn, U. Parzefall, N. Zorzi, "Characterization of 3D-DDTC strip sensors with passing-through columns," Nuclear Instruments and Methods A, vol. 730, pp. 38–43, 2013.
52. M. Bubna, E. Alagoz, M. Cervantes, A. Krzywda, K. Arndt, M. Obertino, A. Solano, G.-F. Dalla Betta, D. Menace, L. Moroni, L. Uplegger, R. Rivera, I. Osipenkov, J. Andresen, G. Bolla, D. Bortoletto, M. Boscardin, J. M. Brom, R. Brosius, J. Chramowicz, J. Cumalat, M. Dinardo, P. Dini, F. Jensen, A. Kumar, S. Kwan, C. M. Lei, M. Povoli, A. Prosser, J. Ngadiuba, L. Perera, I. Shipsey, P. Tan, S. Tentindo, S. Terzo, N. Tran, S. R. Wagner, "Testbeam and laboratory test results of irradiated 3D CMS pixel detectors," Nuclear Instruments and Methods A, vol. 732, pp. 52–56, 2013.
53. A. Krzywda, E. Alagoz, M. Bubna, M. Obertino, A. Solano, K. Arndt, L. Uplegger, G. F. Dalla Betta, M. Boscardin, J. Ngadiuba, R. Rivera, D. Menasce, L. Moroni, S. Terzo, D. Bortoletto, A. Prosser, J. Adreson, S. Kwan, I. Osipenkov, G. Bolla, C. M. Lei, I. Shipsey, P. Tan, N. Tran, J. Chramowicz, J. Cumalat, L. Perera, M. Povoli, R. Mendicino, A. Vilela Pereira, R. Brosius, A. Kumar, S. Wagner, F. Jensen, S. Bose, S. Tentindo, "Pre- and post-irradiation performance of FBK 3D silicon pixel detectors for CMS," Nuclear Instruments and Methods A, vol. 763, pp. 404–411, 2014.
54. M. Bubna, E. Alagoz, M. Cervantes, A. Krzywda, K. Arndt, M.M. Obertino, A. Solano, G. F. Dalla Betta, D. Menace, L. Moroni, L. Uplegger, R. Rivera, I. Osipenkov, J. Andresen, G. Bolla, D. Bortoletto, M. Boscardin, J. M. Brom, R. Brosius, J. Chramowicz, J. Cumalat, M. Dinardo,

P. Dini, F. Jensen, A. Kumar, S. Kwan, C. M. Lei, M. Povoli, A. Prosser, J. Ngadiuba, L. Perera, I. Shipsey, P. Tan, S. Tentindo, S. Terzo, N. Tran, S. R. Wagner, "Testbeam and laboratory characterization of CMS 3D pixel detectors," Journal of Instrumentation, JINST 9, C07019, 2014.

55. M. Povoli, A. Bagolini, M. Boscardin, G. F. Dalla Betta, G. Giacomini, F. Mattedi, E. Vianello, N. Zorzi, "Impact of the layout on the electrical characteristics of double-sided silicon 3D sensors fabricated at FBK," Nuclear Instruments and Methods A, vol. 699, pp. 22–26, 2013.
56. M. Povoli, G.-F. Dalla Betta, A. Bagolini, M. Boscardin, G. Giacomini, F. Mattedi, N. Zorzi, "Layout and Process Improvements to Double-Sided Silicon 3D Detectors Fabricated at FBK," 2012 IEEE Nuclear Science Symposium, Anaheim (USA), October 29–November 3, 2012, Conference Record, paper N14-204.
57. G. Pellegrini, M. Lozano, M. Ullán, R. Bates, C. Fleta, D. Pennicard, "First double sided 3-D detectors fabricated at CNM-IMB Nuclear Instruments and Methods A, vol. 592, pp. 38–43, 2008.
58. D. Pennicard, G. Pellegrini, M. Lozano, C. Fleta, R. Bates, C. Parkes, "Design, simulation, production and initial characterisation of 3D silicon detectors," Nuclear Instruments and Methods A, vol. 598, pp. 67–70, 2009.
59. D. Pennicard, C. Fleta, R. Bates, V. O'Shea, C. Parkes, G. Pellegrini, M. Lozano, J. Marchal, N. Tartoni, "Charge sharing in double-sided 3D Medipix2 detectors," Nuclear Instruments and Methods A, vol. 604, pp. 412–415, 2009.
60. A. Mac Raighne, K. Akiba, L. Alianelli, R. Bates, M. van Beuzekom, J. Buytaert, M. Campbell, P. Collins, M. Crossley, R. Dumps, L. Eklund, C. Fleta, A. Gallas, M. Gersabeck, E.N. Gimenez, V. V. Gligorov, M. John, X. Llopart, M. Lozano, D. Maneuski, J. Marchal, M. Nicol, R. Plackett, C. Parkes, G. Pellegrini, D. Pennicard, E. Rodrigues, G. Stewart, K. J. S. Sawhney, N. Tartoni, L. Tlustos, "Precision scans of the Pixel cell response of double sided 3D Pixel detectors to pion and Xray beams," Journal of Instrumentation, JINST, vol. 6, P05002, 2011.
61. C. J. Kenney, J. Hasi, S. Parker, A. C. Thomson, E. Westbrook, "Use of active-edge silicon detectors as X-ray microbeam monitors," Nuclear Instruments and Methods A, vol. 582, pp. 178–181, 2007.
62. S. I. Parker, A. Kok, C. Kenney, P. Jarron, J. Hasi, M. Despeisse, C. Da Vià, G. Anelli, "Increased speed: 3D silicon sensors; fast current amplifiers," IEEE Transactions on Nuclear Science, vol. 58(2), pp. 404–417, 2011.
63. Z. Li, "New BNL 3D-Trench electrode Si detectors for radiation hard detectors for sLHC and for X-ray applications," Nuclear Instruments and Methods A, vol. 658, pp. 90–97, 2011.
64. A. Montalbano, D. Bassignana, Z. Li, S. Liu, D. Lynn, G. Pellegrini, D. Tsybychev, "A systematic study of BNL's 3D-Trench Electrode detectors," Nuclear Instruments and Methods A, vol. 765, pp. 23–28, 2014.
65. G. Jia, J. Plentz, I. Höger, J. Dellith, A. Dellith, F. Falk, "Core–shell diodes for particle detectors," Journal of Physics D: Applied Physics, vol. 49, 065106, 2016.
66. G. Jia, U. Hübner, J. Dellith, A. Dellith, R. Stolz and G. Andrä, "Core-shell diode array for high performance particle detectors and imaging sensors: status of the development," Journal of Instrumentation, JINST, vol. 12, C02044, 2017.
67. G. Jia, U. Hübner, R. Stolz, J. Plentz, J. Dellith, A. Dellith, G. Andrä, "Compact 3D core-shell diode array for high performance particle detector and imaging sensor applications," presented at IWORID 2017, Cracow, Poland, July 2–6, 2017.
68. A. Aurola, V. Marochkin, T. Tuuva, "A novel 3D detector configuration enabling high quantum efficiency, low crosstalk, and low output capacitance," Journal of Instrumentation, JINST 11, C03040, 2016.
69. C. Da Vià, S. Parker, M. Deile, T.-E. Hansen, J. Hasi, C. Kenney, A. Kok, S. Watts, "Dual readout—strip/pixel systems Instruments and Methods A, vol. 594, pp. 7–12, 2008.
70. S. Parker, C. Da Vià, M. Deile, T.-E. Hansen, J. Hasi, C. Kenney, A. Kok, S. Watts, "Dual readout: 3D direct/induced-signals pixel systems," Instruments and Methods A, vol. 594, pp. 332–338, 2008.

71. C. Da Vià, M. Deile, J. Hasi, C. Kenney, A. Kok, S. Parker, S. Watts, G. Anelli, V. Avati, V. Bassetti, V. Boccone, M. Bozzo, K. Eggert, F. Ferro, A. Inyakin, J. Kaplon, J. Lozano Bahilo, A. Morelli, H. Niewiadomski, E. Noschis, F. Oljemark, M. Oriunno, K. Österberg, G. Ruggiero, W. Snoeys, S. Tapprogge, "3D active edge silicon detector tests with 120 GeV muons," IEEE Transactions on Nuclear Science, vol. 56, pp. 505–554, 2009.
72. L. Adamczyk et al., "Technical design report for the ATLAS forward proton detector," CERN-LHCC-2015-009; ATLAS-TDR-024, May 2015.
73. M. Albrow et al., "CMS-TOTEM precision proton spectrometer," Tech. Rep. CERN LHCC-2014-021. TOTEM-TDR-003. CMS-TDR-13, September 2014.
74. S. Parker, N. V.Mokhov, I. L. Rakhno, I. S. Tropin, C. Da Vià, S. Seidel, M. Hoeferkamp, J. Metcalfe, R. Wang, C. Kenney, J. Hasi, P. Grenier, "Proposed triple-wall, voltage isolating electrodes for multiple-bias-voltage 3D sensors," Nuclear Instruments and Methods A, vol. 685, pp. 98–103, 2012.
75. C. Da Vià, M. Boscardin, G.-F. Dalla Betta, G. Darbo, C. Fleta, C. Gemme, P. Grenier, S. Grinstein, T.-E. Hansen, J. Hasi, C. Kenney, A. Kok, S. Parker, G. Pellegrini, E. Vianello, N. Zorzi, "3D silicon sensors: Design, large area production and quality assurance for the ATLAS IBL pixel detector upgrade," Nuclear Instrumentation and Methods A, vol. 694, pp. 321–330, 2012.
76. M. Povoli, A. Bagolini, M. Boscardin, G.-F. Dalla Betta, G. Giacomini, E. Vianello, N. Zorzi, "Slim edges in double-sided silicon 3D detectors," Journal of Instrumentation, JINST 7, C01015, 2012.
77. G. F. Dalla Betta, M. Povoli, M. Boscardin, G. Kramberger, "Edgeless and slim-edge solutions for silicon pixel sensors", Proceedings of Science—22nd Workshop on Vertex Detectors (Vertex 2013), Paper 042, 2014.
78. M. Povoli, A. Bagolini, M. Boscardin, G. F. Dalla Betta, G. Giacomini, F. Mattedi, R. Mendicino, N. Zorzi, "Design and testing of an innovative slim-edge termination for silicon radiation detectors," Journal of Instrumentation, JINST 8, C11022, 2013.
79. G. F. Dalla Betta, C. Da Vià, M. Povoli, S. Parker, M. Boscardin, G. Darbo, P. Grenier, S. Grinstein, J. Hasi, C. Kenney, A. Kok, C.-H. Lai, G. Pellegrini, S. J. Watts, "Recent developments and future perspectives in 3D silicon radiation sensors," Journal of Instrumentation, JINST 7, C10006, 2012.

CHAPTER 5

Fabrication Technologies

5.1 GENERAL ASPECTS OF SILICON DETECTOR PROCESSING

Although the technology of solid-state integrated circuits (IC) has been continuously and steadily progressing since the 1940s, it was only in the early 1980s that its impact on silicon detectors fabrication became really important. It was in this period that the so-called planar process was successfully applied for the first time, with ion implantation of doped regions to form junctions on silicon substrates by Joseph Kemmer. Kemmer's breakthrough idea also exploited the passivating properties of silicon dioxide to obtain devices with low leakage currents. This was possible, thanks to the substantial reduction of surface currents typical of metal semiconductor (Schottky) junctions, which was the fashionable technology for fabrication of semiconductor sensors at that time [1, 2].

Since then, silicon sensor fabrication technology has continuously improved, following the trends of the microelectronics industry, allowing the production of more complex and reliable devices. However, despite the existence of many similarities among integrated circuits (ICs) and sensor processing technologies, some major differences exist:

- Silicon sensors have spatial resolutions of the order of microns, while the minimum features size of advanced transistors is of the order of tens of nanometers. Moreover, as the overall dimensions of a sensor reaches as many as tens of square centimeters, yield becomes a major concern since it scales with its active area.
- Integrated circuits are usually fabricated on Czochralski (CZ) low-resistivity substrates, whereas sensors typically require high-resistivity and high-purity silicon as starting material to reduce leakage currents and depletion bias voltages (V_{bias}) the latter depending on the junction space charge. Furthermore, the presence of impurities and dopants gives rise to the formation of stable defects after irradiation (as already pointed out in Chapter 3) and should therefore be kept to a minimum.

- The need for low-leakage currents introduces some important limitations in the fabrication process, with respect to the typical IC technology, and so sometimes the adoption of extrinsic or external gettering is required. This important technological concept is discussed in more detail later in this chapter.

5.1.1 Materials

Radiation detectors are normally fabricated on so-called 'detector-grade' silicon wafers. As already pointed out, these substrates are quite different from CZ substrates, which are commonly used for IC fabrication, since high resistivity and high purity are their main characteristics. High resistivity and high purity mean in practice that one can reach totally depleted detectors with a maximum signal-to-noise ratio, since the presence of a strong electric field is essential to ensure fast signals, high charge collection efficiency, high avalanche breakdowns, and reliable performance before and after irradiation. Furthermore, a high-purity material is necessary to obtain high minority carrier lifetimes (defining how long carriers will stay around without recombining), which is extremely important for the fabrication of reliable low-leakage current devices.

"Detector-grade" silicon is usually obtained by float zone (FZ) refinement of hyperpure silicon rods [3]. In this technique, induction heating by a radio frequency (RF) coil allows obtaining a free suspension of the silicon melt without any contact with contaminating materials. The incorporation of impurities (both dopants and contaminants) into the growing crystal is mainly determined by their solid solubility in the silicon melt and their equilibrium distribution (or segregation) coefficient [4]. The equilibrium segregation coefficient $K_0 = C_S/C_L$, represents the concentration of solute in the solid C_S to the one in the liquid C_L in thermal equilibrium. K_0 strongly depends on the size of the impurity atom, which means that large atoms are not easily incorporated into the silicon crystal, thereby resulting in a small (less than 1) segregation coefficient. As this is the case for most of the typical dopants and impurities used in silicon sensors, they preferentially segregate to the melt, which becomes progressively enriched with them as the crystal is being pulled from it. At this point, it is possible to purify the material by "multiple zone refining" where a short region called "zone" is melted and moved slowly through a relatively long rod of the solid, resulting in a smaller but purer solid by phase separation. This process can be repeated at different temperatures for further purification (since solidification at higher temperatures is inversely proportional to the concentration of impurities). An important exception to this rule is boron, whose segregation coefficient (0.8) is too close to unity to provide segregation in the melt, and since the boron evaporation rate from the melt is low, its concentration remains constant during the growth process. For this reason, different approaches are required for detector grade silicon production, depending on the type and concentration of doping needed. As an example, for p-type materials, a starting polycrystalline silicon (polysilicon) rod of a certain boron and phosphorus concentration is used, and then phosphorus is reduced by multiple FZ refinements until the concentration of boron with the desired concentration is reached. For n-type doping, as starting material, one would use polysilicon rods with a boron content that is as low as possible, where again phosphorus is reduced to the desired concentration by multiple

FZ refinements. N-type materials were generally preferred for detector fabrication until the early 2000s, while more recently p-type has become more popular for applications in high-radiation environments due to postirradiation effects such as "type inversion," as discussed in Chapter 3.

FZ silicon substrates with resistivity ranging from 1 kΩ cm to 30 kΩ cm, corresponding to dopant concentrations of ~10^{12} cm^{-3} and less, are usually available at wafer manufacturers, with doping nonuniformities lower than 30%. For such substrates, concentrations of undesirable impurities—among them oxygen and carbon—is typically lower than 10^{16} cm^{-3}, while minority carrier lifetimes are of the order of several milliseconds [5].

If really lower doping nonuniformities are required, neutron transmutation doping (NTD) can be used. This technique is based on exposing high-purity p-type silicon wafers to fast neutrons to form unstable silicon isotopes that finally decay into phosphorus, with the release of beta particles, according to the following nuclear reaction:

$$Si_{14}^{30} + n \rightarrow Si_{14}^{31} \rightarrow P_{15}^{31} + \beta^{-} \tag{5.1}$$

The resulting material is n-type, as the initial p-type material is "overcompensated" by phosphorus atoms, which are donors, with resistivity nonuniformities lower than 5%. This method can be used for silicon substrates with resistivity up to 5 kΩ cm [5].

More recently, other types of substrates have been used to fabricate silicon detectors, among them CZ, magnetic Czochralski (MCZ), and epitaxial substrates. This trend started in the late 1990s, when it was observed that using oxygen-rich silicon was beneficial for devices exposed to heavy irradiation, while carbon was detrimental. In fact, oxygen atoms can capture radiation-induced vacancies in silicon, thus reducing the formation of electrically active defects, particularly double-charged vacancies that act as acceptors, lowering the formation of radiation-induced "effective" space charge after irradiation [6]. (For further information on this topic, see Chapter 3.)

The first oxygen-rich silicon devices used diffusion-oxygenated float zone (DOFZ) wafers, where FZ wafers are oxidized at high temperatures (~1150°C) for 72 hours for optimal oxygen distribution in the volume [6, 7]. This technique resulted in an oxygen concentration of the order of 10^{17} cm^{-3}. However, depending on diffusion times and temperatures, the resulting oxygen profiles were largely nonhomogeneous, with peaks at the surface (where an oxide was deposited and therefore where oxygen generated), and much lower values at mid-wafer of about 1×10^{16} cm^{-3} for 24-hour diffusion time. Much higher and more uniform oxygen concentrations could be achieved in the case of CZ materials, with concentrations up to 1×10^{18} cm^{-3} and in the case of MCZ with several 10^{17} cm^{3} wafers [8]. A relatively high oxygen concentration could also be obtained for thin epitaxial layers due to the in-diffusion of oxygen from the CZ substrate, which acts as a seed for the silicon deposition, while thicker epitaxial layers have been shown to suffer from a largely nonuniform oxygen profile [9]. Today CZ and especially MCZ wafers with sufficiently high purity are available, and therefore with high resistivity; accordingly, they are expected to become increasingly important for radiation detector production, also considering that FZ ingots are limited to 6-inch diameters.

5.1.2 Technological Aspects

As pointed out at the beginning of this chapter, most silicon sensor technologies are based on the process developed by Kemmer [10]. Although this approach is well suited to standard sensors, it constrains the fabrication of more complex structures, requiring, for example, multiple junctions with deeper doping profiles. This is particularly the case for "monolithic detectors" where sensors and front-end electronics are processed on the same wafer. In technical terms, considerable results have been obtained using detector-compatible P-type Metal Oxide Semiconductor (PMOS) and Complementary Metal Oxide Semiconductor (CMOS) technologies, which again exploit "extrinsic gettering" techniques to counteract the detrimental effect of contaminants on carrier lifetimes [11, 12].

The basic processing steps involved in the fabrication of silicon sensors are as follows[1]:

5.1.2.1 Passivation Oxide Deposition

Silicon dioxide is an electrical insulator and a barrier material for diffusion of impurities, the discovery of which is known to be the primary reason why silicon is playing such a dominant role in modern microelectronics. Passivation of silicon surfaces by thermal oxidation is usually performed at the beginning of the fabrication sequence by creating an effective protection against contamination and mechanical damage and by reducing the formation of interface states responsible for surface leakage currents. The oxidation process can be divided into two categories, wet and dry; the wet is obtained in water vapor and the dry in pure oxygen environments.

Wet oxidation is typically used for growing thick isolation layers in integrated circuit (IC) processing since the growth rate is much higher in the presence of water, whereas dry oxidation is preferred for detector processing because it results in lower surface state densities and enhanced oxide quality. In this respect, <100> oriented silicon substrates provide better results than <111> ones in terms of both surface leakage currents and oxide charge density. This is because <111> orientation has a higher concentration of dangling bonds, resulting in a higher density of interface traps and consequently a higher amount of oxide charge.

Thermal oxidation is performed at temperatures ranging from 800°C to 1200°C; warm-up and cool-down steps in nitrogen atmosphere are usually performed before and after each oxidation, respectively, in order to bring the wafers into thermal equilibrium with the oxidation environment. By so doing, uniform oxide layers are obtained with minimal stress. As thermal oxidation has to be performed at high temperatures, only one initial oxidation step is suggested for the fabrication of silicon detectors [10]. If further protective oxide layers are required during the fabrication process, they should be more conveniently obtained by, for example, chemical-vapor-deposition (CVD), which can be performed at lower temperatures [15]. By using CVD techniques, thin films of several materials can be deposited on the substrate's surface by thermal decomposition and/or reaction of gaseous compounds.

If a silicon dioxide film is required over aluminum metallization, the deposition has to be performed at a temperature below the silicon–aluminum eutectic

[1] A detailed treatment can be found in several books like [13, 14, 15]

point (577°C).[2] The following reaction between silane (SiH_4) and oxygen is commonly taking place at temperatures between 300 and 500°C, at atmospheric pressure or in a reduced pressure system like low-pressure chemical vapor deposition (LPCVD), with the following reaction:

$$SiH_4 + O_2 \rightarrow SiO_2 + 2H_2 \tag{5.2}$$

The resulting oxide, known as low-temperature oxide (LTO), can be used as a final passivation layer. Oxide layers can also be deposited using plasma-enhanced CVD (PECVD) at 300°C using the following reaction:

$$SiH_4 + 2N_2O \rightarrow SiO_2 + 2N_2 + 2H_2 \tag{5.3}$$

As previously stated, oxide depositions prior to metallization can be performed at higher temperatures, with better results in terms of uniformity, step coverage, and a wider choice of possible reactions. As an example, the decomposition of the vapor produced from a liquid source, tetra-ethyl-ortho-silicate (TEOS), is used in a LPCVD system at about 800°C:

$$Si(OC_2H_5)_4 + O_2 \rightarrow SiO_2 + by\ products \tag{5.4}$$

Both LTO and TEOS oxides can be deposited either undoped or doped. In the latter case, phosphine, arsine, or diborane gases are typically used as dopant sources.

5.1.2.2 Silicon Nitride and Polysilicon Deposition

Besides silicon dioxide, an additional passivation layer of silicon nitride (Si_3N_4), whose density is higher than that of SiO_2, is considered helpful for protecting devices from external contaminations, as it provides a better barrier to the diffusion of mobile ions such as Na^+ and K^+. Silicon nitride can also be used as a dielectric material, especially when the integration of coupling capacitors is required [16]. In fact, thanks to its high dielectric constant ($\varepsilon_{nit} = 7.4$), silicon nitride allows high capacitances to be obtained while keeping the dielectric layer much thicker than possible for capacitors using oxide alone. This fact is extremely important to increase the process yield.

The use of silicon nitride introduces some complications in the process sequence since, unless properly patterned, it could have negative effects on the sensor noise by increasing the surface leakage current [17]. Silicon nitride is typically deposited in CVD reactors, both LPCVD and PECVD, starting from either silane or dichlorosilane. Thermal growth of silicon nitride is also possible, in principle, but is not practical. A typical reaction between dichlorosilane and ammonia is used in a LPCVD system at about 800°C, with the following reaction:

$$3SiH_2Cl_2 + 4NH_3 \rightarrow Si_3N_4 + 2HCl + 6H_2 \tag{5.5}$$

[2] The eutectic temperature is the lowest possible melting temperature over all of the mixing ratios for the involved component species, in this case silicon, oxygen, and aluminum.

These LPCVD films are typically hydrogen-rich, containing up to 8% hydrogen, and usually have high internal tensile stresses. As an alternative, silicon nitride can be used as a final passivation layer using a low-temperature (300°C) deposition by PECVD by the following reaction:

$$SiH_4 + NH_3 \rightarrow Si_xN_yH_z + H_2 \tag{5.6}$$

Such a nitride layer is nonstoichiometric[3] and is usually very rich in hydrogen.

For detector fabrication, polysilicon is generally used to fabricate bias resistors, or gate electrodes of MOS structures, but its "gettering" properties can also be exploited, as will be explained in detail in the following sections.

Polysilicon is usually deposited in an LPCVD system using thermal decomposition of silane, at a temperature ranging from 600°C to 700°C, as can be seen in the following reaction:

$$SiH_4 + 4NH_3 \rightarrow Si + 2H_2 \tag{5.7}$$

It can also be doped during deposition (in situ doping) by the addition of dopant gases such as phosphine, arsine, or diborane. More often, polysilicon is doped by high-temperature diffusion or ion implantation. Diffusion is normally preferred to obtain lower resistivity, but ion implantation is necessary for selective doping of poly-Si, as required for the realization of both coupling capacitors and high-ohmic resistors.

5.1.2.3 Junction Fabrication

The sensors' fabrication on high-resistivity silicon substrates requires the creation of both p^+ and n^+ doped regions to process junctions and ohmic contacts. To this purpose, ion implantation is the most commonly used method, as it can be performed at low temperature. In this way, lifetime degradation that often results from high-temperature processing can be avoided. Low-temperature processing also allows using a wide variety of barrier materials to mask the implantation, such as photoresist, oxide, nitride, aluminum, and other metal films, thus increasing the flexibility of the process design. Moreover, the control of the dopant dose introduced into the wafer is much better for ion implantation when compared with diffusion and a much wider range of doses can be reproducibly achieved. In addition, the p-n junctions that can be obtained by ion implantation are nearly abrupt and very shallow, reducing surface "dead layers"[4] and allowing for good detection efficiency even for low-energy radiation. For p^+ doping, boron is the most frequently used element since, being very light, it is easily implanted through a thin screen oxide layer up to a few

---

[3] Nonstoichiometric compounds are chemical compounds having an elemental composition whose proportions cannot be represented by integers; sometimes in such materials, a small percentage of atoms are missing, or too many atoms are packed into an otherwise perfect lattice.

[4] A dead layer on the surface of a sensor is the result of either natural or induced passivation, or it is formed during the process of contact fabrication. Charged particles passing through this region produce ionization that is incompletely collected and recorded, which leads to departures from the ideal in both energy deposition and resolution [5.18].

hundred nanometers, using photoresist as a barrier. For n^+ doping, phosphorus and arsenic can be used. Although the first detectors realized by Kemmer implemented arsenic for backside n^+ implantation [10], phosphorus is generally preferred because it is lighter than arsenic and can be implanted through a screen oxide layer, reducing the chance of contamination. Moreover, as will be explained further in the following, phosphorus is known to have much better gettering properties than arsenic [19, 20], thus resulting in lower leakage currents. In fact, the gettering effectiveness is much higher in the case of phosphorus diffusion than for phosphorus implantation [19]. For this reason, phosphorus diffusion is sometimes used to process so-called "n^+-n high-low" junctions. Such junctions are needed to lower the contact resistance between the semiconductor and the metal used to connect the sensor to the readout electronics.

As implantations always damage silicon surfaces to a certain extent, more markedly for heavy ions than for light ions, an annealing step is required to remove the effects of this damage. Therefore, wafers are typically heated to a temperature between 600°C and 1000°C. This thermal cycle also allows dopants to be electrically activated. On the one hand, a low-annealing temperature of about 600°C is usually sufficient for damage recovery, since this is enough for silicon atoms to move back into lattice sites, but it provides only partial dopant activation and results in low conductivity for the implanted regions. On the other hand, a higher annealing temperature would lead to better conductivity, but it could also increase the leakage current. For this reason, an appropriate trade-off between conductivity and leakage current must be found for optimal detector operation.

5.1.2.4 Etching and Metallization

For electrical contacts to the junctions to be possible, openings have to be processed on the oxide surface. This operation becomes possible by using either "wet chemical" or "dry plasma" etchings. Dry etching, in which a high-speed charged or neutral plasma beam is being shot at the interested region, is known to be a cleaner process than wet etching and therefore leads to a reduced chance of contamination. However, if the energy of the impinging plasma is high enough, it might induce displacement damage, with consequent deterioration of the sensor's electrical performance. For this reason, careful evaluation of the effects of these etching techniques on detectors' leakage currents should be performed to define an optimized process sequence.

To create a useful detector, junctions need to be covered with a good conductor to have a point of electrical contact, and aluminum is the most suitable material for this purpose. Aluminum can be deposited by evaporation or by sputtering, where a small percentage of silicon (1%) is added to aluminum to suppress its diffusion into the silicon bulk. This effect is particularly important for radiation detectors since the need for shallow junctions would make the aluminum diffusion into silicon give rise to spikes, causing catastrophic junction shorts. Other techniques based on the use of the so-called diffusion barriers can prevent the formation of spikes. They consist of the deposition of a stack of metal layers, including platinum, palladium, titanium, and tungsten, between silicon and aluminum, the presence of which would also allow for formation of a low contact resistance. However, as diffusion barriers introduce a non-negligible complication into the fabrication process, their use is

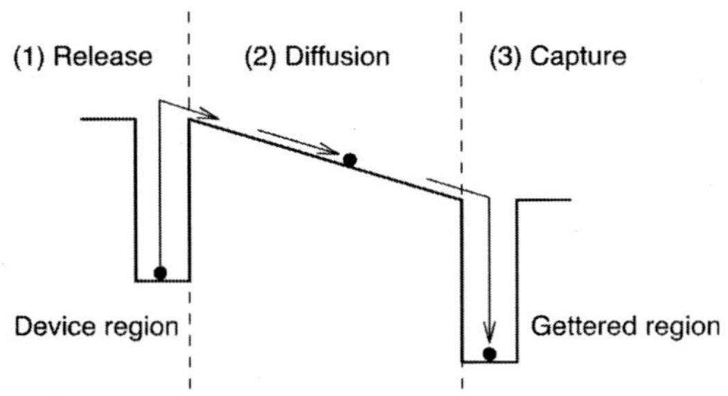

FIGURE 5.1 Concept of the gettering process involving release, diffusion, and capture of defects.

limited to cases where very shallow junctions are required. After aluminum deposition, the electrical contact with silicon would still be rather poor, and so it is normally improved by "alloying" the wafer. This operation, which is also called "contact sintering" consists in heating the wafers at a temperature of about 400°C for a few minutes in a forming gas environment consisting, for example, of 15% H_2/N_2.

5.1.2.5 Gettering

Imperfections in the crystalline structure such as vacancies or self-interstitials or various metallic impurities such as Fe, Au, Ni, Pt, Cu, and the like, are known to have a harmful influence on the semiconductor device operation, as they induce the formation of deep-level centers in the band gap. Such centers usually generate junction or bulk leakage currents and reduce the storage time of MOS capacitors. Impurity atoms can either be present in the crystal "as grown" or be introduced during device processing. Crystal imperfections can be due to processing, for example, during ion implantation, but they can also be caused by irradiation during device operation in highly radioactive environments. Although many efforts are always devoted to maintaining an ultra-clean processing environment,[5] it may not be sufficient to prevent contamination. Several techniques have been developed to counteract the detrimental effects of contaminating impurities during device processing, which are generically referred to as *gettering* [14].

Gettering is defined as the process used to reduce the amount of metallic impurities or crystalline defects in a wafer by confining them in regions away from the device active volume, where their presence would not be detrimental, or by completely removing them from the wafer. Gettering techniques can be performed before processing, or they can be an integral part of the processing sequence, depending on the technology employed. Regardless of the adopted technique, the effectiveness of gettering is strictly dependent on three basic processes [21] (see Fig. 5.1):

1. Release: impurities or defects need to be released (dissolve) to have them in a solid solution rather than in form of stable, immobile precipitates.

[5] Processing cleanrooms have a controlled level of contamination that is specified by the number of particles per cubic meter at a specified particle size. A "class 100" or "class 1000" refers to FED-STD-209E, an American standard that denotes the number of particles of size 0.5 μm or larger permitted per cubic "foot" of air.

2. Diffusion: impurities or defects should diffuse through the silicon substrate to the gettering sites.

3. Capture: impurities or defects should be captured by the gettering sites and should not be released during the subsequent heat treatments.

Gettering techniques can be classified into two groups: "extrinsic," which provides external sinks to the impurities, and "intrinsic," where sinks are created inside the bulk by thermal treatments. The presence of oxygen is fundamental for intrinsic gettering, as heat cycles can cause oxygen to aggregate and form precipitates that act as effective gettering sites [13]. A temperature sequence known as 'high-low-high' is commonly used to dissolve these precipitates, to diffuse them away from the surface, creating the so-called denuded zones, and to nucleate and grow them again in the silicon bulk. As a specific example, it is interesting to know that silicon wafers are used just as a mechanical support for integrated circuit fabrication, so that microelectronic devices can be conveniently processed in their denuded zone, usually on the surface, resulting in higher carriers lifetime and lower leakage currents [22]. However, since silicon sensors' active regions are typically extended to the whole substrate thickness, they would not be able to benefit from this specific effect. Furthermore, the concentration of oxygen in FZ silicon is very low, although, as already discussed, it could be increased by diffusion with the already discussed demonstrated benefits in parameter reproducibility of devices processed in different facilities and their use in highly radioactive environments [23].

Extrinsic gettering, in which the wafer back surface is typically used as a sink for impurities and defects, can, unlike intrinsic gettering, be conveniently applied for detector fabrication. There are several ways to perform backside extrinsic gettering, which are based either on the induction of chemical reactions among impurities and elements introduced at the surface or on the creation of mechanical stresses in the form of crystal damage at the wafer's surface [14].

In addition, heavy ion implantation of argon, or other chemical species like phosphorus and boron, have been shown to provide effective gettering [19, 24]. Another classical method consists of phosphorus diffusion followed by a segregation anneal [25, 26, 27]. Finally, the gettering action of the popular phosphorus can be technically explained as an increased substitutional solubility of impurities due to the formation of phosphorus-metal pairs and by extra crystal defects that are created during silicon phosphide precipitation [28, 29].

The CVD deposition of a silicon nitride film on the wafer backside has also been reported to be an effective gettering technique, thanks to the mechanical stress that it induces [14]. Moreover, according to several authors, the CVD deposition at a temperature of about 650°C of a thick layer of polysilicon of ~1 μm on the wafer backside is an effective gettering technique, due to the high density of defects produced by the grain boundaries and dislocations [14]. In fact, it was successfully proven that a poly-Si layer as a backside gettering sink for the fabrication of CMOS-compatible silicon detectors is possible [11, 30] with in situ doping of polysilicon, with phosphorus and leakage current densities as good as 1 nA/cm^2.

A backside gettering treatment consisting of the deposition of a 500 nm thick poly-Si layer, doped by phosphorus diffusion from a $POCl_3$ source, has also been used for the fabrication of silicon microstrip detectors with integrated JFET-MOSFET-based electronics, showing that good quality transistors can be obtained while preserving low diode leakage currents (~0.5 nA/cm^2) [31].

5.2 DEEP ETCHING TECHNIQUES

5.2.1 Deep Reactive Ion Etching

The fabrication of 3D sensors became possible following the rapid development of micro-electro-mechanical systems (MEMS) technologies during the early 1990s, and, in particular, following the availability of deep reactive ion etchers (DRIE), which allowed columnar openings with high aspect, or depth-to-diameter, ratios to be obtained. Today etching rates higher than 3 μm/min, selectivity to masking materials higher than 70:1, very good etching profiles, and nonuniformities lower than 5% across the wafer can be achieved [32]. The so-called Bosch process is currently a standard for DRIE [33] and consists of the repetition, during several cycles, of a two-part etching phase based on fluorine compounds, alternating nearly isotropic plasma etching steps (by SF_6) and sidewall passivation steps (by C_4F_8). In this way, high conformity, or anisotropy, can be achieved in the overall etching profiles, as can be seen in Fig. 5.2. After patterning the surface masking layer (Fig. 5.2a), a first isotropic etching is carried out by SF_6, with fluorine ions interacting with the unmasked regions of silicon to form SiF_4 gas (Fig. 5.2b); then SF_6 is replaced by C_4F_8, and CF_2 is obtained, forming a Teflon-like protective layer on all surfaces, including the walls of the openings (Fig. 5.2c). Because of the accelerating voltage between the plasma and the wafer, the following fluorine ion bombardment easily removes the protective film from the horizontal surfaces, including the bottom surface of the etched openings, whereas the continuously renewed sidewall protections resist remarkably well through the entire etching cycle (Fig. 5.2d). The repeated cycling of etching and coating causes the characteristic "scalloping" effect, which can be seen as "ripples" in the scanning electron micrograph of Fig. 5.2e.

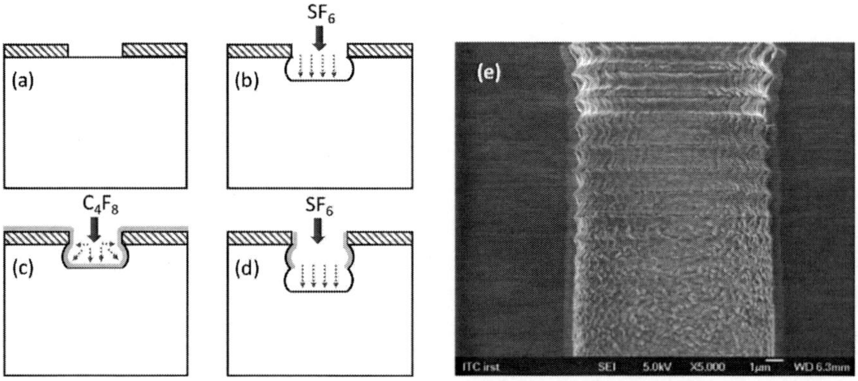

FIGURE 5.2 Schematic representation of the etching sequence by the Bosch process: (a) Patterning of the masking layer; (b) first shallow isotropic etching by SF_6; (c) protective layer deposition by C_4F_8; (d) removal of the protective film from the horizontal surfaces by directional ion bombardment, followed by second shallow isotropic etching. (e) Scanning electron micrograph showing a detail of the column sidewalls after DRIE etching with the characteristic "scalloping" effect (courtesy of FBK, Trento).

Aspect ratios of the order of 30:1 are possible today with this technique. These values should be kept in mind when evaluating the current limits achievable for the more aggressive electrode geometries in 3D technology.

5.2.2 Other Etching Techniques

Although deep reactive ion etching using the Bosch process is by far the most standard and reliable technique for 3D sensor fabrication, two alternative techniques for deep anisotropic etching are worth mentioning: laser drilling and photo-electro-chemical (PEC) etching. Although these techniques have not yet reached the same level of maturity as DRIE, they are interesting approaches for special applications. Comparison among them and DRIE is reported in [34] and, in more detail, in [35]. Laser drilling has the advantage of being material-independent and of providing a high aspect ratio. An example is reported in [36] where an aspect ratio of 100:1 was obtained for cylindrical opening arrays in silicon [36]. This technique can, however, induce non-negligible damage to the sidewalls and is not an easy scalable process since holes are etched one by one. Currently, laser drilling was successfully used in processing 3D diamond detectors, making it the center of several studies, which will be addressed in more detail in Chapter 9. PEC, another promising anisotropic etching option, allows for very high aspect ratios while inducing minor damage to the sidewalls. The etching rate, however, is slower than DRIE, and most of all, it depends on the silicon lattice orientation, with etched holes having square rather than circular shapes, which is not ideal for sensor behavior since sharp corners could always be the cause of spikes, electric field peaks, and consequent sensor breakdown.

5.3 FULL 3D DETECTORS WITH ACTIVE EDGE

The first 3D sensor prototypes were fabricated at the Stanford nanofabrication facility (SNF) in Stanford, California, in the early 1990s. The key aspects relevant to the fabrication technology used at SNF are addressed in [37] and further detailed in [38]. With reference to these papers, the main process steps are summarized in the following with the aid of Fig. 5.3.

a. Oxidation and Wafer Bonding

Oxidation is usually performed as a first step in any silicon detector fabrication, for surface passivation purposes while wafer bonding, which is a relatively recent technology, emerged mainly for MEMS applications (see [39] and the references therein). Wafer bonding consists in the adhesion of two wafers by exploiting hydrogen-bonding forces at room temperature. This process can be strengthened by further high-temperature annealing. For 3D detector fabrication a support wafer, which is afterward removed, is wafer-bonded to the sensor wafer for several reasons:

i. It provides high mechanical resistance, preventing the detector wafer from being damaged by cracks that might have developed during the electrode etching steps and other stress-inducing steps, so the process yield can be increased.

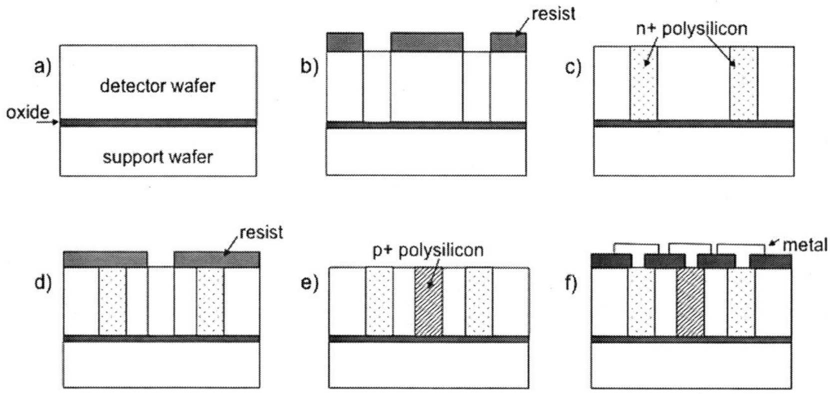

FIGURE 5.3 Schematic representations of the main steps in the fabrication process of 3D detectors at the Stanford nanofabrication facility. For more details, see the discussion in the text.

ii. It is essential for detectors with active edges, so they can hold together after trench etching, as will be seen at the end of this section.

iii. It allows the sensor wafer to be back thinned to the desired thickness without affecting the process yield, thus increasing design flexibility.

b. p^+ Hole Definition and Etching

Geometries of the p^+ columnar electrodes are defined by lithography and oxide etching. Then, deep, narrow openings are etched in the silicon bulk by DRIE, using both oxide and a thick photoresist layer as a protective mask. The photoresist is finally removed at the end of the etching cycle.

In earlier fabrication attempts, the achieved opening aspect ratio was at most 11:1, so that column diameters larger than 20 μm were produced for 220 μm deep electrodes [38]. As already mentioned, new DRIE etchers are now enabling aspect ratios of about 30:1, so that full through electrodes with diameters as low as 10 microns can be processed on 300 micron thick wafers.

c. p^+ Hole Filling and Doping

Electrode openings are fully filled with polysilicon deposited in an LPCVD reactor providing conformal coating by properly setting the temperature and pressure conditions. As for the columns doping, this step is performed by thermal diffusion from a boron source either before the poly-Si deposition or after a first poly-Si deposition step, partially filling the holes, and later followed by a second deposition. The fact that the columnar openings are completely filled with poly-Si offers some advantages. For example, during the lithography steps following the poly-Si deposition, the photoresist can be uniformly deposited or "spinned" on the wafer's surface without getting trapped in the opening voids. Furthermore, it has been shown that when a charged particle traverses the poly-Si electrodes, part of the generated charge could still be collected [40], provided that the carrier lifetime in poly-Si is not too short, since the charge motion in the electrodes is governed by diffusion. In particular, carrier lifetimes in poly-Si are proportional to grain boundaries size and correspond to 0.5 ns for a grains of 1 μm diameter. It has also been shown that thermal treatment after poly-Si deposition can cause recrystallization and increase in grain size [38]. Note that in [37], electrodes filling with crystalline silicon, using an epitaxial process, is also mentioned. This solution would ease charge collection from electrodes due to an enhanced electric field sustained by a doping gradient, but, unfortunately, it is very difficult to implement this solution in practice.

Poly-Si columns filling can cause additional problems, one of which is the coverage of the wafer's surface by a thick poly-Si layer of 10 μm or more. Etching such a thick layer of poly-Si is a difficult task and, as an alternative, chemical mechanical polishing (CMP) can be used [41]. In all cases, this step further increases the processing complexity.

d. n^+ Hole Definition and Etching

Similar to step (b), geometries of the n^+ columnar electrodes are defined by lithography and oxide etching; deep cylindrical apertures are etched by DRIE.

e. n^+ Hole Filling and Doping

Similar to step (c), holes are completely filled by polysilicon and doped by thermal diffusion from a phosphorus source. The thick poly-Si layer has then to be removed from the surface one more time.

f. Metal Deposition and Definition

An oxide layer is deposited, and then, contact holes are defined and etched through the oxide, following which metal is deposited and patterned. A final passivation layer could then be deposited and etched only in the probe/bonding pad regions.

This process sequence does not account for two additional ion implantation steps, which would be necessary in case signals are read out from n^+ electrodes. They are the isolation implantations (p-spray or p-stop) on both the top and bottom sides of the wafer.

Not shown in Fig. 5.3 is the support wafer removal, which could be performed either by etching, mechanical grinding, or CMP [41]. However, handling sensors after support wafer removal would be very difficult, so this step should be properly combined with interconnect processes, especially for pixel sensors requiring bump bonding.

The process shown in Fig. 5.3 requires the addition of a few additional steps to account for implementation of active-edges [42]:

- One of the two DRIE steps used to etch electrode openings can be used to etch border trenches at the same time. Due to the length of the trenches along the sensor's edge, which is at least a few mm, the etching gas and the etching products can enter and leave trenches more easily than they do for columnar openings. Therefore, to

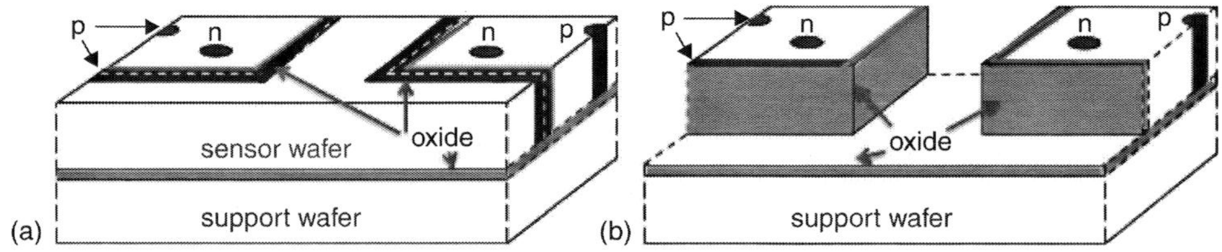

FIGURE 5.4 Schematic representation of the active edge process at the Stanford nanofabrication facility showing two adjacent detectors still bonded to a support wafer: (a) sensors after the holes and trenches for the n and p electrodes have been etched, doped, and filled; (b) sensors after the larger dicing trenches (reaching to the dashed white lines in the top diagram) have been etched to separate the sensors.

maintain comparable trench and electrode vertical etch rates, the trench width should be smaller than the column diameter. Trenches should then be filled with poly-Si and doped like columnar electrodes, becoming in this way "active" (see Fig. 5.4, left);

- An additional etching step is needed at the end of the process to remove all exceeding material surrounding the sensor, while leaving a few microns of poly-Si to protect the sensor's edge from mechanical damage and contamination from impurities (see Fig. 5.4, right).

This fabrication process is long and rather complex, since it involves several nonstandard steps. It should be stressed however, that such a process was developed at SNF, which is a research laboratory, where only a few batches were fabricated. More recently, the SNF 3D processing was transferred to the Norwegian Foundation for Scientific and Industrial Research, SINTEF (in Oslo, Norway) in view of potential industrial production [43]. Figure 5.5 shows

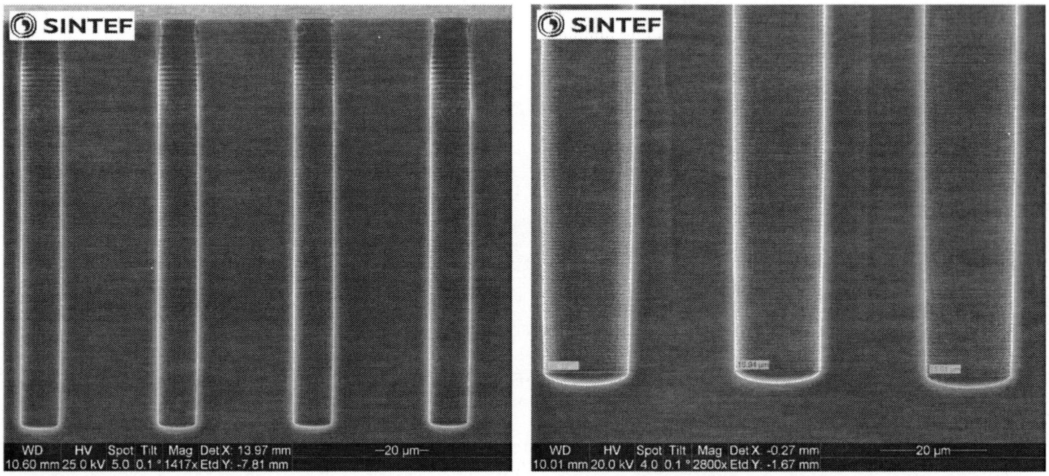

FIGURE 5.5 (Left) Cross section of holes etched at SINTEF with (right) details of the column (courtesy of SINTEF).

cross sections of columns etched at SINTEF obtained with scanning electron microscopy (SEM), showing excellent uniformity and a smooth profile also at the column end—evidence of good control of this fundamental processing step [43].

5.4 ALTERNATIVE APPROACHES

As mentioned in Chapter 4, several 3D architectures have been proposed to simplify the fabrication process. The positive and negative electrical and functional aspects of the different approaches have already been discussed. Here we will focus on the technological details of each design.

The main steps in the fabrication sequence of the "single-type-column" 3D-STC made at FBK [44] are briefly summarized in the following, with reference to the device schematic cross section and SEM micrograph of Fig. 5.6:

a. Boron implantation was used to obtain the ohmic contact on the backside and the isolation regions (p-spray or p-stop) between n^+ electrodes on the front-side.

b. Circular columns with a diameter of 6–10 μm were DRIE etched using thick oxide and photoresist layers as a mask.

c. Column n^+ doping was performed by phosphorus diffusion from a solid source. It extended to a toroidal region around the top of the openings to allow formation of contacts to the metal.

d. After doping, columns were only partially filled (passivated) with an oxide layer.

e. Contact openings were defined inside the surface region only, out of the column opening.

f. Aluminium sputtering was used for metal layer deposition.

The fabrication process of the VTT "Semi 3D" approach [45] was very similar. A difference, however, is present in the fact that the columns are hollow for FBK, while they are filled with in situ doped poly-Si for VTT.

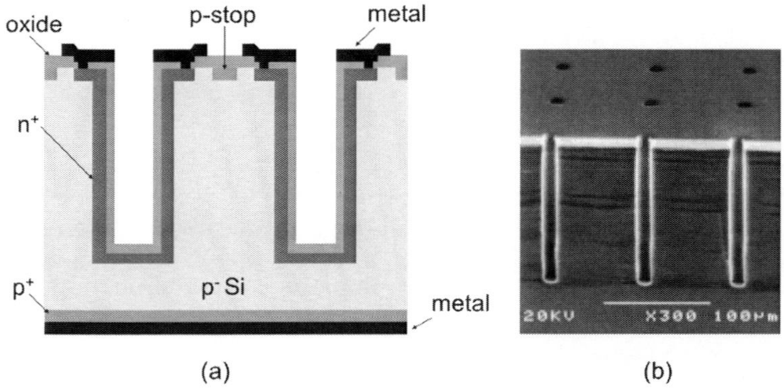

FIGURE 5.6 3D-STC detectors fabricated at FBK: (a) Schematic cross section (not to scale), and (b) SEM cross section of columnar electrodes (courtesy of FBK).

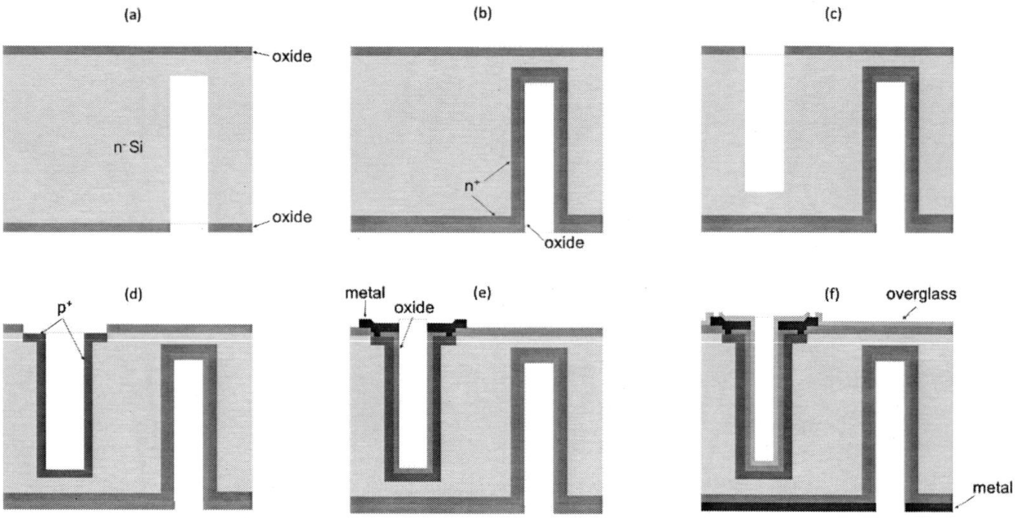

FIGURE 5.7 Main steps of the fabrication of 3D-DDTC detectors on n-type substrates at FBK. See text for discussion.

The fabrication technology was more complex in the case of double-sided, double-type-column 3Ds (3D-DDTC), but it still provided several advantages over single-sided full 3Ds from SNF, in particular since the support wafer was not necessary.

As an example, Fig. 5.7 illustrates the main processing steps for fabrication of 3D-DDTCs on n-type substrates at FBK [46]:

a. A thick oxide was grown and patterned on the backside to be used as a mask for the first DRIE process.

b. The thick oxide was etched from the backside while phosphorus was diffused from a solid source into the columns and at the surface. Later on, a thin oxide layer was grown to prevent the dopant from outdiffusing.

c. The oxide layer on the frontside was patterned, and the DRIE step was performed on the top surface defining the readout electrodes.

d. The thick oxide was removed from a circular region around all columns to allow boron to be diffused inside the columns and on the surface to ease contact formation.

e. A thin oxide layer was grown to prevent dopants from outdiffusion. Contact holes were defined and etched through the oxide. Then, aluminum was sputtered and patterned.

f. A final passivation layer was deposited and patterned to define the access to the metal layer, while on the backside aluminum was sputtered to produce a metal contact to the back electrode.

The fabrication steps for 3D-DDTC sensors made on p-type substrate are basically the same, apart from the obvious inversion of the column doping types and one additional

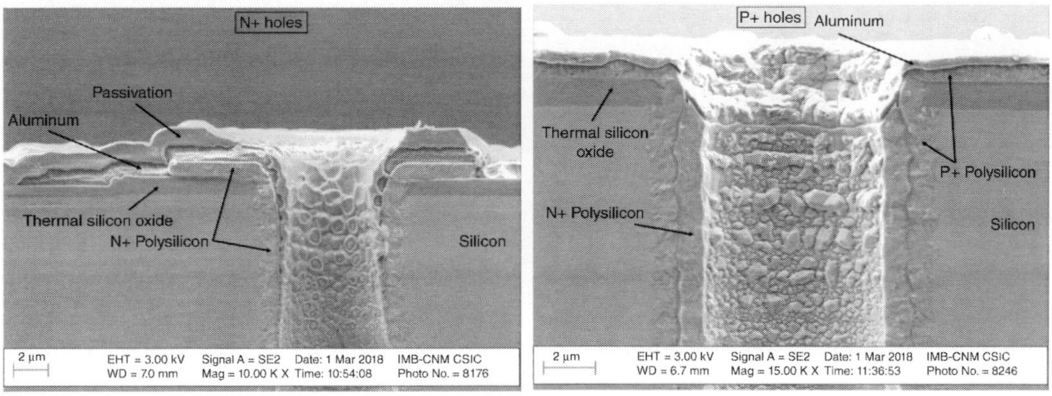

FIGURE 5.8 Cross section of (left) n^+ and (right) p^+ columnar electrodes in p-type silicon. The images show the top part of the columns with detailed views of the different layers (courtesy of IMB-CNM-CSIC).

process step to implant a p-spray surface isolation between n-columns on the frontside. If p-stops or a combination of p-stops and p-spray (moderated p-spray) are employed, an additional mask for the p-stop patterning is needed.

The fabrication process for double-sided 3D sensors proposed by CNM was similar to the one just described, apart from some significant details [47, 48]. In particular, the openings were partially filled by the deposition of ~1 μm of polysilicon, which was afterward doped with either a boron nitride solid source to form p^+ electrodes or $POCl_3$ to form n^+ electrodes. Unlike FBK, boron was only diffused into the columns and not on the backside surface. The p^+ columns were "passivated" by a thin, wet oxide grown directly on the polysilicon, while the n^+ columns were passivated by depositing 1 μm of TEOS inside the columnar openings. Surface isolation of the n^+ columns on the front side was obtained by p-stops. A metal layer was finally deposited on both wafer sides to produce contacts to the polysilicon. Detailed cross sections of device regions including n^+ and p^+ columns are shown in Fig. 5.8.

As already mentioned, for the ATLAS IBL production, FBK developed a variant of its 3D-DDTC process (called 3D-DDTC$^+$) where passing-through columns of both doping types could be made [49]. A schematic cross section of a device is shown in Fig. 5.9. Since

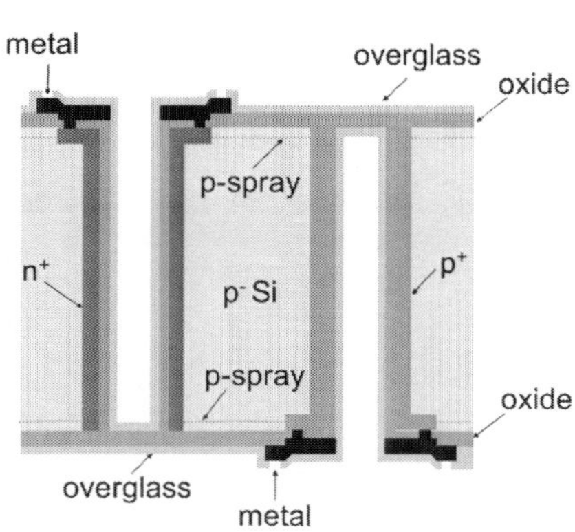

FIGURE 5.9 Schematic cross section of a 3D-DDTC detector with passing-through columns fabricated at FBK on a p-type substrate (not to scale).

the junction (n^+) columns are passing through, p-spray layers were processed on both wafer sides to ensure proper n^+ electrodes isolation. Passing-through columns also require patterning all the frontside and backside layers, making this process fully double-sided. Without a support wafer, mechanical fragility can be an issue. Special care should be taken to avoid accidental damage that could result in the wafers becoming very brittle, leading to breakage during processing. Furthermore, the DRIE steps were more critical than in full 3Ds, since the etch stop for DRIE, which in full-3Ds is provided by the thick bonding oxide layer between the sensor and support wafers, is only provided by a relatively thin stack of passivation layers.

Figure 5.10 shows a SEM micrograph of a set of columns, with those etched from the front-side (n^+) alternating with those etched from the backside (p^+). As can be seen, all columns penetrate all the way through the substrate, and although their shape is not perfectly cylindrical, but rather tends to shrink as the etching progresses through the substrate, the nonuniformity is limited. The maximum column diameter of 11 μm is slightly lower than the mask design value of 12 μm, while the minimum one is about 7 μm. A detail of the end of a frontside (n^+) column is also shown in the inset of Fig. 5.10, which shows that the shape is very uniform and no sign of damage appears on the membrane used as an etch stop.

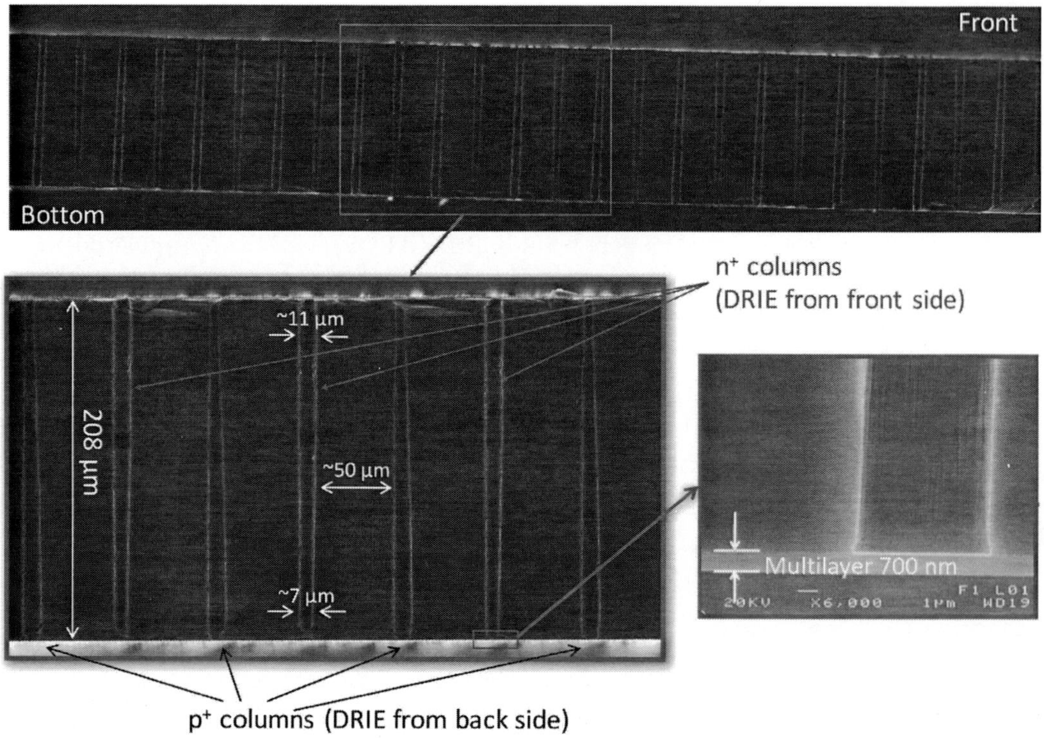

FIGURE 5.10 SEM micrographs of etched columns, with increasing magnification. The bottom right inset shows a detail at the end of a column with no sign of damage on the etch-stop membrane (courtesy of FBK).

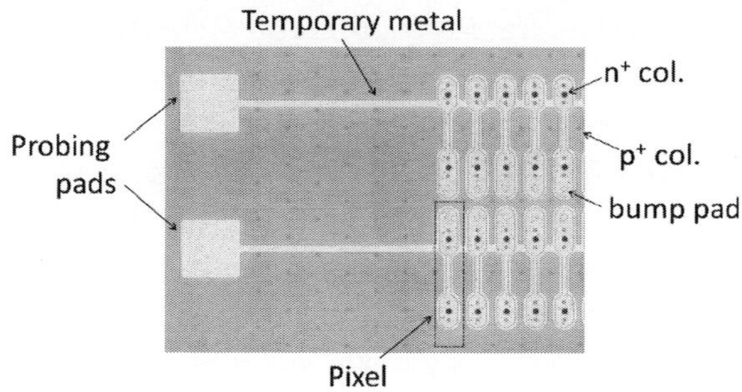

FIGURE 5.11 Micrograph of a detail of a FE-I4 compatible 3D pixel sensor, showing the temporary metal lines shorting rows of pixels and the probing test pads (courtesy of FBK).

Following an idea developed at SNF for the original full 3Ds, FBK introduced an additional step in the 3D-DDTC$^+$ processing useful for electrical characterizations of pixel sensors before bump bonding. A temporary metal was deposited over the passivation layer and patterned in strips that contacted the underlying pixels through the passivation openings at the bump-bonding sites, with probing pads at the end of each strip (see Fig. 5.11). The I-V curves of all strips were then measured and summed to obtain the total current of an entire pixel sensor. To mitigate the risk of mechanical damage during testing, the p-side of the wafers was protected by a passivation layer, which was patterned after completion of the measurements, along with removal of the temporary metal.

While the absence of a support wafer makes it impossible to fabricate active edges in any double-sided 3D detector technology, it allows the processing of slim edges. As mentioned in Chapter 4, for the IBL sensors, FBK implemented a slim edge based on a multiple fence of p^+ columns, without any process modification. Other geometries, like segmented trenches, can be used to improve the slim-edge effectiveness [50], at the expense of a further tuning of the etching step. In fact, etching segmented trenches and columnar openings with DRIE at the same time could result in nonuniform depths, since the etching rate could be different for different geometries. For this reason, the layout of the segmented trenches should be optimized to obtain similar etching times of the cylindrical electrodes openings.

5.5 RECENT DEVELOPMENTS

3D sensors have shown to be radiation tolerant by design and are therefore a promising technology for particle tracking at the high luminosity LHC (HL-LHC) [51]. The very high luminosity of 5×10^{34} cm^{-2} s^{-1} will cause event pile-up as high as 200 events/bunch crossing, thus requiring very high hit-rate capabilities, increased pixel granularity, extreme radiation hardness (up to a fluence of 2×10^{16} n_{eq}/cm^2), and reduced material budget. For the above reasons, significant improvements are required in both pixel sensors and readout electronics to be used in the innermost pixel layers of LHC tracking detectors. Since 2013, the CERN RD53 Collaboration has been working on the design of a new readout chip in

65 nm technology [52], while several R&D programs were launched to optimize different sensor technologies.

Compared to the existing ones, future 3D sensors will have significantly downscaled features, with pitches as small as 50 × 50 or 25 × 100 μm^2, interelectrode spacing as short as ~30 μm, electrodes as narrow as ~5 μm, and an active thickness as thin as ~100 μm [53]. Obtaining these challenging parameters requires dedicated efforts. The main effort is mainly on the technological side, and it can be tackled in two different ways.

An improved double-sided process shown in Fig. 5.12a was pursued by CNM for small-pitch sensors [54]. Due to mechanical fragility considerations, a minimum wafer thickness of ~200 μm is required, which is compatible with 4-inch diameter wafers, but too thin for 6-inch diameter ones. Moreover, since electrodes are dead regions, their impact on the geometrical inefficiency could become more important in small-pitch sensors, unless their volume is minimized. This demands aspect ratios as high as ~30:1, which can be obtained using cryogenic DRIE techniques [55]. As already mentioned in this chapter, besides being already a well-established process, the double-sided approach offers several important advantages: (1) the sensor bias can be applied from the backside, significantly simplifying the sensor assembly within a pixel detector system; (2) the front-side layout is less dense, which is a feature of paramount importance in small-pitch devices since it allows for a comfortable accommodation of bump pads; (3) with ~200 μm active thickness, the signal charge generated by a minimum ionizing particle remains high enough with high efficiency at a moderated tilted angle. But some disadvantages are also present: (1) the mechanical fragility (yield); (2) the limited accuracy (a few μm) of the alignment between the front- and backside of the wafers, which could be non-negligible for small-pitch sensors; (3) the larger geometrical capacitance, and (4) the larger pixel clusters for high-tilted particles, due to the ~200 μm active thickness.

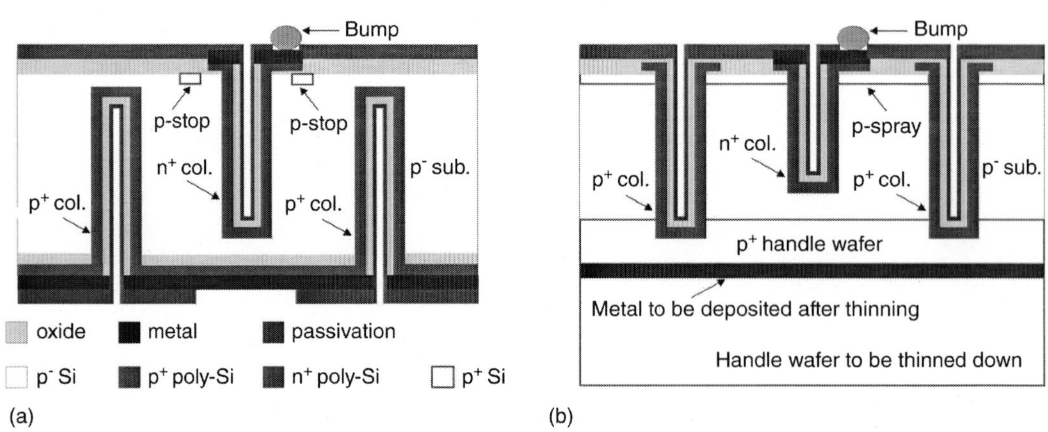

FIGURE 5.12 Schematic cross sections of small-pitch 3D sensors: (a) double-sided and (b) single-sided. (From Dalla Betta et al., Vertex 2016, Paper 42, 2017. [56] Copyright owned by the authors under the terms of the Creative Commons Attribution-NonCommercial-ShareAlike Licence. http://creativecommons.org/licenses/by-nc-sa/3.0/.)

As an alternative, FBK proposed a single-sided 3D process with support wafer, with modifications of the original SNF technology, mainly to ensure homogeneous backside sensor biases [53, 57, 58]. As an example, Fig. 5.12b shows the schematic cross section of a device fabricated on silicon–silicon direct wafer bonded (Si-Si DWB) substrates from IceMOS Technology Ltd. (Belfast, North Ireland). These substrates consist of a float zone high-resistivity layer of the required thickness directly bonded, therefore without an oxide layer in between, to a thick low-resistivity handle wafer. The p^+ (ohmic) columns are etched deep enough to reach the highly doped handle wafer, so that a low-resistance ohmic contact is achieved on the sensor backside. The latter can eventually be partially thinned with a postprocess, and a uniform metal layer can be deposited on it to improve the quality of the sensor bias contact. The high doping of the handle wafer also requires etching of the n^+ (readout) columnar electrodes to be stopped at a distance of ~20 μm from it in order to prevent early breakdowns. This technology would also be feasible using epitaxial wafers, and it was also demonstrated at FBK for silicon on insulator (SOI) wafers. In fact, it was proven that the p^+ columns could be etched through the wafer-bonding oxide to reach the heavily doped handle wafer [59].

Among the advantages of the single-sided solutions are the mechanical robustness provided by the thick handle wafer and the possibility of easily tailoring the active layer thickness to the desired value. With a thin active layer (~100 μm), the column width can be kept narrow, even though the etching aspect ratio is not improved, and all the device dimensions can be more easily downscaled. Among the disadvantages are the extra effort and cost for postprocessing the handle wafer thinning and for backside metal deposition. Moreover, the frontside layout can become quite dense for small-pitch sensors, making bump bonding demanding.

An interesting modified single-sided technology was proposed by CNM, based on a previous development of ultrathin 3D sensors [60]. The technology is very similar to the original one from Stanford and uses SOI wafers. However, as can be seen in Fig. 5.13, the process is completed by a local wet etching of the support wafer, patterned with a mask, visible on the bottom in the figure, and a metal deposition to access the backside of the active layer for sensor biasing [61].

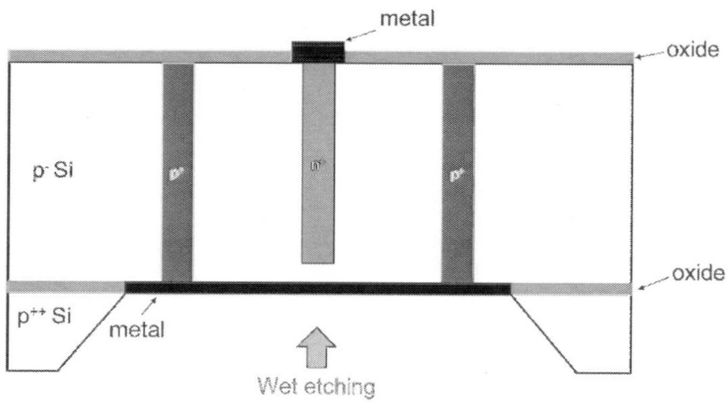

FIGURE 5.13 Schematic cross section of a small-pitch 3D sensor on SOI wafer proposed by CNM, with local wet etching of the handle wafer and a metal deposition on the backside.

REFERENCES

1. J. Kemmer, "Fabrication of low noise silicon radiation detectors by the planar process," Nuclear Instruments and Methods, vol. 169, pp. 499–502, 1980.
2. J. Kemmer, P. Burger, R. Henck, E. Heijne, "Performance and applications of passivated ion-implanted silicon detectors," IEEE Transactions on Nuclear Science, vol. 29(1), pp. 733–737, 1982.
3. W. von Ammon, H. Herzer, "The production and availability of high resistivity silicon for detector application," Nuclear Instruments and Methods, vol. 226, pp. 94–102, 1984.
4. R. Hull, "Properties of crystalline silicon," INSPEC, The Institution of Electrical Engineers, London, UK, 1999.
5. P. Dreier, "High resistivity silicon for detector applications," Nuclear Instruments and Methods A, vol. 288, pp. 272–277, 1990.
6. G. Lindstroem et al. (The RD48 collaboration), "Radiation hard silicon detectors—developments by the RD48 (ROSE) collaboration," Nuclear Instruments and Methods A, vol. 466, pp. 308–326, 2001.
7. Z. Li, W. Chen, L. Dou, V. Eremin, H. W. Kraner, C. J. Li, G. Lindstroem, E. Spiriti, "Study of the long term stability of the effective concentration of ionized space charges (Neff) of neutron irradiated silicon detectors fabricated by various thermal oxidation processes," IEEE Transactions on Nuclear Science, vol. 42(4), pp. 219–223, 1995.
8. M. Moll, E. Fretwurst, G. Lindstroem, "Investigation on the improved radiation hardness of silicon detectors with high oxygen concentration," Nuclear Instruments and Methods A, vol. 439, pp. 282–292, 2000.
9. E. Fretwurst, F. Hoenniger, G. Kramberger, G. Lindstroem, I. Pintilie, R. Roeder, "Radiation damage studies on MCz and standard and oxygen enriched epitaxial silicon devices," Nuclear Instruments and Methods A, vol. 583, pp. 58–63, 2007.
10. J. Kemmer, "Improvement of detector fabrication by the planar process," Nuclear Instruments and Methods, vol. 226, pp. 89–93, 1984.
11. S. Holland, "An IC compatible detector process," IEEE Transactions on Nuclear Science, vol. 36(1), pp. 283–289, 1989.
12. W. Snoeys, J. D. Plummer, S. Parker, C. Kenney, "PIN detector arrays and integrated read-out circuitry on high-resistivity float-zone silicon," IEEE Transactions on Electron Devices, vol. 41(6), pp. 903–912, 1994.
13. S. M. Sze, "VLSI technology," McGraw-Hill, New York, 1988.
14. W. R. Runyan, K. E. Bean, "Semiconductor integrated circuit processing technology," Addison-Wesley, New York, 1990.
15. S. Wolf, R. N. Tauber, "Silicon processing for the VLSI era," vol. 1, Lattice Press, 1986.
16. S. Okuno, H. Ikeda, T. Akamine, Y. Saitoh, K. Kadoi, Y. Kojima, "A stacked dielectric film for a silicon microstrip detector," Nuclear Instruments and Methods A, vol. 361, pp. 91–96, 1995.
17. S. Holland, "An oxide-nitride-oxide capacitor dielectric film for silicon strip detectors," IEEE Transactions on Nuclear Science, vol. 42(4), pp. 423–427, 1995.
18. B. L. Wall, J. F. Amsbaugh, A. Beglarian, T. Bergmann, H. C. Bichsel, L. I. Bodine, N. M. Boyd, T. H. Burritt, Z. Chaoui, T. J. Corona, P. J. Doe, S. Enomoto, F. Harms, G. C. Harper, M. A. Howe, E. L. Martin, D. S. Parno, D. A. Peterson, L. Petzold, P. Renschler, R. G. H. Robertson, J. Schwarz, M. Steidl, T. D. Van Wechel, B. A. VanDevender, S. Wüstling, K. J. Wierman, J. F. Wilkerson, "Dead layer on silicon p-i-n diode charged-particle detectors," Nuclear Instruments and Methods A, vol. 744, pp. 73–79, 2014.
19. T. E. Seidel, R. L. Meek, A. G. Cullis, "Direct comparison of ion-damage gettering and phosphorus-diffusion gettering of Au in Si," Journal of Applied Physics, vol. 46(2), pp. 600–609, 1975.
20. M. L. Polignano, G. F. Cerofolini, H. Bender, C. Claeys, "Gettering mechanisms in silicon," Journal of Applied Physics, vol. 64(2), pp. 869–876, 1988.

21. J. S. Kang, D. K. Schroeder, "Gettering in silicon," Journal of Applied Physics, vol. 65(8), pp. 2974–2985, 1989.
22. M. L. Polignano, G. F. Cerofolini, H. Bender, C. Claeys, "Lifetime engineering by oxygen precipitates in silicon," in "Solid State Devices,"G. Soncini and P. U. Calzolari (Editors), Elsevier Science Publishers B.V., pp. 455–458, 1988.
23. Z. Li, H.W. Kraner, "Gettering in high resisitive float zone silicon wafers for silicon detector applications," IEEE Transactions on Nuclear Science, vol. 36(1), pp. 290–294, 1989.
24. T. M. Buck, K. A. Pickar, J. M. Poate, C. M. Hsieh, "Gettering rates of various fast-diffusing metal impurities at ion-damaged layers of silicon," Applied Physics Letters, vol. 21(10), pp. 485–487, 1972.
25. R. Falster, "Platinum gettering in silicon by phosphorus," Applied Physics Letters, vol. 46(8), pp. 737–739, 1985.
26. D. Lecrosnier, J. Paugam, G. Pelous, F. Richou, M. Salvi, "Gold gettering in silicon by phosphorus diffusion and argon implantation: mechanisms and limitations," Journal of Applied Physics, vol. 52(8), pp. 5090–5097, 1981.
27. G. F. Cerofolini, M. L. Polignano, H. Bender, C. Claeys, "The role of dopant and segregation annealing in silicon p-n junction gettering," Physica Status Solidi, vol. a103, pp. 643–654, 1987.
28. W. F. Tseng, T. Kolj, J. W. Mayer, T. E. Seidel, "Simultaneous gettering of Au in silicon by phosphorus and dislocations," Applied Physics Letters, vol. 33(5), pp. 442–444, 1978.
29. A. Ourmazd, W. Schroter, "Phosphorus gettering and intrinsic gettering of nickel in silicon," Applied Physics Letters, vol. 45(7), pp. 781–783, 1985.
30. S. Holland, "Fabrication of detectors and transistors on high-resistivity silicon," Nuclear Instruments and Methods A, vol. 275, pp. 537–541, 1989.
31. G. F. Dalla Betta, M. Boscardin, P. Gregori, N. Zorzi, G. U. Pignatel, G. Batignani, M. Giorgi, L. Bosisio, L. Ratti, V. Speziali, V. Re, "A fabrication process for silicon microstrip detectors with integrated electronics," IEEE Transactions on Nuclear Science, vol. 49(3), pp. 1022–1026, 2002.
32. A. A. Ayón, R. L. Bayt, K. S. Bruner, "Deep reactive ion etching: A promising technology for micro- and nanosatellites," Smart Materials and Structures, vol. 10, pp. 1135–1144, 2001.
33. F. Laermer, A. Urban, "Milestones in deep reactive ion etching," Proceedings of the 13th International Conference on Solid-State Sensors, Actuators and Microsystems (TRANSDUCERS '05), Seoul, Korea, June 5–9, 2005, pp. 1118–1121.
34. G. Pellegrini, P. Roy, R. Bates, D. Jones, K. Mathieson, J. Melone, V. O'shea, K.M. Smith, I. Thayne, P. Thornton, J. Linnros, W. Rodden, M. Rahman, "Technology development of 3D detectors for high-energy physics and imaging," Nuclear Instruments and Methods A, vol. 487, pp. 19–26, 2002.
35. G. Pellegrini, "Technology development of 3D detectors for high energy physics and medical imaging," PhD Thesis, University of Glasgow (UK), 2002.
36. M. Christophersen, B. Phlips, "Laser-micromachining for 3D silicon detectors," 2010 IEEE Nuclear Science Symposium, Knoxville (USA), October 30–November 6, 2010, Conference Record, paper N15-2.
37. S. I. Parker, C. J. Kenney, J. Segal, "3D–A proposed new architecture for solid-state silicon detectors," Nuclear Instruments and Methods A, vol. 395, pp. 328–343, 1997.
38. C. J. Kenney, S. I. Parker, J. Segal, C. Storment, "Silicon detectors with 3-D electrode arrays: fabrication and initial test results," IEEE Transactions on Nuclear Science, vol. 46(4), pp. 1224–1236, 1999.
39. S. H. Christiansen, R. Singh, U. Gösele, "Wafer direct bonding: from advanced substrate engineering to future applications in micro/nanoelectronics," Proceedings of the IEEE, vol. 94(12), pp. 2060–2106, 2006.
40. J. Hasi, "3D—The Next Step to Silicon Particle Detection," PhD Thesis, Brunel University, London, UK, 2004.

41. S. Franssila, Chapter 16, “CMP: Chemical-Mechanical Polishing” in “Introduction to Microfabrication,” John Wiley & Sons, Chichester, UK, pp. 165–172, 2004.
42. C. J. Kenney, S. I. Parker, E. Walckiers, “Results from 3-D silicon sensors with wall electrodes: Near-cell-edge sensitivity measurements as a preview of active-edge sensors,” IEEE Transactions on Nuclear Science, vol. 48(6), pp. 2405–2410, 2001.
43. T.-E. Hansen, A. Kok, T. A. Hansen, N. Lietaer, M. Mielnik, P. Storas, C. Da Vià, J. Hasi, C. Kenney, S. Parker, “First fabrication of full 3D-detectors at SINTEF,” Journal of Instrumentation, JINST 4, P03010, 2009.
44. C. Piemonte, M. Boscardin, G. F. Dalla Betta, S. Ronchin, N. Zorzi, “Development of 3D detectors featuring columnar electrodes of the same doping type,” Nuclear Instruments and Methods A, vol. A541, pp. 441–448, 2005.
45. S. Eränen, T. Virolainen, I. Luusua, J. Kalliopuska, K. Kurvinen, M. Eräluoto, J. Härkönen, K. Leinonen, M. Palvianen, M. Koski, “Silicon semi 3D radiation detectors, 2004 IEEE Nuclear Science Symposium, Rome (Italy), October 16–22, 2004, Conference Record, Paper N28-3.
46. A. Zoboli, M. Boscardin, L. Bosisio, G.-F. Dalla Betta, C. Piemonte, S. Ronchin, N. Zorzi, “Double-sided, double-type-column 3D detectors at FBK: Design, fabrication and technology evaluation,” IEEE Transactions on Nuclear Science, vol. 55(5), pp. 2775–2784, 2008.
47. G. Pellegrini, M. Lozano, M. Ullán, R. Bates, C. Fleta, D. Pennicard, “First double sided 3-D detectors fabricated at CNM-IMB,” Nuclear Instruments and Methods A, vol. 592, pp. 38–43, 2008.
48. G. Pellegrini, J. P. Balbuena, D. Bassignana, E. Cabruja, C. Fleta, C, Guardiola, M. Lozano, “3D double sided detector fabrication at IMB-CNM,” Nuclear Instruments and Methods A, vol. 699, pp. 27–30, 2013.
49. G. Giacomini, A. Bagolini, M. Boscardin, G.-F. Dalla Betta, F. Mattedi, M. Povoli, E. Vianello, N. Zorzi, “Development of double-sided full-passing-column 3D Sensors at FBK,” IEEE Transactions on Nuclear Science, vol. 60(3), pp. 2357–2366, 2013.
50. M. Povoli, A. Bagolini, M. Boscardin, G. F. Dalla Betta, G. Giacomini, F. Mattedi, R. Mendicino, N. Zorzi, “Design and testing of an innovative slim-edge termination for silicon radiation detectors,” Journal of Instrumentation, JINST 8, C11022, 2013.
51. F. Gianotti, M. L. Mangano, T. Virdee, S. Abdullin, G. Azuelos, A. Ball, et al., “Physics potential and experimental challenges of the LHC luminosity upgrade,” European Physical Journal, vol. C39, pp. 293 Eur. Phys. J 333, 2005.
52. M. Garcia-Sciveres, J. Christiansen, Development of pixel readout integrated circuits for extreme rate and radiation, CERN-LHCC-2013-002, LHCC-I-024. http://rd53.web.cern.ch/RD53.
53. G. F. Dalla Betta, “3D silicon detectors,” Proceedings of Science—1st INFN Workshop on Future Detectors for HL-LHC (IFD 2014), Paper 013, 2015.
54. D. Vázquez Furelos, M. Carulla, E. Cavallaro, F. Förster, S. Grinstein, J. Lange, I. López Paz, M. Manna, G. Pellegrini, D. Quirion, S. Terzo, “3D sensors for the HL-LHC,” Journal of Instrumentation, JINST 12 (2017) C01026.
55. B. Wu, A. Kumar, S. Pamarthy, “High aspect ratio silicon etch: A review,” Journal of Applied Physics, vol. 108, 051101, 2010.
56. G.-F. Dalla Betta, R. Mendicino, D. M. S. Sultan, M. Boscardin, G. Giacomini, S. Ronchin, N. Zorzi, G. Darbo, M. Meschini, A. Messineo, “Small-pitch 3D devices,” Proceedings of Science—25th Workshop on Vertex Detectors (Vertex 2016), Paper 028, 2017.
57. G. F. Dalla Betta, M. Boscardin, G. Darbo, R. Mendicino, M. Meschini, A. Messineo, S. Ronchin, D. M. S. Sultan, N. Zorzi, “Development of a new generation of 3D pixel sensors for HL-LHC,” Nuclear Instruments and Methods A, vol. 824, pp. 386–387, 2016
58. G. F. Dalla Betta, M. Boscardin, R. Mendicino, S. Ronchin, D. M. S. Sultan, N. Zorzi, “Development of new 3D pixel sensors for Phase 2 upgrades at LHC,” IEEE Nuclear Science Symposium and Medical Imaging Conference (NSS—MIC’15), Conference Record, San Diego (USA), October 31–November 7, 2015, paper N3C3-5

59. G. F. Dalla Betta, M. Boscardin, M. Bomben, M. Brianzi, G. Calderini, G. Darbo, R. Mendicino, M. Meschini, A. Messineo, S. Ronchin, D.M.S. Sultan, N. Zorzi, "The INFN-FBK 'Phase-2' R&D Program," Nuclear Instruments and Methods A, vol. 824, pp. 388–391, 2016.
60. G. Pellegrini, F. Garcia, J. P. Balbuena, E. Cabruja, M. Lozano, R. Orava, M. Ullan, "Fabrication and simulation of novel ultra-thin 3D silicon detectors," Nuclear Instruments and Methods A, vol. 604, pp. 115–118, 2009.
61. J. Lange, Recent progress on 3D silicon detectors, Proceedings of Science—24th Workshop on Vertex Detectors (Vertex 2015), Paper 026, 2016.

CHAPTER 6

Radiation Hardness in 3D Sensors

6.1 INTRODUCTION

3Ds are sometimes referred to as "ultra" radiation hard sensors. For the sake of clarity, radiation hardness is defined here as the residual signal that is left after irradiation compared to the signal before irradiation, measured in the same conditions. This was also referred to as signal efficiency (SE) in earlier chapters. Sometimes radiation hardness is referred to as the maximum signal-to-noise ratio (SNR) after irradiation compared to the one before irradiation, meaning that the system, including the readout electronics, will have to withstand high-radiation levels and that the noise of the system, comprising the readout electronic noise due to sensor capacitance, will be low. In systems where the readout is "binary" and where a threshold is used to select the particle signal, the minimal threshold the electronics can tolerate without any noise is normally quoted. Binary systems are extensively used today in high-energy physics pixel detectors with so-called time over threshold (ToT) readout [1]. In ToT, the electronics records how long the signal is above a selected threshold within a selected time window. ToT is directly proportional to the amplitude of the signal and therefore is an indication of the energy released by the impinging particle. The risetime of the signal above threshold is inversely proportional to the capacitance of the sensors. A fast rise is an indication of a lower sensor capacitance. Sometimes SNR is used in this chapter to clarify concepts related to design choices. The radiation effects on p-type or n-type silicon substrates are basically the same for planar or for 3D sensors. This means that silicon, as a material, will react to radiation in the same way independently from the processing choice, so radiation-induced defects causing carrier trapping, increased leakage current, and space charge changes will follow the same behavior for both designs. However, 3D silicon sensors have a higher SE after irradiation compared to planar sensors with the same substrate thickness (but also for diamond or silicon carbide!) because of their geometrical configuration and the way the signal forms (see Chapters 2 and 3 for more details on Ramo's theorem and signal formation before and after irradiation).

To better understand these concepts, let's look again at how 3D sensors are operated when compared to traditional planar silicon sensors. Figure 6.1 shows a sketch (not to scale) of 3D (top) and planar (bottom) sensors, both having the same substrate thickness Δ traversed by a minimum ionizing particle (MIP). In 3Ds, different electrode configurations can be designed to adapt to the same readout pitch, as was shown in Chapters 4 and 5, meaning that if a pixel has a dimension of, for example, $250 \times 50\ \mu m^2$ one could use one, two, or more electrodes connected together to cover it, provided that the distribution of p-n electrodes is symmetrical to maintain as homogeneous as possible electric field in the volume. This is what is sometimes referred to as a "by-design" approach and is the main characteristics of microfabricated devices.

For planar sensors, the interelectrode distance is approximately the substrate thickness $L = \Delta$. As pointed out in Chapter 3, the signal generated by an MIP in an irradiated planar silicon detector depends on SE, which is inversely proportional to the number of traps generated by radiation in silicon, assuming that the electric field always traverses all the volume, and to the number of electron-hole pairs generated per micron of traversed material, *N*. The most probable signal generated over Δ microns in planar

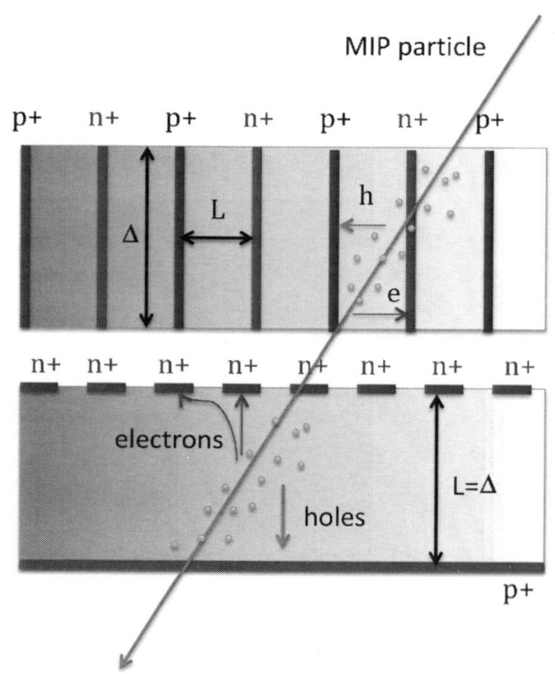

FIGURE 6.1 Schematic representation (not to scale) of the charge collection process in planar and 3D sensors after a minimum ionizing particle has traversed them. In the sketch, L = interelectrode spacing and Δ = substrate thickness. In planar structures L = Δ.

silicon devices is therefore $N \sim 75 \times \Delta$ for an MIP before irradiation. After heavy irradiation, SE ~ λ/L; thus:

$$S_{MIP\ planar} = SE \times N \sim 75\ \frac{\lambda}{L} \times \Delta\ \sim 75\lambda$$

which is independent of thickness since $\Delta = L$. The data shown in Fig. 6.2 clearly illustrate this point since the signal amplitude for all considered planar detectors tends to be ~2500 electrons at high fluences, hinting an effective drift length λ of about 33 microns at about 1.6×10^{16} n_{eq} cm^{-2}. In the 3D geometry, the decoupling between the interelectrode distance L and thickness Δ allows one to exploit the improved signal efficiency given by shorter interelectrode distances and the possibility of using the thickness as a "signal generator" also after irradiation. In this case, the signal generated by a MIP is approximately $S_{MIP3D} \sim 75\lambda \times \left(\frac{\Delta}{L}\right)$

So, by using a suitable substrate thickness Δ and reducing the interelectrode distance L, one can still improve S_{MIP}3D even if λ is strongly reduced by heavy irradiation. A 3D sensor's signal estimate for a 230 micron substrate thickness Δ, a 33 micron λ and 71 microns L gives S ~ 8000 electrons at 1.6×10^{16} n_{eq} cm^{-2}. The 3D curve in Fig. 6.2 obtained with a device with the above parameters shows a signal charge of ~5000 electrons and consequently a λ ~ 20 microns at 1.6×10^{16} n_{eq} cm^{-2} rather than 33 microns as for planar. This is a good indication that the electric field used was far from the one allowing for drift velocity saturation across the junction. Details on the experimental data later on will in

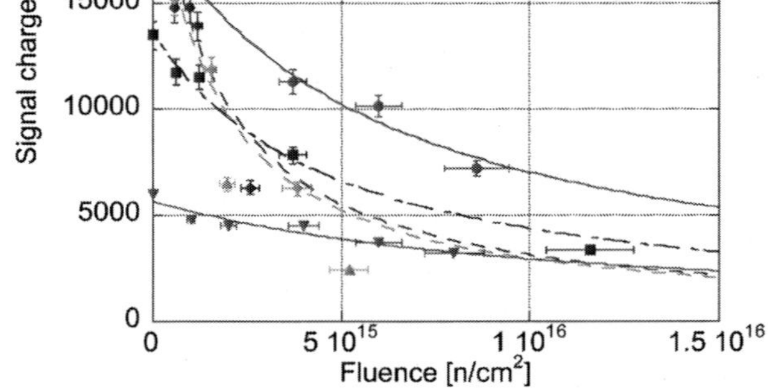

FIGURE 6.2 Signal charge data compilation for planar, 3D, and diamond sensors versus fluence (high-energy protons top scale and 1MeV equivalent neutron bottom scale). Data from [2, 3, 4, 5, 6, 7].

fact confirm this hypothesis. Projecting this estimate further for future devices, for example, with thicknesses Δ of 150 microns (to reduce multiple scattering) and pixel pitch of 50 microns, one would get $L = 35$ microns and, assuming $\lambda = 33$ microns, one would expect a signal after 1.6×10^{16} n_{eq} cm^{-2} of $75 \times 33 \times 150/35 \sim 10600$ e^-. For planar sensors, one would again expect $75 \times 33 \sim 2500$ e^- without charge multiplication.

When thinking about the design of 3D sensors for radiation-harsh environments, one should not be tempted to go for the extreme available technological conditions and increase the substrate thickness as much as possible while choosing the smallest interelectrode spacing L available. As will be seen shortly, the best results are obtained with optimization of the design to a specific application. Thus, the interelectrode spacing and the substrate thickness should be carefully selected depending on the expected radiation levels. This consideration is necessary because parameters like the sensor's capacitance, the electrode aspect ratio,[1] and the amplifier's input current need to be closely examined for the best system operation. For example, in 3D sensors the capacitance, and therefore the noise, increases with thickness contrary to planar sensors. Also, the electrode dimension increases with thickness, adding extra capacitance and reducing the overall sensitive volume, since the electrodes themselves are not efficient and most processes actually leave the electrode's center empty. The noise also increases with smaller interelectrode spacing. So thick substrates with an aggressive interelectrode spacing are only useful at extreme fluences where the noise increase due to the extra capacitance and the active volume reduction due to the larger electrode diameter (for a fixed aspect ratio) is counterbalanced by the preservation of the signal such that SNR is always maximized.

6.2 SOME HISTORY: INITIAL IRRADIATION TESTS

The first 3D detector prototypes were processed at the SNF in 1994–1995 on 121 μm thick p-type wafers, with active edges and a wafer-bonded support for mechanical robustness. These first prototypes had electrodes pitch of 100 μm, and the induced current of a low-intensity defocused pulsed infrared light-emitting diode to cover the entire sensor's surface was manually recorded and compared before and after low-energy proton irradiation of 1×10^{15} n_{eq} cm^{-2}. Figure 6.3 shows the signal height as a function of the applied bias voltages recorded after irradiation. The data, collected at room temperature, were plotted in arbitrary units due to the difficulty in controlling the measurement conditions in time. It was therefore impossible to extract the net signal charge. However, the shape of the signal heights gave already crucial information about the behavior of the sensors and in particular the full depletion voltage before and after irradiation. The nonirradiated detector (not shown) rapidly depleted at ~10 V, where the current value reached a plateau after a rapid increase. After irradiation, the full depletion voltage is ~70 V, as can be extracted from the knee point before plateau of the plot on the right. The two datasets visible in the

[1] Defined as the thickness versus electrode diameter in this case. The value is fixed and depends on the etching machine used. At present, it is common to get a value of ~25:1 for most etchers.

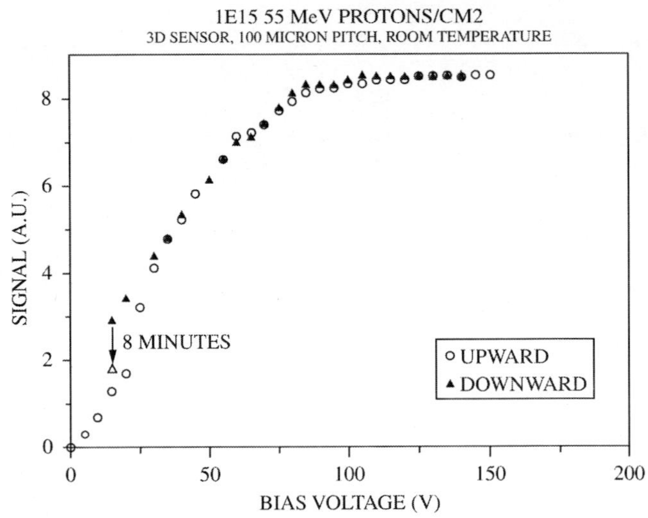

FIGURE 6.3 Signal height as a function of applied bias voltage in response to an infrared defocused laser beam after $1 \times 10^{15}\mathrm{cm}^{-2}$, 55-MeV proton irradiation (From Parker et al., IEEE Trans. Nucl. Sci., 2001. With permission of IEEE.) [8]

plot represent signals that were collected with increased (upward) and decreased (downward) applied bias voltage to check the reproducibility of the results and the absence of unwanted perturbing effects like hysteresis [8]. These bias voltages are much lower than those that would be necessary for the corresponding-thickness planar sensors irradiated at the same fluence. These promising results encouraged a few groups to pursue the investigation of this technology, despite the complexity of the fabrication and the skepticism that it would ever be possible to fabricate large sensors on a large scale with reproducible behavior.

To better explain what follows, it should again be stressed that 3D geometry has shown great flexibility in the use of readout electronics. Since both electrode types are accessible from the front and backside of the wafer, it is possible to process the readout electrodes to be compatible with both pixel and microstrip readout chips (alone or together) for both input polarities and on either side of the wafer. This approach has proven very handy in testing segmented structures since, on the one hand, single-channel (normally known as diodes) or strip structures can be better connected to the readout electronics after irradiation. Pixelated structures, on the other hand, because of their two-dimensional layout, require the connection process called bump bonding to be readout. Bump bonding can be performed with different soldering metals that require high-temperature steps at about 100°C or beyond. At this temperature, some radiation-induced defects are known to anneal, making any test obsolete. A picture of one of the devices used for irradiation tests supporting this concept is shown in Fig. 6.4: It is a test device processed on n-type high-resistivity silicon, with p-type active edges surrounding the perimeter of the device. As can be seen, the test structure has 10 rows of n-electrodes with a pitch of

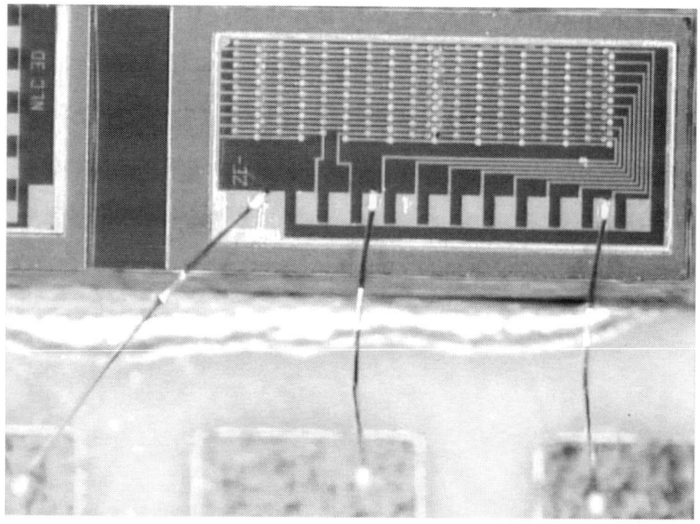

FIGURE 6.4 Picture of one of the 3D detectors with active edge devices used in irradiation experiments. Electrodes of the same type have been tied together to allow the entire sensor to be tested using only two probes. A third probe is being used for redundancy. In this present configuration, with 3 n + signal electrodes in each 50 × 400 μm^2 rectangle, the interelectrode spacing was 71 μm, the thickness was 0.23 mm, and the total volume was 0.367 mm^3. The pixel active volume was $(10 \times 0.05) \times (4 \times 0.40) \times 0.23\ mm^3 = 0.18\ mm^3$. The n+ strips end at the bonding pads at the bottom of the picture.

50 μm and a p-n interelectrode distance of 71 μm tied together with aluminum strips. Also, the p-electrodes are tied together with aluminum strips. The total area of the sensors is 1.6 mm^2, and the thickness was measured to be between 225 and 235 μm, giving an average volume of 0.367 mm^3. The active pixel volume was finally calculated to be $(10 \times 0.05) \times (4 \times 0.40) \times 0.23\ mm^3 = 0.18\ mm^3$ and was used to establish the radiation-induced damage parameter $\alpha = I/(V \times \phi)$.

The same strategy is used to form pixels, compatible, for example, with the ATLAS or CMS pixel readout electronics [9, 10, 11], where electrodes of the same type (n-type in this case) are tied together by aluminum strips to cover the pixel area of 50 × 400 μm^2 or 50 × 250 μm^2 or 100 × 150 μm^2 readouts. In Fig. 6.4, there are two bias pads organized vertically (for redundancy) on the bottom left of the figure. All the other pads are joined by an aluminum strip that is only visible under a microscope with a particular focus. These strips can be easily opened using a scalpel, allowing, if necessary, separate readout of each strip.

Several identical sensors to the one in Fig. 6.4 were irradiated with neutrons at three main 1 MeV equivalent fluences: $(3.7, 6.0, 8.6) \times 10^{15}\ n_{eq}\ cm^{-2\,u}$. Again, a low-intensity defocused and pulsed infrared 1060 nm laser with 700 ps rise time and 5 ns pulse width was used to generate the signal, which was then displayed and averaged 1000 times and then stored, after amplification, on a fast oscilloscope. The penetration length of this wavelength in silicon larger than 500 micron, so it is reasonable to use low-intensity IR light to

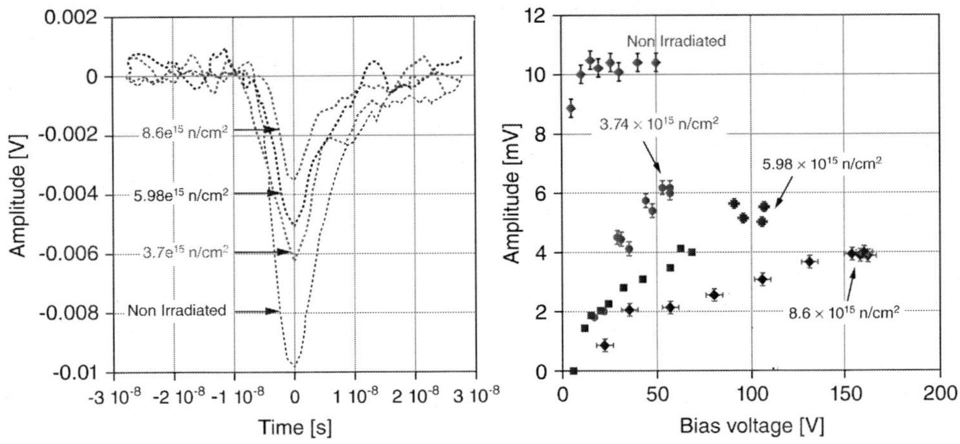

FIGURE 6.5 (Left) Averaged oscilloscope traces of the IR 1060 laser pulses from samples irradiated at different fluences. A response from a nonirradiated sample is inserted for comparison. Each trace in the figure is the average of 1000 pulses. (Right) Peak amplitude versus bias voltage scan of the measured samples. The nonirradiated sample is visible at the top of the plot and shows a full depletion voltage of 15 V, while the plateau amplitude is reached at 160 V for the most highly irradiated sample. (Copyright Elsevier 2008, Da Vià, C., et al., NIMA, 587, 2008. With permission.) [2]

simulate minimum ionizing particles. The laser intensity was calibrated, using a 300 μm parallel-plate silicon sensor, to inject a charge of 14 fC. This charge corresponds to ~3.5 minimum ionizing particles and should not perturb the charge status of the radiation-induced defects. Figure 6.5 (left) shows the entire set of averaged oscilloscope traces of the pulses collected at −10°C of the samples irradiated at different fluences, overlapped, and compared with the trace of the corresponding averaged pulses from a nonirradiated reference sample taken at the full-depletion bias voltage of the data shown in Fig. 6.5 (right). For the larger irradiation fluence, the highest applied voltage corresponds to 160 V. The maximum amplitude obtained for all fluences is probably underestimated, as pointed out previously when discussing the compilation of Fig. 6.2. This figure included the same 3D dataset, since the sensors could not be biased to plateau due to a wrong surface passivation that induced an early breakdown (see Chapter 3 for details on the mitigation of surface radiation damage to segmented sensors).

However, the results obtained already give important hints on the different performance of 3D sensors. Another interesting consideration should be reserved for the shape of the oscilloscope signals, which, as can be seen from the noise in Figure 6.5 (left), have not been filtered. They are all symmetrical (with some exception at the highest fluences, again due to the small bias voltage leading to a softer electric field), and mostly do not show the characteristic slow return time in segmented planar sensors due to the much longer hole drift times. This specific property is the consequence of the electric field distribution, which is identical throughout the thickness of the sensor and is pointing to potentially unique speed properties of this technology, as will be described in more detail in chapter 9.

6.3 DEVICES WITH A DIFFERENT ELECTRODE CONFIGURATION

Returning to irradiation results, more samples were fabricated on 210 micron thick p-type silicon substrate with the same perimeter but different interelectrode spacing to study the geometrical dependence of the signal efficiency. Devices were named 2E, 3E, 4E, and later 1E and 5E to identify how many electrodes had to be tied together to form pixels compatible with the 400 micron long ATLAS FE-I3 pixel readout.

A microscopic picture of the sensors and their layouts are shown in Fig. 6.6 (left and center), while averaged signals measured at 0°C versus bias voltage after irradiation are shown on the right. The irradiation was performed with neutrons at fluences of 7.55×10^{14}, 2.00×10^{15}, and 8.81×10^{15} n_{eq} cm^{-2}. The last-named was already a remarkable fluence for silicon devices! The picture at the top of Fig. 6.6 (left) shows a peculiar structure in which some pixels are connected together with vertical strips. These "ganged" pixels were implemented to completely cover the perimeter of the front-end chip, which needs an extra ~100 μm in the perimeter region, so the pixels at the edge of the sensors are extended by approximately the same amount. An extra electrode is used to cover this extended length, as can be seen in the picture. This arrangement allows one to "tile" assemblies together with virtually zero dead space.

The main difference among these configurations, as can be seen in the sketch, is the decreasing interelectrode distance and increased electrode density. The increasing number of electrodes leads to a smaller bias voltage for full depletion (from ~20 to ~5 V) and an increased volume with reduced, but when properly processed nonzero, efficiency corresponding to the sum of the central volume of each electrode. Tests with X-rays and 120 GeV muons (see Chapter 4) have shown that this inefficiency corresponds to a reduction of ~40% and ~60% of the bulk signal for the n^+ and p^+ electrodes, respectively [13]. As already shown, these voids would not be good for a traditional imaging detector, but in high-energy physics it is possible to mitigate this inconvenience by tilting the sensors to a 15° angle, with respect to an incident minimum ionizing particle beam, where, it was observed that this inefficiency is completely recovered (99.9%).

The data plots in Fig. 6.6 (right) show that the maximum pulse amplitude for the nonirradiated sensors is slightly different for the three different configurations. This can be explained by the fact that, depending on the electrode density, a fraction of the photons impinging the transparent region at an angle could be converted in the electrodes inside the bulk. The corresponding signal would then be averaged with the lower intensity electrode signals and therefore would be lower. Moreover, the fraction of pixel area covered by aluminum strips, which is 57%, 61%, and 65% for the 2E, 3E, and 4E configurations, respectively, would reflect the IR light. A maximum bias voltage of ~150 V was applied to the most heavily irradiated samples to reach the plateau. This corresponds to an average field along the line between the p and n electrodes of 1.4, 2.1, and 2.6 V/μm. It should be noted that after $\sim 2 \times 10^{15}$ n_{eq} cm^{-2}, the bias voltage needed for full depletion is ~100 V for the 3E and 4E configurations, with an available average signal efficiency SE of ~80%.

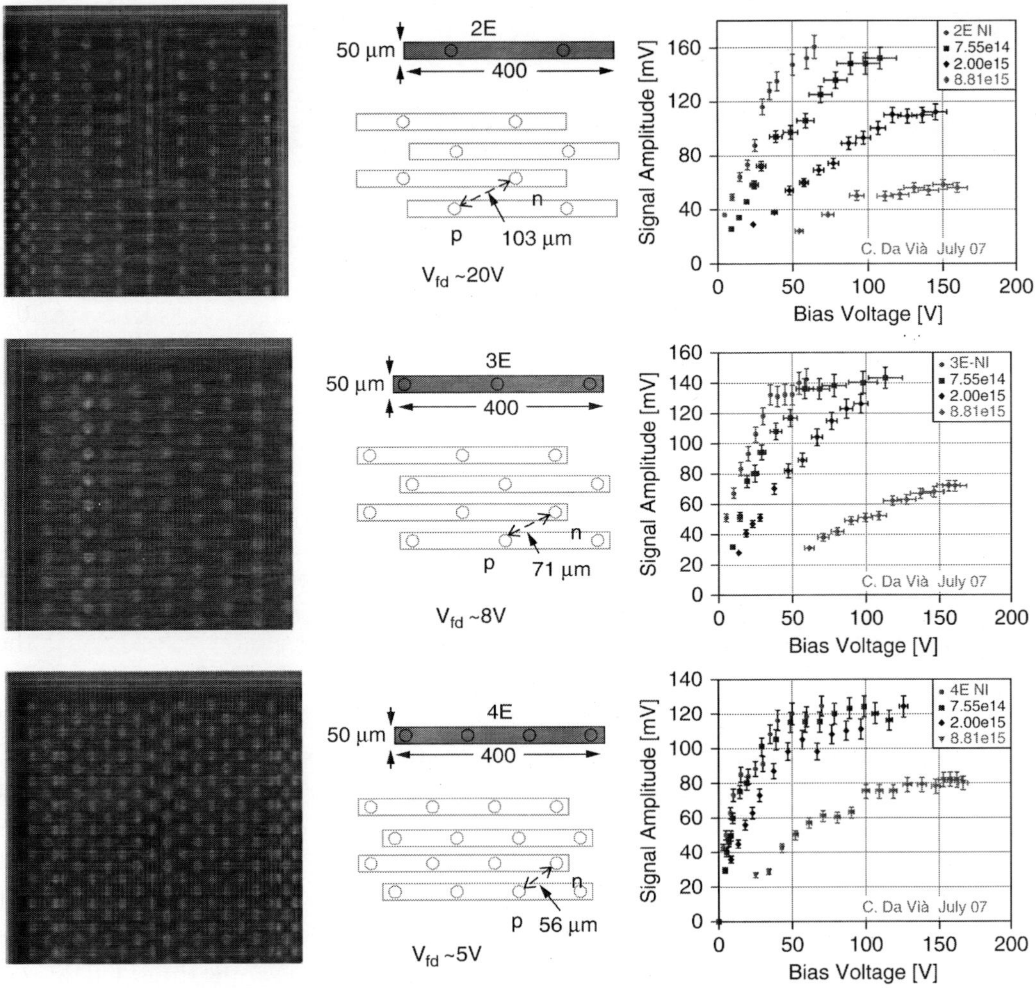

FIGURE 6.6 3D silicon sensors with 2E, 3E, and 4E electrode configurations. The number of electrodes used here is needed to cover the 400 μm length of an ATLAS pixel compatible with the so-called FE-I3 readout electronics chip. This results in a different p–n interelectrode distance of 103, 71, and 56 microns and a consequent different bias voltage needed to deplete the diode. On the right: peak amplitude versus bias voltage for the 2E (top), 3E (middle), and 4E (bottom) samples measured at 0°C. The nonirradiated sample's curves give the maximum signal amplitude in each plot. The nonirradiated full-depletion voltage should theoretically be 20, 8, and 5 V for the 2E, 3E, and 4E samples, respectively, while the plateau amplitude is reached at ~40 V for all three configurations. This anomalous behavior was shown to depend on the different oxide treatments during processing. About 150 V bias voltage is applied to the most heavily irradiated samples of each configuration to reach the plateau. The different maximum amplitude for the nonirradiated samples can be explained by a combination of inclined photons converting in the electrodes and reflections of the IR light from the aluminum strips. (Copyright Elsevier 2009, Da Vià et al., NIMA, 604, 2009. With permission.) [12]

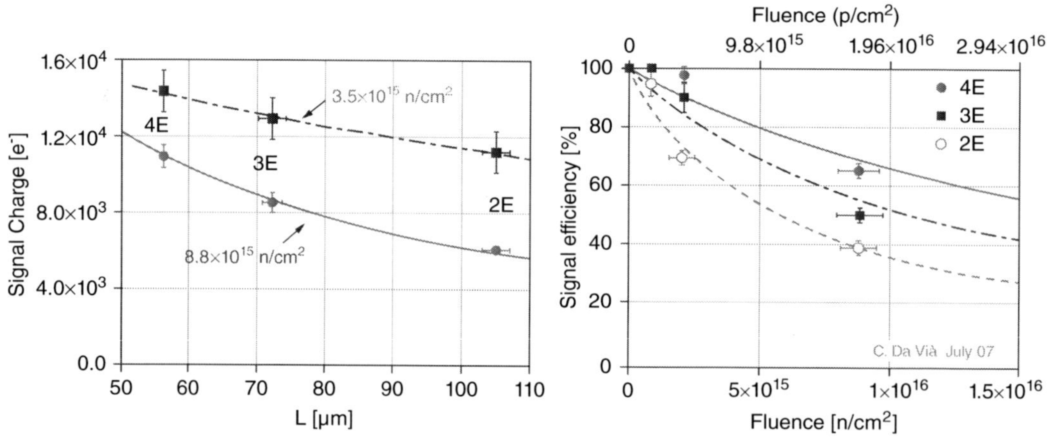

FIGURE 6.7 Left: Signal charge versus interelectrode distance L after 3.5×10^{15} and 8.8×10^{15} n_{eq} cm^{-2}. The curves show the improvement of the signal with reduced L for the same fluence. Right: Signal efficiency versus 1 MeV equivalent neutron (bottom legend) and 24 GeV/c proton fluence (top legend) per cm^2. The data are from the 2E, 3E, and 4E 3D samples. The fits extrapolate the signal efficiency up to 1.5×10^{16} n_{eq} cm^{-2}. (Copyright Elsevier 2009, Da Vià et al., NIMA, 604, 2009. With permission.) [12]

The normalized peak signal of the three configurations versus interelectrode distance L with the corresponding SE versus irradiation fluence can be seen in Fig. 6.7 (left and right, respectively). A study of the noise on the same set of 3D configurations (2E, 3E, and 4E) connected, in this case, to the ATLAS pixel FE-I3 electronics with pixel dimension 50×400 um^2, highlights the impact of the capacitance on the noise, since, as pointed out previously, the capacitance increases with the number of electrodes.

The equivalent noise charge ENC[2] of the entire pixel sensor was measured by injecting a fixed amount of charge into each pixel front-end chip and looking at the threshold dispersion over the entire matrix. This operation is possible since each front-end electronics chip is equipped with a test input capacitance. Noise versus bias voltage for 3D pixel configurations fabricated at Stanford with a 19:1 aspect ratio is shown in Fig. 6.8 [12]. The noise measured before irradiation for the various configurations corresponds to 200 e (2E), 275 e (3E), and 290 e (4E). These values are assumed not to change significantly after irradiation, so an extrapolation of the SNR is possible at the considered fluences, as summarized in Table 6.1.

TABLE 6.1 Summary of SNR values measured on sensors of different geometries after irradiation

Fluence (n_{eq} cm^{-2})	2E	3E	4E
3.5×10^{15}	56	47	49
8.8×10^{15}	30	31	38

[2] The equivalent noise charge (ENC) equals the number of electrons one would have to collect from a silicon sensor in order to create a signal equivalent to the noise of this sensor.

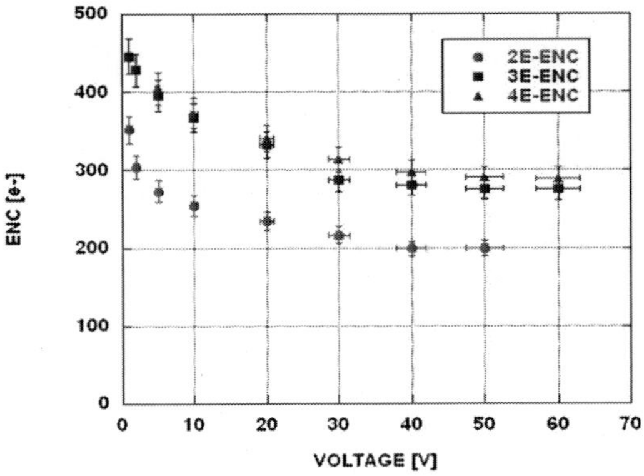

FIGURE 6.8 Equivalent noise charge (ENC) of the 2E, 3E, and 4E 3D sensors after bump bonding with the FE-I3 ATLAS pixel readout chip, as measured before irradiation. (Copyright Elsevier 2009, Da Vià, C., et al., NIMA, 604, 2009. With permission.) [12]

6.4 RADIATION HARDNESS OF 3D-STC (OR SEMI-3D) DETECTORS (FBK, VTT)

As already mentioned, single-type column (STC), where a single columnar readout electrode is partially penetrating the silicon substrate and the bias electrode is a planar implantation, represented the first step in 3D detector developments at FBK and VTT [14, 15]. As noted earlier, while this approach is simple from the technological point of view, since it does not require a support wafer and the number of processing steps is greatly reduced, it is not as radiation tolerant as fully penetrating columns 3D are. This is because the electric field is not as strong and homogeneously distributed (see Chapter 4 for more details).

To verify this concept, STC sensors produced at FBK were irradiated with neutrons up to fluences of 5×10^{15} n_{eq} cm^{-2} and the induced signal by a fast pulsed infrared laser (TCT) on one readout strip was measured for different incident points of the laser spot [16]. Figure 6.9

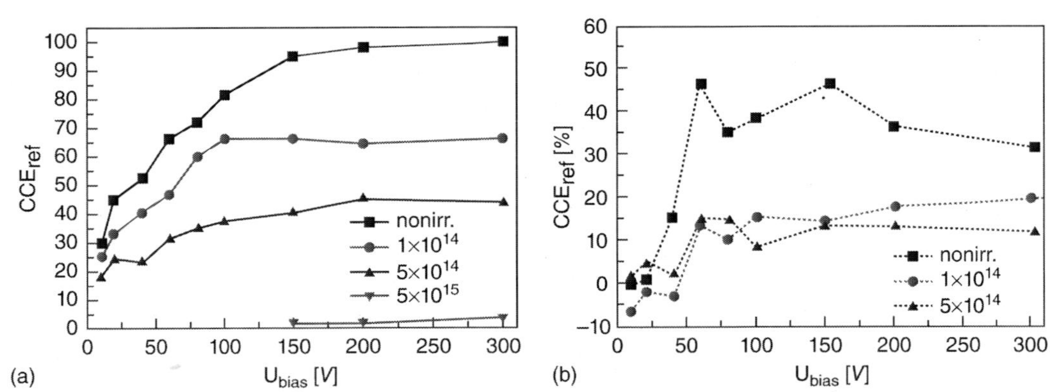

FIGURE 6.9 Relative charge collection efficiency versus reverse bias in 3D-STC detectors irradiated with neutrons at different fluences for different incidence points of the laser beam: (a) close to the readout electrode and (b) at the center of the cell. Data were extracted by TCT measurements, normalizing the signals after irradiation to the signals before irradiation. (From Zavrtanik et al., 2007 IEEE Nuclear Science Symposium. With permission of IEEE.) [16]

shows results for just two points: the first (point 1) is at few microns from the readout electrode (Fig. 6.9a), and the second (point 3) is at the center of a four-column cell[3] (Fig. 6.9b). Both plots refer to the induced signal integrated over 25 ns. As can be seen, the SE reaches 40% after $5 \times 10^{14}\ n_{eq}\ cm^{-2}$ for point (1), whereas for point (3) the SE is only 10% at the same fluence. This is a confirmation of the nonuniform response of the 3D-STC configuration even at large applied bias voltages. At the largest fluence, the signal efficiency is less than 5% also for point 1, further proving that 3D-STC detectors are not the best choice for extreme radiation environments.

Similar measurements were performed at the University of Freiburg with a 980 nm infrared laser and where 3D-STC strips were connected to the ATLAS ABCD3T readout chip [17]. Results after irradiation with 25-MeV protons at $1 \times 10^{15}\ n_{eq}\ cm^{-2}$ are shown in Fig. 6.10 [18]. As can be seen from the images, the induced signals are not uniform on the studied region delimited by four columns. The low signals, in particular in the central

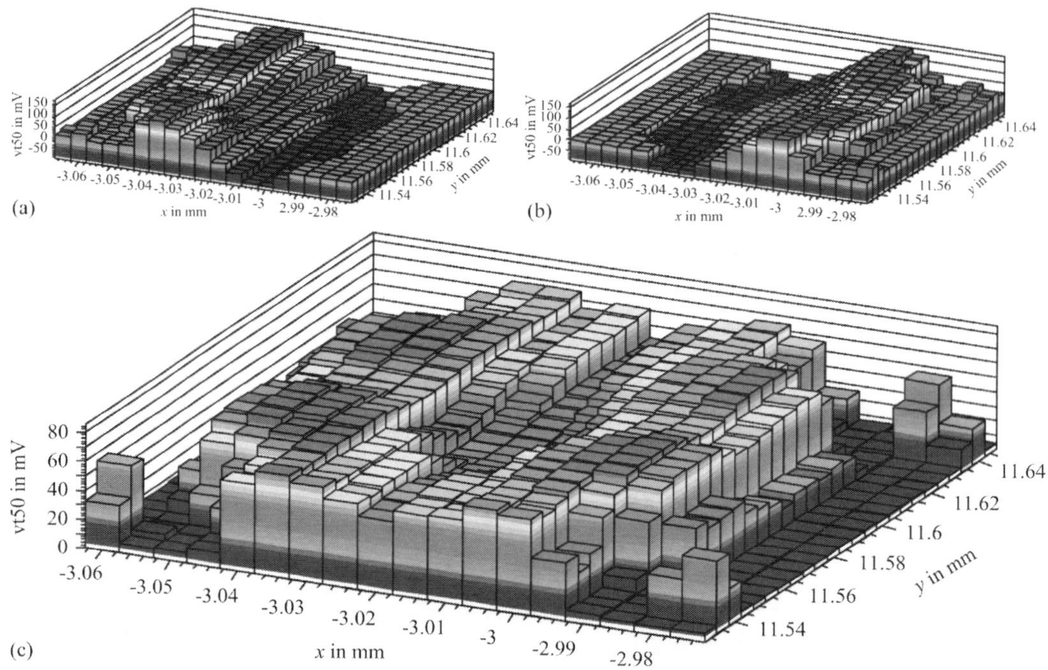

FIGURE 6.10 Results of a position-resolved scan over a square cell at a bias voltage of 110V after proton irradiation at $1 \times 10^{15}\ cm^{-2}$: (a) and (b) represent the signals induced on each single strip, whereas (c) shows the sum from both strips. (Copyright Elsevier 2007, Eckert et al., NIMA, 581, 2007. With permission.) [18]

[3] This is the region where the electric field is the weakest. See Chapter 4 for more details.

region visible in Fig. 6.10c, are probably a combination of charge trapping, together with the above-mentioned low electric field.

Complementary studies by the same group with MIP radioactive sources [19] and in a 180 GeV pion beam [20] further confirmed the behavior of such structures.

With regard to 3D-STC sensors developed at VTT Finland (Semi 3D), samples were irradiated with protons at CERN up to 1.0×10^{16} p/cm^2 showing a depletion voltage below 100 V even at a fluence 6.0×10^{15} p/cm^2, but they could not be measured at the highest fluence because of very high leakage currents [15].

6.5 RADIATION HARDNESS OF 3D-DDTC DETECTORS (FBK, CNM)

As already shown in the previous chapters, 3D double-sided double-type column (3D-DDTC) detectors were independently proposed by FBK [22] and CNM [23] in slightly different versions. The idea behind a 3D sensor with a double-side process was motivated by the possibility of achieving performance comparable to full 3D detectors while reducing the process complexity. The success of this approach proved to be a fundamental advantage in the case of the first medium-volume production of these devices and finally their use in an experiment.

In the first 3D-DDTC, as already stated, both columns did not penetrate fully the wafer substrate leaving a region, ideally of about 30 microns between the column tip and the surface. This region would therefore have a weak electric field across it, which would require the application of substantial bias voltages to be sensitive, and even more after irradiation.

An example is shown in Fig. 6.11. [24, 25], where signals before and after irradiation, versus bias voltage, are shown for devices fabricated with full-through (FBK) or partially

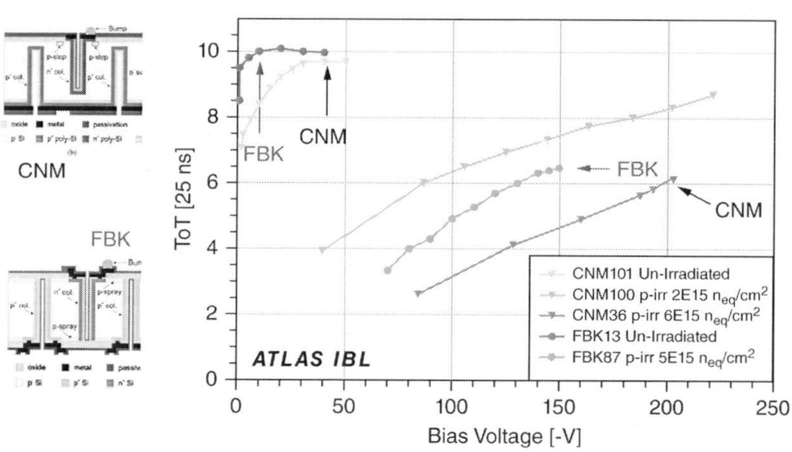

FIGURE 6.11 Signals from 3D sensors with full-through FBK electrodes compared with partially through CNM electrodes before and after irradiation where the different electric field distribution is clearly visible from the bias voltage required for full response to minimum ionizing particles. (Copyright Elsevier 2014, Da Vià, C., NIMA, 765, 2014. With permission.) [24]

through (CNM) geometries (sketches on the left) with identical interelectrode distance. As can be seen from the nonirradiated signals at the top left of the plot, there is already a significant difference between the two designs. FBK fully depletes at about 10 V, while CNM requires about 30 V to get a full signal. After irradiation (bottom curves), the difference is even more evident, since similar signals are reached by both devices after the same fluence, but FBK reaches it with 150 V while CNM requires about 200 V to allow a strong electric field to reach the regions between tips and surface.

The data shown in Fig. 6.11 suggested that a possible use could be made of the low electric field volumes in partially through electrodes. With a proper bias voltage, one could select only the p-n column overlapping regions while "suppressing" the low-electric field ones increasing the intrinsic resolution inside the sensor volume [26]. Figure 6.12 shows simulations of this concept.

The partially through column design has shown other potential benefits such as improved breakdown voltages before and after irradiation [27] and also charge multiplication due to "tip" effects. This can be seen in Fig. 6.13, where an excess of signal amplitudes, above the original nonirradiated signals, after irradiation above 200 V is clearly visible. Data in

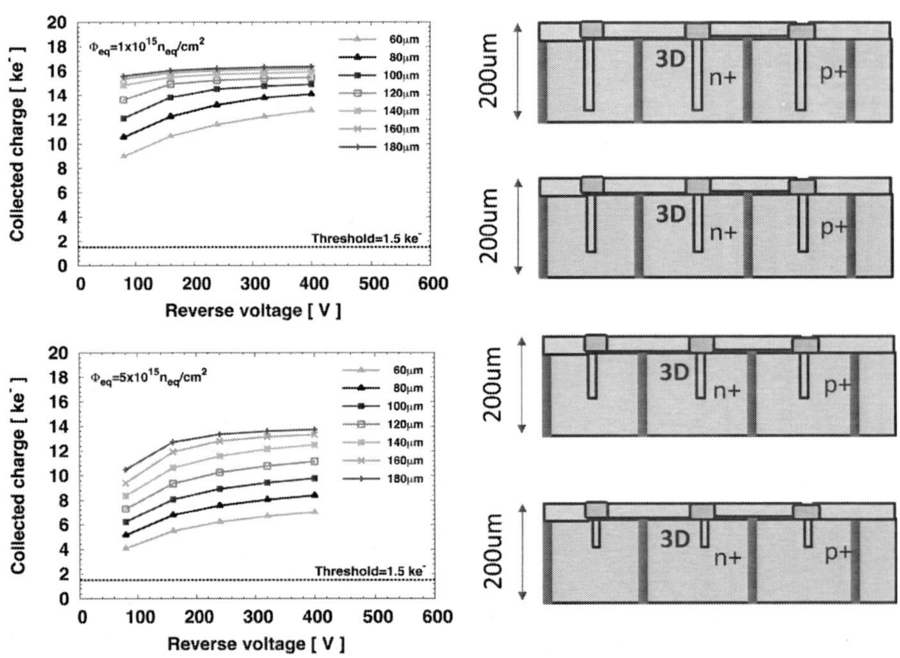

FIGURE 6.12 On the right: sketches of devices with 200 micron thick substrate, full-through p+ electrodes and n+ collecting electrodes of different depths ranging from 60 microns (bottom) to 180 microns (top). On the left: collected charge versus voltage simulations of devices with different collecting n+ electrode depths after $1 \times 10^{15}\ n_{eq}\ \mathrm{cm}^{-2}$ (top) and $5 \times 10^{15}\ n_{eq}\ \mathrm{cm}^{-2}$ (bottom) irradiation. The high-fluence simulation clearly shows the difference among signals, depending on the p-n electrode overlap. The figures also indicate the threshold one should expect to apply to suppress the noise in the current/future detector systems. (Da Vià et al., JINST 10, C04020, 2015. With permission]. [26]

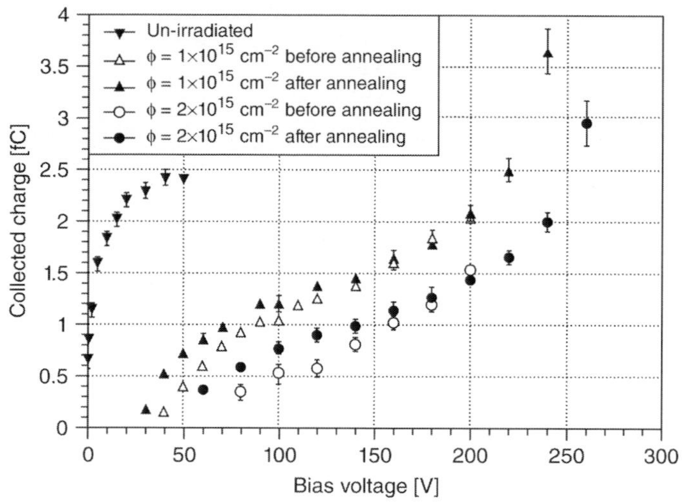

FIGURE 6.13 Signal charge of 3D-DDTC detectors irradiated with 25-MeV protons at 1×10^{15} n_{eq} cm^{-2} and 2×10^{15} n_{eq} cm^{-2}. The figure includes data measured before and after irradiation and an annealing step at 60° for 80 minutes for the irradiated samples to allow short-term annealing. Both irradiated samples show charge multiplication above 200 V, probably due to the tip effect. [Zoboli et al., IEEE Nuclear Science Symposium, 2008. With permission of IEEE. [29]

Fig. 6.13 refer to 3D-DDTC strip sensors from the first batch fabricated at FBK on n-type substrates, where the column overlap was limited [28]. The samples were irradiated with 25-MeV protons at fluences up to 2×10^{15} n_{eq} cm^{-2} and annealed for 80 minutes at 60°C to allow beneficial annealing. After irradiation, all sensors were tested at the University of Freiburg with minimum ionizing particles at −10°C so that the sensors could be biased up to 200 V without excessive noise [29]. TCAD simulations incorporating radiation damage and impact ionization effects confirmed the observed charge multiplication to be due to the high electric field at the columns tips [30].

Similar conclusions were obtained from TCAD simulations performed at the University of Glasgow for 3D-DDTC sensors fabricated at CNM in [31], where the electrical and signal properties together with breakdown voltage and the possible use of column tips for charge enhancement are discussed. Further simulations are also available in [32]. It should be mentioned that initial CNM 3D detectors were fabricated with double-sided technology with 250 μm deep electrodes on about 300 μm thick, n-type silicon wafers, and they also showed encouraging results after irradiation.

In particular, p-in-n and n-in-p configurations, with an interelectrode distance of about 56 μm, strip sensors connected to the Alibava readout electronics, were tested after irradiation up to 2×10^{16} n_{eq} cm^{-2} [33, 34]. Results for the two configurations are visible in Fig. 6.14, which shows that both sensor types are fully efficient after such fluence. From the plots one can notice that for both types charge multiplication is clearly visible above 200 V, where signal efficiencies well exceed 100% after 2×10^{15} n_{eq} cm^{-2} while they reach 70% above 350 V after 2×10^{16} n_{eq} cm^{-2} [34]. For the p-in-n configuration (6.14b, right) there is an indication

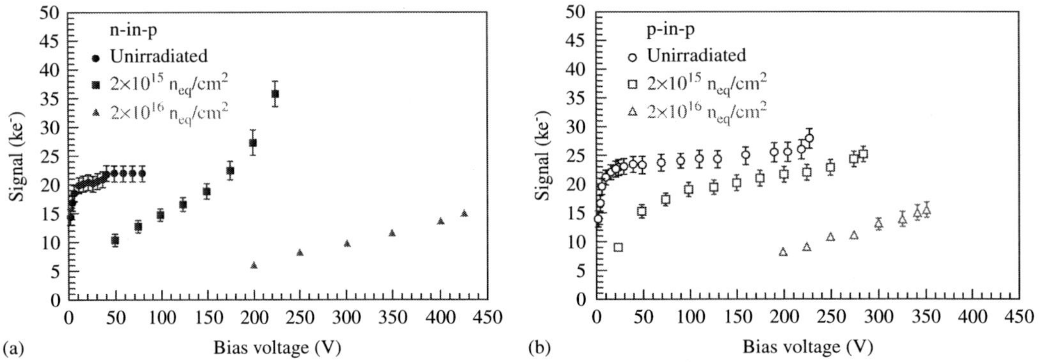

FIGURE 6.14 Signal versus reverse bias voltage for different irradiation fluences measured with (a) n-in-p sensors and (b) p-in-n sensors. All tests were performed at different temperatures, depending on fluence: $T = -16°C$ (before irradiation), $T = -40°C$ (2×10^{15} n_{eq} cm^{-2}), and $T = -43°C$ (2×10^{16} n_{eq} cm^{-2}). (Copyright Elsevier 2011. Koehler et al., Nucl. Instrum. Methods A, 659, 2011. With permission.) [34]

that charge multiplication also appears before irradiation at high voltages. This was not observed previously probably owing to early breakdown of full-through devices of FBK's early double-sided processing.

The results shown in Figure 6.14 address a very interesting point: is the symmetry typical of 3D sensors design saying more about their behavior? We have already seen that the timing characteristics of a signal collected from a 3D sensor without shaping shows a symmetric shape before and after irradiation, hinting that the contribution of holes to the signal seems to be equally important than the one of electrons, despite their lower mobility and lower trapping time after irradiation. This is not the case in the planar sensor. But what is probably even more surprising is the confirmation of this behavior when the electron collection and hole collection are viewed independently, as shown in Fig. 6.14. Even if charge multiplication is playing a role, it is nevertheless interesting to see that it does it independently on the electric field polarity which, even after type inversion for the n^+ substrate, is opposite if the readout is performed on the p^+ or n^+ electrode.

To further analyze this fact, it is useful to look at the electric and weighting fields and potential lines simulated for the above-mentioned 3D sensors shown in Fig. 6.15 [34]. Figure 6.15a shows the simulated electric field inside a cell where the junction electrode in the center is surrounded by four ohmic electrodes and four other junction electrodes are visible on the corners. From Fig. 6.15b, it is possible to appreciate the electric field symmetry between two electrodes of opposite polarity, where the simulation shows that for a substrate effective doping concentration of 3×10^{11} cm^{-3} and an applied bias of 225V, the field reaches a strength of about 12 V/um at the contacts throughout the entire column length, with the exception of the tips where the field strength is probably much higher. The weighting potential distribution for an extension of the same cell geometry along one of the strips direction is shown in Fig. 6.15c, where the intensity of the equipotential

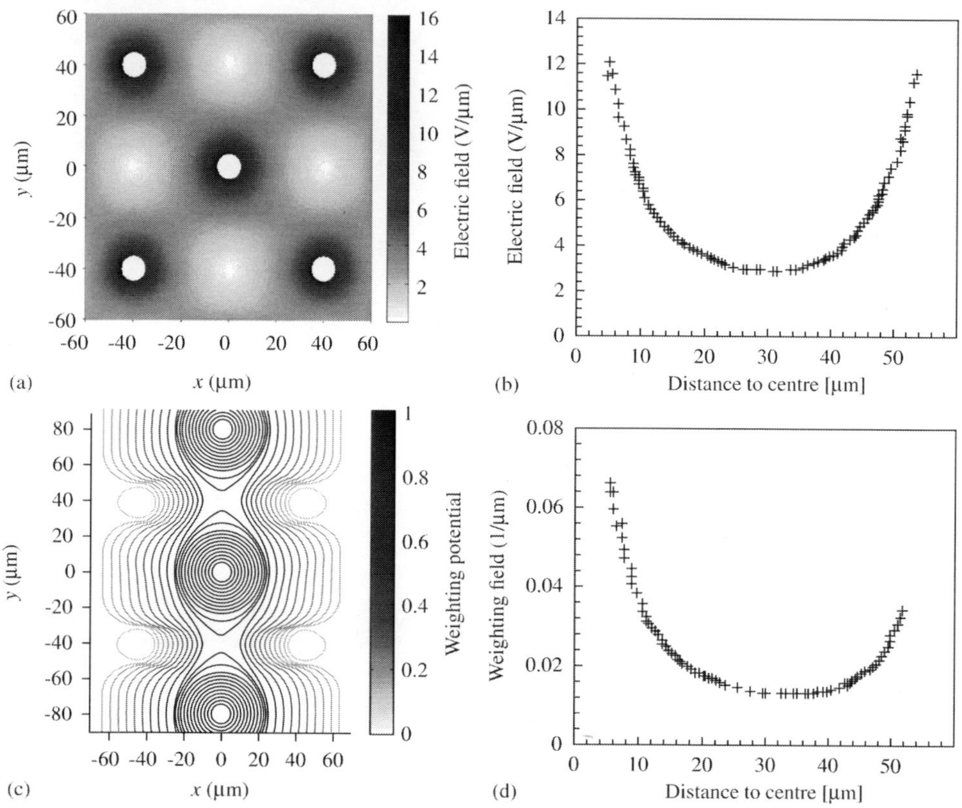

FIGURE 6.15 (a) Electric field distribution within a cell with a central collecting (junction) electrode surrounded by four ohmic ones. (b) Electric field strength in the region connecting two opposite polarity electrodes. (c) Weighting potential lines in an extension of one cell showing the equivalent density of the equipotential lines around both polarity electrodes. (d) Weighting field between two opposite polarity electrodes stressing the symmetric nature of the distribution, which looks like the one of two planar segmented geometry back to back. (Copyright Elsevier 2011. Koehler et al., Nucl. Instrum. Methods A, 659, 2011. With permission.) [34]

lines around electrodes of both polarities is clearly visible. The weighting filed between two opposite polarity electrodes is shown in Fig. 6.15d. On the basis of the shape of the weighting field, which differs substantially from the one of a segmented planar geometry that goes from maximum peaking at the collected electrode to zero at the opposite side, it is possible to appreciate the potential advantage of a 3D geometry before and after irradiation since both carriers keep contributing to the signal despite the collecting electrode polarity.

Other batches of 3D-DDTC were processed on 200–220 μm thick p-type substrates at FBK, with a larger column overlap (from 100 to 150 μm). Tests on devices coming from these batches confirmed the expected improvement on the signal amplitude due to the improved electric field distribution while maintaining similar electrical characteristics [35, 36].

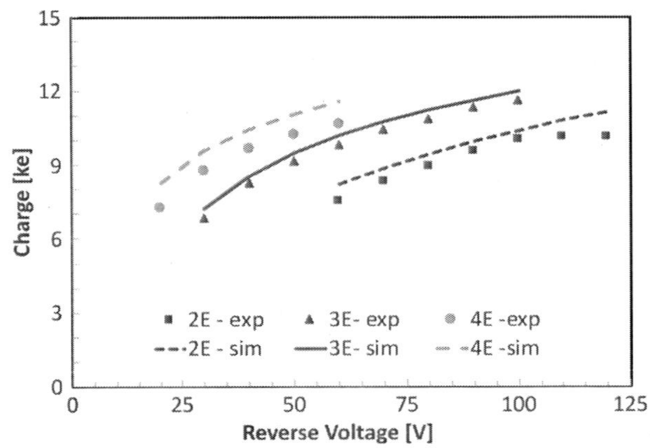

FIGURE 6.16 Comparison between measured and simulated signals as a function of reverse bias voltage 3D-DDTC FBK sensors with different electrode configurations (2E, 3E, 4E) irradiated with 25-MeV protons at 1×10^{15} n_{eq} cm^{-2}.

Results of sensors compatible with the ATLAS FEI3 readout chip with different electrode configurations are available in [37, 38], and signal amplitudes versus bias voltage are shown in Figure 6.16 for devices irradiated with protons up to 1×10^{15} $n_{eq}cm^{-2}$ [39]. Tests on the noise confirmed previous studies and ranged between 200 and 240 electrons rms at full depletion, depending on the electrodes' configuration. Tests with γ- and β-sources as well as high-energy pion beams at CERN, with and without a 1.6T magnetic field, showed regular behaviors [38, 40].

In summary, the performance of 3D-DDTC confirmed once more to be comparable to that of full 3D sensors. In particular, the test beam data with and without magnetic fields did not show substantial differences between the two processing technologies. In test beam data, some loss of tracking efficiency is observed for normal incident tracks due to the electrodes being insensitive, but full efficiency is recovered when the sensor is tilted to 15° or more, as can be seen in Fig. 6.17, which shows the efficiency maps for a proton-irradiated and a neutron-irradiated sample of 3E type. Note that the overall hit efficiency at 15° particle incidence angle is 99.9% and 98.1% for the proton- and neutron-irradiated sample, respectively [40].

Similar tests were performed on sensors from the batches processed at CNM and FBK for the ATLAS IBL project and assembled with the FEI4 chip [11]. Irradiations were performed with 25-MeV protons at the KIT cyclotron (in Karlsruhe, Germany) and with neutrons at the TRIGA reactor (in Ljubljana, Slovenia), up to very large fluences. After irradiations, all assemblies were annealed for 120 minutes at 60°C before testing them with 120 GeV pions at CERN SPS and with 4 GeV positrons at DESY. The high-resolution EUDET telescope was used in beam tests, and measurements were performed with both particle tracks perpendicular to the sensors and nonperpendicular incidence tracks to emulate the IBL geometry. More details on the measurement setups can be found in [26].

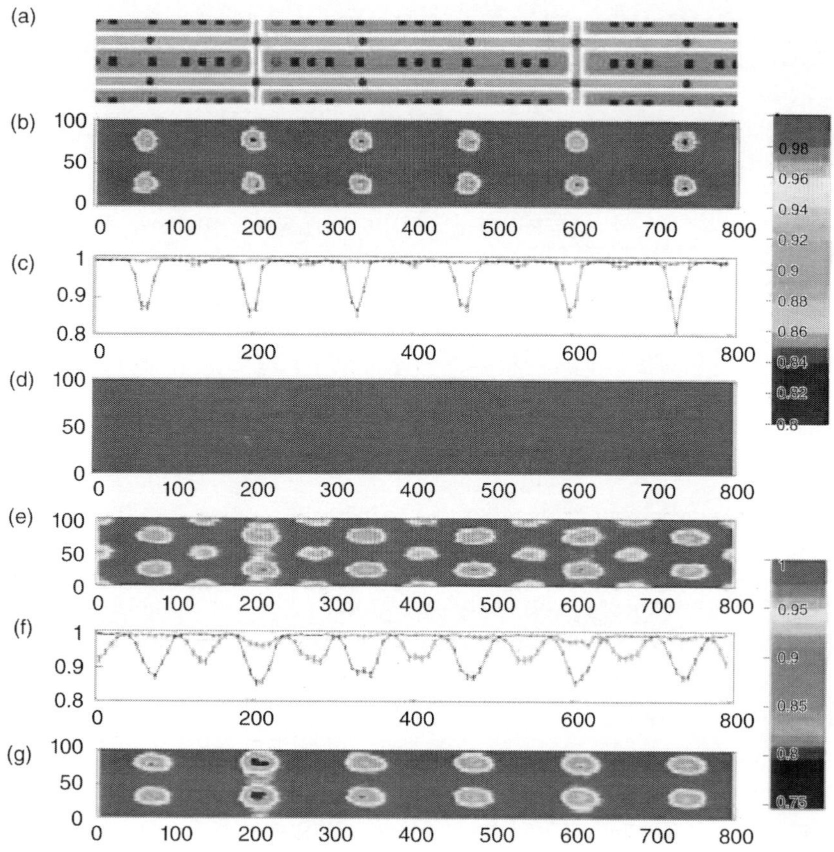

FIGURE 6.17 Efficiency maps for FBK DDTC 3E sensors after 25-MeV proton and reactor neutron irradiation up to 1×10^{15} n_{eq} cm^{-2}. From top to bottom: (a) mask details showing the sensor region centered on one cell and extending to half a cell in both directions; (b) 2D efficiency map at 0° particle incidence angle, orthogonal to the surface, for the proton-irradiated sensor; (c) 1D efficiency projections of (b) along planes including the readout (blue curve) and bias (red curve) electrode regions at 0°; (d) same as (b) at 15° sample rotation with respect to the incident particle; (e) 2D efficiency map at 0° for the neutron-irradiated sensor; (f) 1D efficiency projections of (e) along planes including the readout (blue curve) and bias (red curve) electrode regions at 0°; (g) same as (e) at 15° rotation with respect to the impinging particle. (Copyright Elsevier 2011. Micelli et al., Nucl. Instrum. Methods A, 650, 2011. With permission.) [40]

As an example of the results obtained, Table 6.2 summarizes the data from beam tests performed at CERN in 2011, including nonirradiated and irradiated sensors measured in different conditions. These results highlight the very good performance and high-radiation tolerance of 3D sensors, also addressing the importance of several parameters in optimizing the hit efficiency, and particularly the particle tilt angle, the threshold value and the bias voltage [41].

In recent years, many irradiation tests have been performed on new and optimized 3D devices, with applications to upgraded high-energy physics experiments at CERN [42, 43, 44].

TABLE 6.2 Summary of the main results from the 2011 beam tests at CERN for the ATLAS IBL project

Sensor ID	Bias (V)	Tilt Angle (°)	Irradiation	Fluence (n_{eq} cm^{-2})	Threshold (e^-)	Hit Efficiency (%)
CNM55	20	0	No	–	1600	99.6
FBK13	20	0	No	–	1500	98.8
CNM34	140	0	25 MeV p	5×10^{15}	1500	97.6
CNM34	160	0	25 MeV p	5×10^{15}	1500	98.1
CNM97	140	15	25 MeV p	5×10^{15}	1800	96.6
CNM34	160	15	25 MeV p	5×10^{15}	1500	99.0
CNM81	160	0	Reactor n	5×10^{15}	1500	97.5
FBK90	60	15	25 MeV p	2×10^{15}	3200	99.2
FBK11	140	15	25 MeV p	5×10^{15}	2000	95.6
FBK87(*)	160	15	25 MeV p	5×10^{15}	1500	98.2

(*) Results relevant to FBK87 sensor have been obtained from a beam test at DESY in April 2012. All results have been obtained with a tuning of the FEI4 chip with 10 ToT at 20 ke- signal.

These include both the innermost pixels and the forward regions where the sensors are positioned vertically with respect to the beam direction as close as possible to the beam inside the beam pipe at about 200 meters from the interaction point using special vacuum-tight enclosures called roman pots (yes, they were developed for the first time by a group of physicists from ROME) [45, 46]. In this configuration where the objective is to detect protons, which escaped intact from the interaction point and lost some of their momentum in the process, the signal is collected mainly in the region close to the beam, making irradiation very inhomogeneous. The robustness of the 3D design proves to be another crucial asset for this experiment since the same bias voltage can be applied equivalently to the irradiated and nonirradiated part of the sensor without causing breakdown [47, 48, 49]. For these applications, 3D sensors also provide significant advantages in terms of edge sensitivity. While this would be maximized in the event active edges are implemented, slim-edge-based sensors have also shown very good performance, with full efficiency recorded up to the last pixel (CNM) and even beyond it (FBK), due to the different terminations adopted [50].

Finally, after many processes at different facilities and several tests performed independently by several groups, it is now clear that the predictions on the radiation tolerance of 3D sensors can be made by simple geometrical considerations [51]. This is confirmed by the plots shown in Figure 6.18 where data from devices processed in different facilities with 71 microns interelectrode spacing (bottom curves) remarkably overlap compared with a simulated 35 microns interelectrode spacing (top curve) [52]. Recent experimental results on new designs with more aggressive pitches and interelectrode spacing and at fluences reaching 3×10^{16} n_{eq} cm^{-2} have been shown to follow the predicted trends, as can be seen in Fig. 6.19, confirming that, indeed, "ultra" radiation hard is a good definition of 3D sensors [53, 54, 55].

Moreover, such a remarkable performance is obtained at low voltage, and hence low-power consumption. As an example, Fig. 6.19 shows the bias voltage necessary to obtain 97% hit efficiency (V97%) in a beam test as a function of fluence at 0° tilt (i.e., worst case scenario for 3D sensors!) for different devices. New, small-pitch 50×50 μm^2 1E devices are

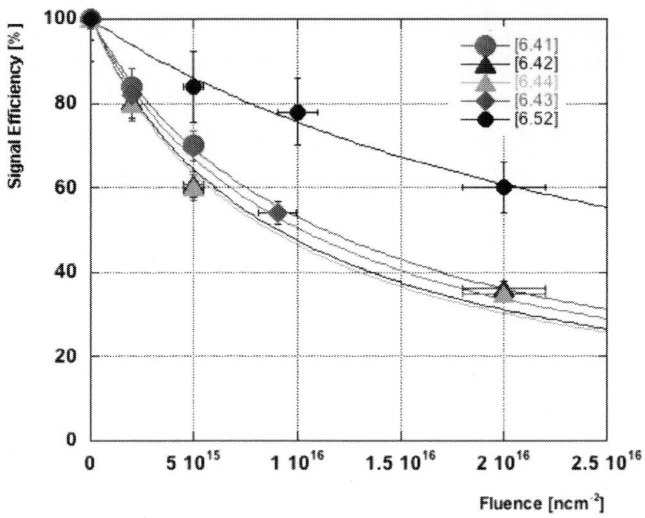

FIGURE 6.18 Signal efficiency compilation of 3D sensors fabricated at different processing facilities and techniques. The overlapping group of curves corresponds to devices with 71 micron interelectrode spacing, and the top curve has an interelectrode spacing of 35 microns.

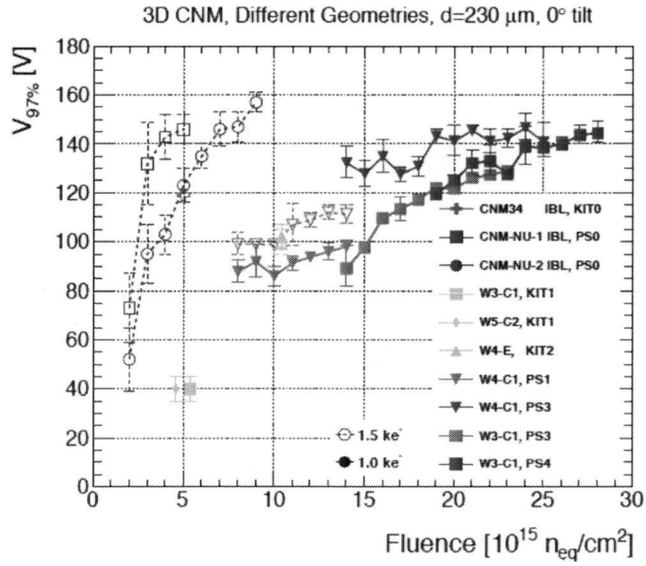

FIGURE 6.19 Bias voltage necessary to obtain 97% hit efficiency (V97%) as a function of fluence at 0° tilt for different devices. The small-pitch $50 \times 50\ \mu m^2$ 1E devices (red/orange) are compared to the large-pitch IBL $50 \times 250\ \mu m^2$ 2E generation (blue/magenta). Open markers refer to a threshold of 1.5 ke-, full ones to 1.0 ke-. (Lange et al., JINST 13, P09009, 2018. With permission.) [55]

compared to the larger pitch IBL 50 × 250 μm^2 2E from a previous generation. As expected, a largely enhanced radiation tolerance can be observed for small-pitch sensors. Notably, only 100 V are required to reach 97% efficiency at the ATLAS ITk target fluence of 1.3 × 1 0^{16} n_{eq}/cm^2, and even at 2.5 × 10^{16} n_{eq}/cm^2, that is, almost two times the maximum fluence of interest for ITk, the necessary voltage is less than 150 V [55].

As will be seen in Chapter 9, other groups are using 3D geometry on different substrates such as diamond, which will potentially improve even further the reach of radiation tolerance in the future.

REFERENCES

1. D. Nygren, "Converting vice to virtue: can time-walk be used as a measure of deposited charge in silicon detectors?", Internal LBL note, May 1991.
2. C. Da Vià, J. Hasi, C. Kenney, V. Linhart, S. Parker, T. Slavicek, S. J. Watts, P. Bem, T. Horazdovsky, S. Pospisil, "Radiation hardness properties of full-3D active edge silicon sensors," Nuclear Instruments and Methods A, vol. 587, pp. 243–249, 2008.
3. T. Rohe, D. Bortoletto, V. Chiochia, L. M. Cremaldi, S. Cucciarelli, A. Dorokhovc, C. Hörmann, D. Kim, M. Konecki, D. Kotliński, K. Prokofiev, C. Regenfus, D. A. Sanders, S. Son, T. Speer, M. Swartz, "Fluence dependence of charge collection of irradiated pixel sensors," Nuclear Instruments and Methods A, vol. 552, pp. 232–238, 2005.
4. P. P. Allport, G. Casse, M. Lozano, P. Sutcliffe, J. J. Velthuis, J. Vossebeld, "Performance of P-type micro-strip detectors after irradiation to 7.5 × 1015 p/cm2", IEEE Transactions on Nuclear Science, vol. 52(5), pp. 1903–1906, 2005.
5. W. Adam, W. de Boer, E. Borchi, M. Bruzzi, C. Colledani, P. D'Angelo, V. Dabrowski, W. Dulinski, B. van Eijk, V. Eremin, F. Fizzotti, H. Frais Kölbl, C. Furetta, K. K. Gan, A. Gorisek, E. Griesmayer, E. Grigoriev, F. Hartjes, J. Hrubec, F. Huegging, H. Kagan, J. Kaplon, R. Kass, K. T. Knöpfle, M. Krammer, W. Lange, A. Logiudice, C. Manfredotti, M. Mathes, D. Menichelli, M. Mishina, L. Moroni, J. Noomen, A.Oh, H. Pernegger, M. Pernicka, R. Potenza, J. L. Riester, A. Rudge, S. Sala, S. Schnetzer, S. Sciortino, R. Stone, C. Sutera, W. Trischuk, J. J. Velthuis, B. Vincenzo, P. Weilhammer, J. Weingarten, N. Wermes, W. Zeuner, "Radiation hard diamond sensors for future tracking applications," Nuclear Instruments and Methods A, vol. 565, pp. 278–283, 2006.
6. G. Kramberger, V. Cindro, I. Dolenc, E. Fretwurst, G. Lindström, I. Mandić, M. Mikuž, M. Zavrtanik, "Charge collection properties of heavily irradiated epitaxial silicon detectors," Nuclear Instruments and Methods A, vol. 554, pp. 212–219, 2005.
7. H. Hoedlmoser, M. Moll, H. Nordlund, "CCE/CV measurements with irradiated p-type MCz diodes," 8th RD50 Workshop, Prague, June 2006.
8. S. I. Parker, C. J. Kenney, "Performance of 3-D architecture silicon sensors after intense proton irradiation," IEEE Transactions on Nuclear Science, vol. 48(5), pp. 1629–1638, 2001.
9. I. Perić, L. Blanquart, G. Comes, P. Denes, K. Einsweiler, P. Fischer, E. Mandelli, G. Meddeler, "The FEI3 readout chip for the ATLAS pixel detector," Nuclear Instruments and Methods A, vol. 565, pp. 178–187, 2006.
10. M. Garcia-Sciveres, D. Arutinov, M. Barbero, R. Beccherle, S. Dube, D. Elledge, J. Fleury, D. Fougeron, F. Gensolen, D. Gnani, V. Gromov, T. Hemperek, M. Karagounis, R. Kluit, A. Kruth, A. Mekkaoui, M. Menouni, J.-D. Schipper, "The FE-I4 pixel readout integrated circuit," Nuclear Instruments and Methods A, vol. 636, pp. S155–S159, 2011.
11. H. Chr. Kästli, M. Barbero, W. Erdmann, Ch. Hörmann, R. Horisberger, D. Kotlinski, B. Meier, "Design and performance of the CMS pixel detector readout chip," Nuclear Instruments and Methods A, vol. 565, pp. 188–194, 2006.

12. C. Da Vià, E. Bolle, K. Einsweiler, M. Garcia-Sciveres, J. Hasi, C. Kenney, V. Linhart, S. Parker, S. Pospisil, O. Rohne, T. Slavicek, S. Watts, N. Wermes, "3D active edge silicon sensors with different electrode configurations: radiation hardness and noise performance," Nuclear Instruments and Methods A, vol. 604, pp. 505–511, 2009.
13. C. Da Vià, M. Deile, J. Hasi, C. Kenney, A. Kok, S. Parker, S. Watts, G. Anelli, V. Avati, V. Bassetti, V. Boccone, M. Bozzo, K. Eggert, F. Ferro, A. Inyakin, J. Kaplon, J. Lozano Bahilo, A. Morelli, H. Niewiadomski, E. Noschis, F. Oljemark, M. Oriunno, K. Österberg, G. Ruggiero, W. Snoeys, S. Tapprogge, "3D active edge silicon detector tests with 120 GeV muons," IEEE Transactions on Nuclear Science, vol. 56, pp. 505–554, 2009.
14. C. Piemonte, M. Boscardin, G. F. Dalla Betta, S. Ronchin, N. Zorzi, "Development of 3D detectors featuring columnar electrodes of the same doping type," Nuclear Instruments and Methods A, vol. A541, pp. 441–448, 2005.
15. S. Eränen, T. Virolainen, I. Luusua, J. Kalliopuska, K. Kurvinen, M. Eräluoto, J. Härkönen, K. Leinonen, M. Palvianen, M. Koski, "Silicon semi 3D radiation detectors," 2004 IEEE Nuclear Science Symposium, Rome (Italy), October 16–22, 2004, Conference Record, Paper N28-3.
16. M. Zavrtanik, V. Cindro, G. Kramberger, J. Langus, I. Mandic, M. Mikuz, M. Boscardin, G.-F. Dalla Betta, C. Piemonte, A. Pozza, S. Ronchin, N. Zorzi, "Position sensitive TCT evaluation of irradiated 3d-stc detectors," 2007 IEEE Nuclear Science Symposium, Honolulu (USA), October 28–November 3, 2007, Conference Record, paper N24-150.
17. F. Campabadal, C. Fleta, M. Key, M. Lozano, C. Martinez, G. Pellegrini, et al., "Design and performance of the ABCD3TA ASIC for readout of silicon strip detectors in the ATLAS semiconductor tracker," Nuclear Instruments and Methods A, vol. 552, pp. 292–328, 2005.
18. S. Eckert, T. Ehrich, K. Jakobs, S. Kühn, U. Parzefall, M. Boscardin, C. Piemonte, S. Ronchin, "Signal and charge collection efficiency of a p-type STC 3D-detector irradiated to sLHC-fluences, read out with 40 MHz," Nuclear Instruments and Methods A, vol. 581, pp. 322–325, 2007.
19. S. Kühn, G. F. Dalla Betta, S. Eckert, K. Jacobs, U. Parzefall, A. Zoboli, N. Zorzi, "Short strips for the sLHC: A P-type silicon microstrip detector in 3-D technology," IEEE Transactions on Nuclear Science, vol. 55(6), pp. 3638–3642, 2008.
20. G. Pahn, R. Bates, M. Boscardin, G. F. Dalla Betta, S. Eckert, L. Eklund, C. Fleta, K. Jacobs, M. Koehler, S. Kuehn, C. Parkes, U. Parzefall, D. Pennicard, T. Szumlak, A. Zoboli, N. Zorzi, "First Beam Test Characterisation of a 3D-stc Silicon Short Strip Detector," IEEE Transactions on Nuclear Science, vol. 56(6), pp. 3834–3839, 2009.
21. L. Tlustos, J. Kalliopuska, R. Ballabriga, M. Campbell, S. Eranen, X. Llopart, "Characterisation of a semi 3-D sensor coupled to Medipix2," Nuclear Instruments and Methods A, vol. 580, pp. 897–901, 2007.
22. G. F. Dalla Betta, M. Boscardin, L. Bosisio, C. Piemonte, S. Ronchin, A. Zoboli, N. Zorzi, "New developments on 3D detectors at IRST," 2007 IEEE Nuclear Science Symposium, Honolulu (USA), October 28–November 3, 2007, Conference Record, paper N18-3.
23. C. Fleta, D. Pennicard, R. Bates, C. Parkes, G. Pellegrini, M. Lozano, V. Wright, M. Boscardin, G. F. Dalla Betta, C. Piemonte, A. Pozza, S. Ronchin, N. Zorzi, "Simulation and test of novel 3D silicon radiation detectors," Nuclear Instruments and Methods A, vol. 579, pp. 642–647, 2007.
24. C. Da Vià, "3D sensors and micro-fabricated detector systems," Nuclear Instruments and Methods A, vol. 765, pp. 151–154, 2014.
25. J. Albert et al. (The ATLAS IBL Collaboration), "Prototype ATLAS IBL modules using the FE-I4A front-end readout chip," Journal of Instrumentation, JINST 7, P11010, 2012.
26. C. Da Vià, M. Borri, G. Dalla Betta, I. Haughton, J. Hasi, C. Kenney, M. Povoli, R. Mendicino, "3D silicon sensors with variable electrode depth for radiation hard high resolution particle tracking," Journal of Instrumentation, JINST 10, C04020, 2015.

27. G.-F. Dalla Betta, N. Ayllon, M. Boscardin, M. Hoeferkamp, S. Mattiazzo, H. McDuff, R. Mendicino, M. Povoli, S. Seidel, D. M. S. Sultan, N. Zorzi, "Investigation of leakage current and breakdown voltage in irradiated double-sided 3D silicon sensors," Journal of Instrumentation, JINST 11, P09006, 2016.
28. A. Zoboli, M. Boscardin, L. Bosisio, G.-F. Dalla Betta, C. Piemonte, S. Ronchin, N. Zorzi, "Double-Sided, Double-Type-Column 3D detectors at FBK: Design, Fabrication and Technology Evaluation," IEEE Transactions on Nuclear Science, vol. 55(5), pp. 2775–2784, 2008.
29. A. Zoboli, M. Boscardin, L. Bosisio, G.-F. Dalla Betta, S. Eckert, S. Kühn, et al., "Functional characterization of 3D-DDTC detectors fabricated at FBK-irst," 2008 IEEE Nuclear Science Symposium, Dresden (Germany), October 19–25, 2008, Conference Record, Paper N34-4.
30. G. Giacomini, C. Piemonte, G. F. Dalla Betta, M. Povoli, "Simulations of 3D detectors", Proc. of Science—20th Workshop on Vertex Detectors (Vertex 2011), Paper 025, 2012.
31. D. Pennicard, G. Pellegrini, M. Lozano, R. Bates, C. Parkes, V.O'Shea, V. Wright, "Simulation results from double-sided 3-D detectors," IEEE Transactions on Nuclear Science, vol. 54(4), pp. 1435–1443, 2007.
32. D. Pennicard, G. Pellegrini, C. Fleta, R. Bates, V.O'Shea, C. Parkes, N. Tartoni, "Simulation of radiation-damaged 3D detectors for the Super-LHC", Nuclear Instruments and Methods A, vol. 592, pp. 16–25, 2008.
33. R. L. Bates, C. Parkes, B. Rakotomiaramanana, C. Fleta, G. Pellegrini, M. Lozano, J. P. Balbuena, U. Parzefall, M. Koehler, M. Breindl, X. Blot, "Charge collection studies and electrical measurements of heavily irradiated 3D double-sided sensors and comparison to planar strip detectors," IEEE Transactions on Nuclear Science, vol. 58(6), pp. 3370–3383, 2011.
34. M. Koehler, R. Bates, C. Fleta, K. Jakobs, M. Lozano, C. Parkes, U. Parzefall, G. Pellegrini, J. Preiss, "Comparative measurements of highly irradiated n-in-p and p-in-n 3D silicon strip detectors," Nuclear Instruments and Methods, vol. 659, pp. 272–281, 2011.
35. A. Zoboli, M. Boscardin, L. Bosisio, G. F. Dalla Betta, C. Piemonte, S. Ronchin, N. Zorzi, "Initial results from 3D-DDTC detectors on p-type substrates," Nuclear Instruments and Methods A, vol. 612, pp. 521–524, 2010.
36. G.-F. Dalla Betta M. Boscardin, L. Bosisio, M. Koehler, U. Parzefall, S. Ronchin, L. Wiik, A. Zoboli, N. Zorzi, "Performance evaluation of 3D-DDTC detectors on p-type substrates," Nuclear Instruments and Methods A, vol. 624, pp. 459–464, 2010.
37. G.-F. Dalla Betta, M. Boscardin, G. Darbo, C. Gemme, A. La Rosa, H. Pernegger, C. Piemonte, M. Povoli, S. Ronchin, A. Zoboli, N. Zorzi, "Development of 3D-DDTC pixel detectors for the ATLAS upgrade," Nuclear Instruments and Methods A, vol. 638-S1, pp. S15–S23, 2011.
38. P. Grenier, G. Alimonti, M. Barbero, R. Bates, E. Bolle, M. Borri, M. Boscardin, C. Buttar, M. Capua, M. Cavalli-Sforza, M. Cobal, A. Cristofoli, G.F. Dalla Betta, G. Darbo, C. Da Vià, E. Devetak, B. DeWilde, B. Di Girolamo, D. Dobos, K. Einsweiler, D. Esseni, S. Fazio, C. Fleta, J. Freestone, C. Gallrapp, M. Garcia-Sciveres, G. Gariano, C. Gemme, M-P. Giordani, H. Gjersdal, S. Grinstein, T. Hansen, T-E. Hansen, P. Hansson, J. Hasi, K. Helle, M.Hoeferkamp, F. Huegging, P. Jackson, K. Jakobs, J. Kalliopuska, M. Karagounis, C. Kenney, M. Koehler, M. Kocian, A. Kok, S. Kolya, I. Korokolov, V. Kostyukhin, H. Krugger, A. La Rosa, C.H. Lai, N. Lietaer, M. Lozano, A. Mastroberardino, A. Micelli, C. Nellist, A. Oja, V. Oshea, C. Padilla, P. Palestri, S. Parker, U. Parzefall, J. Pater, G. Pellegrini, H. Pernegger, C. Piemonte, S. Pospisil, M. Povoli, S. Roe, O. Rohne, S. Ronchin, A. Rovani, E. Ruscino, H. Sandaker, S. Seidel, L. Selmi, D. Silverstein, K. Sjbk, T. Slavicek, S. Stapnes, B. Stugu, J. Stupak, D. Su, G. Susinno, R. Thompson, J-W. Tsung, D. Tsybychev, S.J. Watts, N. Wermes, C. Young, N. Zorzi, "Test beam results of 3D silicon pixel sensors for the ATLAS upgrade," Nuclear Instruments and Methods A, vol. 638, pp. 33–40, 2011.

39. A. La Rosa, M. Boscardin, M. Cobal, G. F. Dalla Betta, C. Da Vià, G. Darbo, C. Gallrapp, C. Gemme, F. Huegging, J. Janssen, A. Micelli, H. Pernegger, M. Povoli, N. Wermes, N. Zorzi, "Characterization of proton irradiated 3D-DDTC pixel sensor prototypes fabricated at FBK," Nuclear Instruments and Methods A, vol. 681, pp. 25–33, 2012.
40. A. Micelli, K. Helle, H. Sandaker, B. Stugu, M. Barbero, F. Huegging, M. Karagounis, V. Kostyukhin, H. Krueuger, J.-W. Tsung, N. Wermes, M. Capua, S. Fazio, G. Susinno, B. Di Girolamo, D. Dobos, A. La Rosa, H. Pernegger, S. Roe, T. Slavicek, S. Pospisil, K. Jakobs, M. Koehler, U. Parzefall, G. Darbo, G. Gariano, C. Gemme, A. Rovani, E. Ruscino, C. Butter, R. Bates, V. Oshea, S. Parker, M. Cavalli-Sforza, S. Grinstein, I. Korokolov, C. Pradilla, K. Einsweiler, M. Garcia- Sciveres, M. Borri, C. Da Vià, J. Freestone, S. Kolya, C. H. Lai, C. Nellist, J. Pater, R. Thompson, S. J. Watts, M. Hoeferkamp, S. Seidel, E. Bolle, H. Gjersdal, K.-N. Sjoebaek, S. Stapnes, O. Rohne, D. Su, C. Young, P. Hansson, P. Grenier, J. Hasi, C. Kenney, M. Kocian, P. Jackson, D. Silverstein, H. Davetak, B. DeWilde, D. Tsybychev, G.-F. Dalla Betta, P. Gabos, M. Povoli, M. Cobal, M.-P. Giordani, L. Selmi, A. Cristofoli, D. Esseni, P. Palestri, C. Fleta, M. Lozano, G. Pellegrini, M. Boscardin, A. Bagolini, C. Piemonte, S. Ronchin, N. Zorzi, T.-E. Hansen, T. Hansen, A. Kok, N. Lietaer, J. Kalliopuska, A. Oja, "3D-FBK Pixel sensors: recent beam tests results with irradiated devices," Nuclear Instruments and Methods A, vol. 650, pp. 150–157, 2011.
41. G.-F. Dalla Betta, M. Povoli, C. Da Vià, S. Grinstein, A. Micelli, S. Tsiskaridze, P. Grenier, G. Darbo, C. Gemme, M. Boscardin, G. Pellegrini, S. Parker, "3D silicon sensors: irradiation results," Proceedings of Science—21st Workshop on Vertex Detectors (Vertex 2012), Paper 14, 2013.
42. G. F. Dalla Betta, C. Betancourt, M. Boscardin, G. Giacomini, K. Jakobs, S. Kuehn, B. Lecini, R. Mendicino, R. Mori, U. Parzefall, M. Povoli, M. Thomas, N. Zorzi, "Radiation hardness tests of double-sided 3D strip sensors with passing-through columns," Nuclear Instruments and Methods A, vol. 765, pp. 155–160, 2014.
43. I. Haughton, C. Da Vià, S. Watts, "Radiation hardness tests of highly irradiated full-3D sensors," Nuclear Instruments and Methods A, vol. 806, pp. 423–431, 2016.
44. M. Fernandez, R. Jaramillo, M. Lozano, F. J. Munoz, G. Pellegrini, D. Quirion, T. Rohe, I. Vila, "Radiation resistance of double-type double-sided 3D pixel sensors," Nuclear Instruments and Methods A, vol. 732, pp. 137–140, 2013.
45. L. Adamczyk et al., "Technical design report for the ATLAS forward proton detector," CERN-LHCC-2015-009; ATLAS-TDR-024, May 2015.
46. M. Albrow et al., "CMS-TOTEM precision proton spectrometer," Technical Report CERN LHCC-2014-021. TOTEM-TDR-003. CMS-TDR-13, September 2014.
47. S. Grinstein, M. Baselga, M. Boscardin, M. Christophersen, C. Da Vià, G.-F. Dalla Betta, G. Darbo, V. Fadeyev, C. Fleta, C. Gemme, P. Grenier, A. Jimenez, I. Lopez, A. Micelli, C. Nellist, S. Parker, G. Pellegrini, B. Phlips, D.-L. Pohl, H. F.-W. Sadrozinski, M. Christophersen, P. Sicho, "Beam test studies of 3D pixel sensors irradiated non-uniformly for the ATLAS forward physics detector," Nuclear Instruments and Methods A, vol. 730, pp. 28–32, 2013.
48. F. Ravera, on behalf of the CMS and TOTEM collaborations, "The CT-PPS tracking system with 3D pixel detectors," Journal of Instrumentation, JINST 11, C11027, 2016.
49. J. Lange, L. Adamczyk, G. Avoni, E. Banas, A. Brandt, M. Bruschi, P. Buglewicz, E. Cavallaro, D. Caforio, G. Chiodini, L. Chytka, K. Ciesla, P. M. Davis, M. Dyndal, S. Grinstein, K. Janas, K. Jirakova, M. Kocian, K. Korcyl, I. Lopez Paz, D. Northacker, L. Nozka, M. Rijssenbeek, L. Seabra, R. Staszewski, P. Swierskad, T. Sykora, "Beam tests of an integrated prototype of the ATLAS Forward Proton detector," Journal of Instrumentation, JINST 11, P09005, 2016.
50. J. Lange, "Recent progress on 3D silicon detectors," Proceedings of Science—24th Workshop on Vertex Detectors (Vertex 2015), Paper 26, 2016.

51. C. Da Vià, S. Watts, "The geometrical dependence of radiation hardness in planar and 3D silicon detectors," Nuclear Instruments and Methods A, vol. 603, pp. 319–324, 2009.
52. G. F. Dalla Betta, M. Boscardin, R. Mendicino, S. Ronchin, D.M.S. Sultan, N. Zorzi, "Development of new 3D pixel sensors for Phase 2 upgrades at LHC," 2015 IEEE Nuclear Science Symposium, San Diego (USA), October 31–November 7, 2015, Conference Record, paper N3C3-5.
53. J. Lange, M. Carulla Areste, E. Cavallaro, F. Förster, S. Grinstein, I. López Paz, M. Manna, G. Pellegrini, D. Quirion, S. Terzo, D. Vázquez, "3D silicon pixel detectors for the High-Luminosity LHC," Journal of Instrumentation, JINST 11, C11024, 2016.
54. D. Vázquez Furelos, M. Carulla, E. Cavallaro, F. Förster, S. Grinstein, J. Lange, I. López Paz, M. Manna, G. Pellegrini, D. Quirion, S. Terzo, "3D sensors for the HL-LHC," Journal of Instrumentation, JINST 12, C01026, 2017.
55. J. Lange, G. Giannini, S. Grinstein, M. Manna, G. Pellegrini, D. Quirion, S. Terzo, D. Vázquez Furelos, "Radiation hardness of small-pitch 3D pixel sensors up to a fluence of 3×10^{16} n_{eq}/cm^2," JINST 13, P09009, 2018.

CHAPTER 7

The Industrialization Phase

7.1 INTRODUCTION

For about a decade after their invention (1997–2007), 3D sensors went through a relatively slow research and development phase. For several years, only a small number of working prototypes were fabricated by the original proponents at Stanford. But starting in 2004, some preliminary, nonoptimized test devices were processed at other facilities using simplified technologies. Those devices made it possible to obtain key, promising results that helped confirm the outstanding performance of these sensors, especially from the radiation hardness point of view. However, the lack of common specific objectives and the absence of coordination among the few groups involved in different R&D programs significantly delayed progress toward producing reproducible devices. Despite the rising interest in this new type of sensors, most of the high-energy physics community, where many of the involved scientists, including the original proponents of 3D, were coming from, was skeptical about the possibility of producing sensors in volumes adequate to their use in an experiment due to research fragmentation and the complexity of the fabrication process.

This scenario drastically changed in 2007, when the ATLAS 3D Sensor Collaboration was formed with the aim of developing, testing, and industrializing 3D silicon sensors in different research institutes and processing facilities in Europe and the United States [1]. The collaboration included 18 institutions and 4 processing facilities: SNF (Stanford, California), SINTEF (Oslo, Norway), CNM (Barcelona, Spain), and FBK (Trento, Italy). Later, VTT (Helsinki, Finland) joined the collaboration. The 3D Sensor Collaboration was initially determined to perform accurate and systematic comparative tests of the available 3D samples from different processing facilities to speed up the transition between the previous fragmented test programs to a true industrialization. Common beam tests and irradiation campaigns were organized, as well as regular update meetings to monitor progress. Test results obtained in the period 2007–2009 proved, as hoped, a comparable performance among different 3D sensor technologies both before and after irradiation. While this effort was still under way, in 2009, the focus shifted to the ATLAS pixel IBL, a new pixel layer that was proposed to surround the beam pipe as close as 3.4 cm from the proton beams.

With this extra layer, the primary vertex determination of the ATLAS experiment would considerably improve as would the overall physics performance [2]. However, the detector proximity to the interaction point required new and extremely radiation-hard technologies for both sensors and front-end chips, none of which was available at the time.

The IBL project provided the first real opportunity for 3D sensors to be considered for an experiment. Within the ATLAS 3D Collaboration it became clear that meeting this challenge could represent a turning point for the technology. Following a topical meeting in Venice, Italy, in June 2009, a nonconventional collaboration model was adopted, and four 3D silicon-processing facilities agreed to share their expertise to speed up production of the required volume of sensors needed for the ATLAS IBL. It was agreed to adopt a common layout design and to pursue a joint processing effort aiming at full mechanical compatibility and equivalent functional performance of the 3D sensors produced, while maintaining some specific intrinsic differences of each processing design [3]. Two different 3D processing options existed at that time, as we explained in detail in Chapter 5: full 3D with active edges, available at SNF and SINTEF, and double-sided 3D with slim edges, available at CNM and FBK.

From this point on in this chapter, the principal milestones of the common design effort agreed to within the 3D ATLAS Collaboration for the IBL project will be explained, followed by the production strategy, the 3D sensor specifications, and analysis of a few key selected results from the electrical characterization of the common production.

7.2 DESIGN SPECIFICATIONS AND COMMON WAFER LAYOUT

The ATLAS IBL 3D pixel sensor design was required to match the geometry of a newly conceived radiation-tolerant front-end chip (FEC), the so-called FEI4, fabricated at IBM in 130 nm CMOS technology [4]. The FEI4 was and still is the biggest FEC ever made for high-energy physics (HEP) applications, with a total area of about 4 cm^2. The chip contains an 80 × 336 pixel array, with a 250 μm × 50 μm single-pixel size. From previous experience with the currently installed ATLAS FEI3 pixel chip, with pixel size 400 × 50 μm^2, coupled to 3D pixel sensors [5] and taking into account the maximum particle fluence foreseen for the IBL (which was predicted to be as high as 5×10^{15} n_{eq} cm^{-2}), a so-called 2E column configuration, with two 3D readout columns per pixel to cover the 250 microns pixel length (therefore at a pitch of 125 μm), was chosen. A wafer thickness of 230 μm was agreed to be a best trade-off between the SNR and the mechanical robustness of the wafers for double-side processing. (More details on the signals and noises after irradiation for 2E and their comparison with other 3D electrode configurations can be found in Chapter 6.) It should be stressed again that single-side processing requires a support wafer during the etching steps. The support wafer needs to be removed before the sensors are connected to the readout chip.

Therefore, to assure full compatibility among the different technological options, the sensor design had to address several issues:

a. Sensor edges: SNF/SINTEF sensors, processed with support wafer, have active edge, while the double processing used at CNM and FBK sensors do not. The IBL specifications required at most 200 μm dead region at the edge between sensors in the

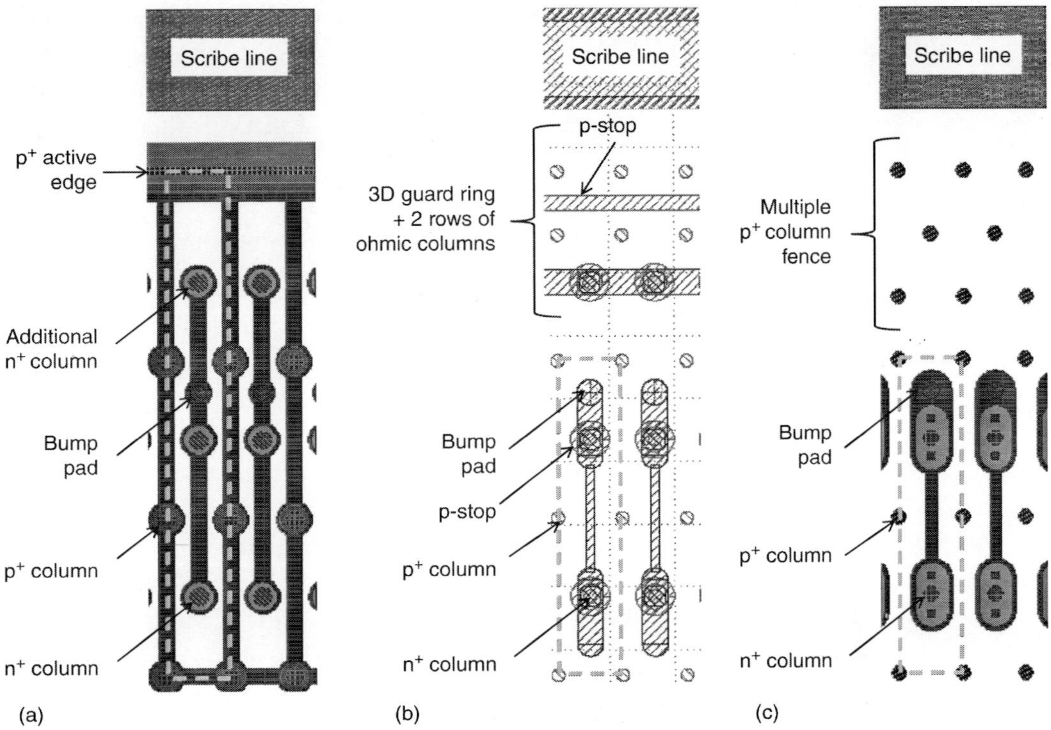

FIGURE 7.1 Layout details of the edge region in the three 3D pixel designs for the ATLAS IBL: (a) SNF/SINTEF, (b) CNM, and (c) FBK.

z direction parallel to the beam, so that active edges were not strictly necessary. Thus, it was decided to go for a slim edge of 200 µm, which could be obtained by all manufacturers. The different solutions adopted for this purpose are sketched in Fig. 7.1, which shows pixel layouts at the sensor edge along Z. For SNF/SINTEF, an additional n^+ column is present, facing the active edge and extending the edge pixel size to 375 µm × 50 µm. As the sketch shows, beyond the active edge there is a standard scribe line for sensor dicing. In the FBK and CNM designs, edge pixels have standard size, and the extra room beyond them is used to host terminating structures, which allow for a "slim" edge. For FBK, the slim edge consists of a multiple ohmic column fence that prevents the depletion region from spreading from the outermost junction column and reaching the cut line forming a conductive path at the edge [6]. For CNM, the slim edge consists of a guard ring of 3D electrodes connected together, able to sink the leakage current originating from the cut line and surrounded by two rows of ohmic columns.

b. Substrate bias: in SNF/SINTEF front-processed sensors, the substrate bias is supplied from a pad placed on the frontside, which is also the side where the bump bonding is made, since the backside is not accessible because of the presence of the support wafer. Another possible option was to perform a selective etching of the backside layer and then deposit a metal pad for wire bonding, but these steps were not fully engineered at that time. In CNM and FBK sensors, the substrate bias can be applied directly from the backside, making these sensors compatible with standard planar sensors without extra processing. To bias the SNF/SINTEF sensors, their layout was extended on one side by ~1.5 mm, permitting a bias tab to be accessible beyond the physical edge of the front-end chip after bump bonding. To ensure compatibility, the same 1.5 mm extension was used for CNM and FBK sensors with a bias tab on the backside. On the sensor side opposite to the bias tab, a 400 μm wide edge was implemented since a slim edge was not required.

c. On-wafer selection electrical tests: bump bonding is a complex and expensive process, so it was important to identify bad sensors at an early stage. It is common for this purpose to perform electrical tests on-wafer before bump bonding to select good sensors. SNF/SINTEF and FBK adopted a deposited temporary metal layer, which allows current-voltage (I-V) tests to be performed in each sensor. This metal layer is deposited at the end of the fabrication process and is removed after its use. The total current can be measured with either just one probing pad or with multiple probing pads connected separately to each column using a probe-card for a more accurate monitoring of defects. Figure 7.2a shows details of the temporary metal layout on an FBK wafer.

For CNM sensors, the use of temporary metal was not compatible with the available testing equipment, so the sensors' quality was measured with the guard ring current. A detail of a CNM sensor and its guard ring-probing pad visible on the left corner is shown in Fig. 7.2b. While this method is useful for monitoring defects at the sensor periphery, it was shown to be not fully reliable since the guard ring current is not representative of the full leakage current drawn by the sensor [7].

The large area of the required FE-I4 compatible 3D sensor was considered a major concern for the production yield since the defect density scales with it. It was therefore agreed

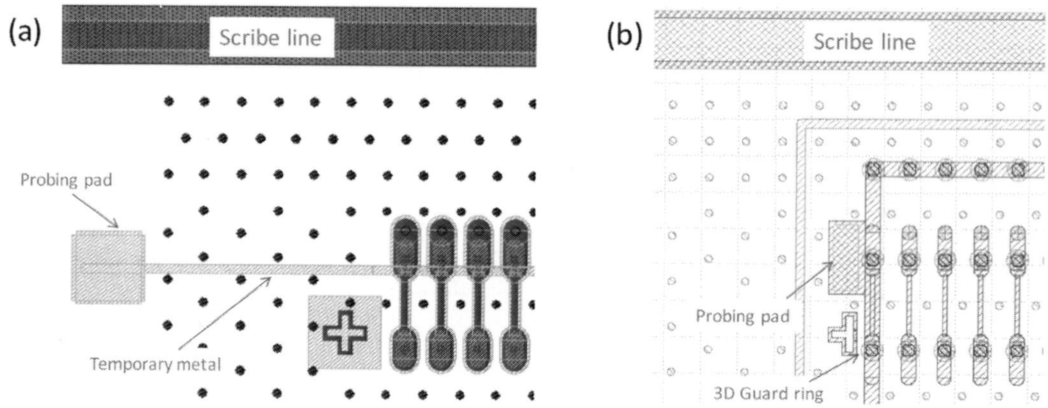

FIGURE 7.2 Layout details of (a) temporary metal used for sensor selection on wafer at FBK and (b) guard ring pad used for sensor selection on wafer at CNM.

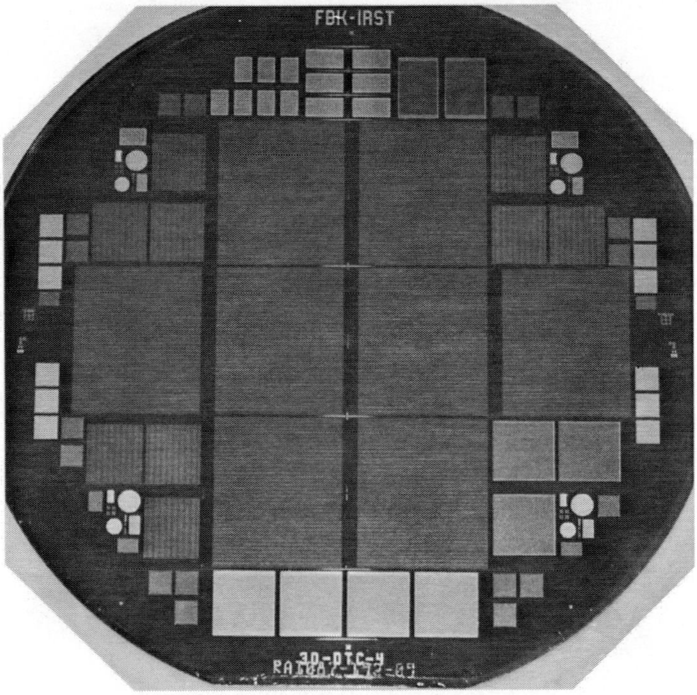

FIGURE 7.3 Photograph of a production wafer from FBK (Courtesy of FBK).

to mitigate this risk by making single-sensor tiles, rather than the required double ones, which would be later mounted in pairs on the modules.

At that time, the fabrication standard for 3D process was 4-inch-diameter wafers, on which eight sensors of the FEI4 dimension could be placed on the floor plan while maintaining a relatively wide safety margin (~1 cm) from the wafer's edge. Figure 7.3 shows the photograph of a wafer processed at FBK. Identical floor plans were produced in all involved facilities.

As can be seen in Fig. 7.3, the core of the wafer layout contains eight single tiles matching the ~4 cm^2 size of the FEI4 chip; nine single-chip sensors compatible with the currently installed ATLAS FE-I3 front-end chip [8], at all corners but the bottom right; and three pixel sensors compatible with the PSI146 readout chip of the CMS-LHC experiment [9], at the bottom right corner. At the wafer periphery, there are test structures used by the different foundries to monitor technological parameters and to perform electrical and functional tests without the need for bump bonding. Among these test structures are 3D strip sensors and 3D diodes.

7.3 SENSOR ELECTRICAL SPECIFICATIONS

The quality assurance electrical specifications, which are summarized in Table 7.1, were extracted from Technology Computer Aided Design (TCAD) simulations and from experimental results. These specifications were then applied to the sensor selection criteria by looking at electrical parameters, which are derived from I-V and/or C-V curves measured

TABLE 7.1 Summary of the sensor electrical specifications for wafer-level tests

Parameter	Symbol	Value
Range of operation temperature	ΔT_{op}	20–24°C
Depletion voltage	V_{depl}	<15V
Operation voltage	V_{op}	$\geq V_{depl}$ + 10V
Leakage current at operation voltage	$I(V_{op})$	<2μA (full sensor) <200nA (guard ring)
Breakdown voltage	V_{bd}	>25V
Slope of the I-V curves at operation voltage	$I(V_{op}) / I(V_{op} - 5V)$	<2

on wafers before bump bonding. As an example, the depletion voltage can be estimated either from C-V curves or from a knee in the I-V curves (see, e.g., the plots in Section 7.5). Previous results from R&D and qualification batches indicated a depletion voltage of about 7-8 V for the chosen 2E configuration and substrate type, so the depletion voltage specification in Table 7.1 includes a wide safety margin. The leakage current, breakdown voltage, and slope of the I-V characteristics are all suitable to detect the presence of defects inside the bulk or at the surface; therefore, the definition of extremes in their values allows a safe selection of good sensors. For CNM devices, where the guard ring current does not represent the entire sensor area, the current limit was scaled according to the number of columnar electrodes involved. Note that, for the slope of the I-V curve, no particular justification to use a different criterion was found for 3D, so the same specification of existing ATLAS planar pixel sensors was used [10].

So, finally, the specifications listed in Table 7.1 were used to select good 3D sensors for bump bonding. Given that a non-negligible fraction of the bump-bonding cost is related to the deposition of an under-bump metallization (UBM), which scales with the number of wafers, it was decided that to be eligible for bump bonding, each wafer should have at least three good 3D sensors matching the FE-I4 ROC.

7.4 PROTOTYPE FABRICATION AND IBL SENSOR PRODUCTION STRATEGY

All four processing facilities fabricated a first batch of 3D sensors with the common wafer layout in 2010. This run was intended as a qualification batch for the IBL [3]. Silicon substrates from the same ingot grown at the company TOPSIL [11] were used by all facilities. These wafers were float zone, double side polished, p-type, with a 100 mm diameter, <100> crystal orientation, 230 μm thickness, and a very high resistivity in the range of 10–30 kΩ cm. All fabrication batches started in the Spring of 2010, but processing times varied for different 3D technologies. As already mentioned in Chapter 5, double-sided 3D sensors provide some advantages since they require a lower number of process steps, so FBK and CNM batches were completed earlier than the SINTEF and SNF ones.

While the 3D processing qualification runs were still ongoing, the IBL schedule was decided with an anticipated installation during the first long shutdown of the LHC in 2013–2014. It was necessary at this point to agree on a new strategy to meet this challenging

TABLE 7.2 Yield of all 3D batches fabricated for the IBL

Batch	Tested wafers	Selected wafers	Good sensors	Yield on selected wafers
FBK-A10	20	12	58	60.4%
FBK-A11	11	4	14	43.8%
FBK-A12	16	13	63	60.6%
FBK-A13	11	4	15	46.9%
CNM-1	19	18	86	59.7%
CNM-2	17	15	85	70.8%
CNM-3	24	15	60	50.0%
OVERALL	118	81	381	58.8%

schedule. This resulted in the choice of the safer and faster processing double-sided 3D from FBK and CNM for the IBL production, while the more aggressive single-side design proposed by SNF and SINTEF with active edges was left for future upgrades.

The sensor design was reviewed, and some modifications to the layout were made to include compatible alignment marks and bias pads for the assembly. After that, the sensor preproduction started at CNM and FBK at the beginning of 2011.

In July 2011, a sensor technology review was made at CERN for the IBL project, based on the available results and involving external experts. At that time, besides results from beam tests of irradiated samples from the qualification batches, the first preproduction batch had been completed at FBK with a relatively good yield, thus contributing to an overall positive evaluation for 3D sensors. At the end of this process, the ATLAS management decided to recommend a mixed planar-3D (75%–25%) scenario as the baseline option for the IBL, with four single-chip 3D sensors to populate each end (forward and backward) of a stave. For the first time, 3D sensors were to be used in a HEP experiment. The production continued at FBK and CNM until May 2012 with a total yield on selected wafers of 58.8%. A total number of 381 sensors were delivered, exceeding the required number of 224, including a 2x safety margin, for the IBL mixed scenario, plus a large number of spares. The yield for all the production batches at CNM and FBK is reported in Table 7.2.

7.5 EXPERIMENTAL RESULTS

This section reports selected results from the electrical characterization of qualification and production batches. I-V measurements were performed on each sensor on wafers using a probe station with a dedicated probe card and standard semiconductor test equipment. Measurements were carried out at room temperature and in the dark to avoid light contamination, using the temporary metal technique at FBK, SNF, and SINTEF and guard-ring current probing at CNM.

Figure 7.4 shows the I-V curves of eight FE-I4 sensors from one wafer from the SINTEF qualification batch. Three sensors show early breakdown with abrupt current rise at low voltage, whereas the other five sensors have total leakage current of a few μA and breakdown voltage in the range of 55 to 70 V. From the knees in the I-V curves of good sensors, a depletion voltage lower than 10 V can be inferred. The leakage current at the operation

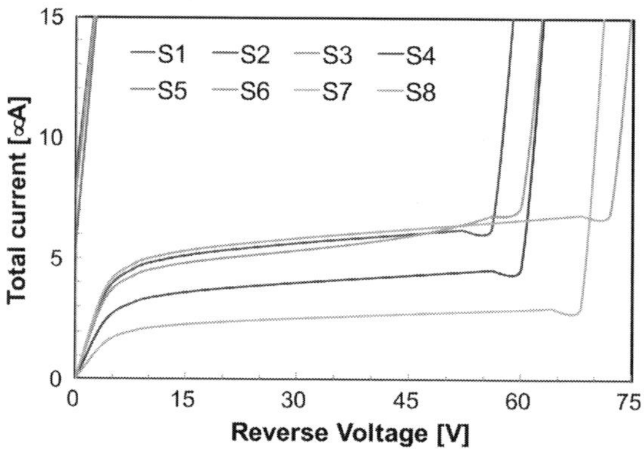

FIGURE 7.4 Current versus voltage measurements of eight 3D sensors from one wafer of the qualification batches from SINTEF, using the temporary metal selection method. Note that three sensors have very early breakdown and that their curves lie on top of each other.

voltage is slightly higher than the specification value, while the breakdown voltage is close to the intrinsic value due to the p-spray; it is well beyond the specification value. Further results relevant to the SINTEF sensors can be found in [12].

Sensors from the SNF qualification batch were properly working. As an example, Fig. 7.5 shows the I-V curves of eight sensors from one wafer. Compared to SINTEF sensors, the leakage current and the I-V curve slope are slightly larger, whereas the intrinsic breakdown voltage is slightly smaller. Moreover, a larger number of sensors had defects, causing early breakdown.

The I-V curves from all 80 columns in two FBK sensors from one wafer of the qualification batch (A09-W14) are shown in Fig. 7.6. The sum of the 80 columns I-V curves gives then the total sensor I-V curve. The detailed mapping performed using temporary metal probing allows one to identify the presence of a defect in a single column of 336 pixels, as in the case of sensor S6. Such a defect is responsible for the early breakdown of the entire sensor. For good sensors, the leakage current at the operation voltage is in the order of 100 nA, a factor of 10 below the specification value. The current slope, as defined in Table 7.1, is in

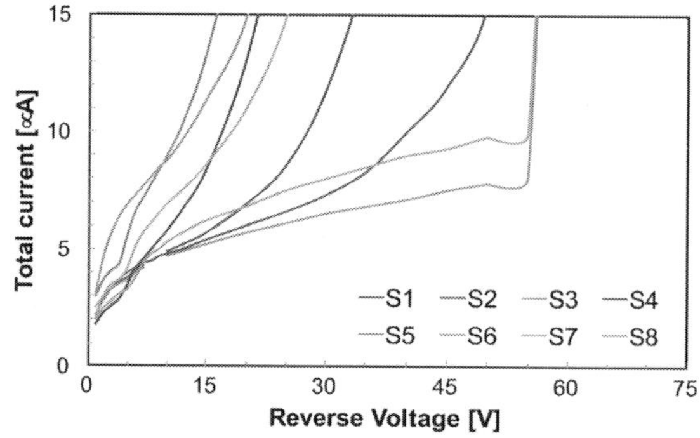

FIGURE 7.5 Current versus voltage measurements of eight 3D sensors from one wafer of the qualification batches from SNF, using the temporary metal selection method.

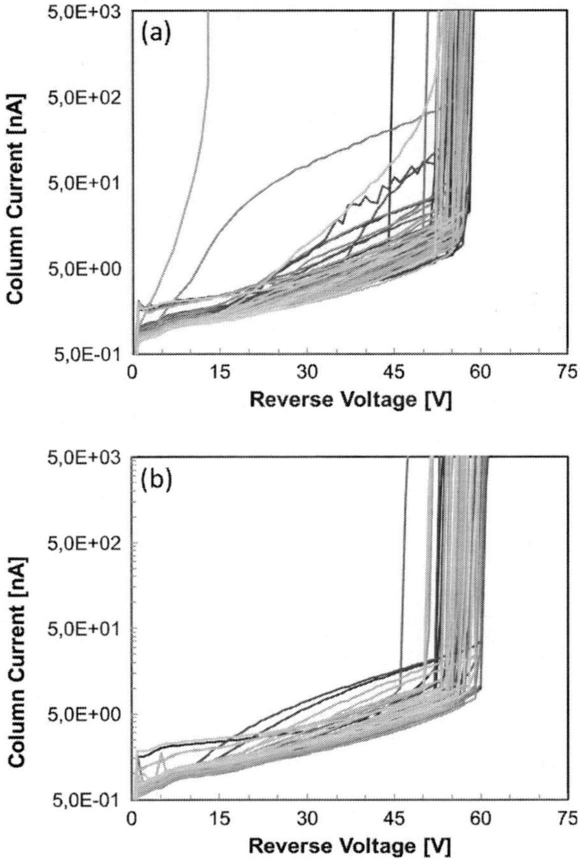

FIGURE 7.6 Current versus voltage measurements of two 3D sensors from one FBK wafer (A09-W14) using the temporary metal selection method: (a) S6, and (b) S5. In this method, the 336 pixels of each of the 80 columns of the matrix composing the sensor are joined together using an aluminium strip, so each I-V probes defects at column level. This can be clearly seen for sensor S6, where one column is responsible for the early breakdown of the sensor. This is not the case of sensor S5, where all columns are recording similar current values and breakdown points.

the range from 1.22 to 1.34, well below the specification. A breakdown voltage of about 45 V, close to the intrinsic value measured on small area devices (3D diodes), demonstrated that the three good sensors on this wafer were free from defects.

The reliability of the temporary metal technique was further validated by comparison of the I-V curves measured on wafer to the ones measured after bump bonding. As an example, Fig. 7.7 shows results relevant to two sensors from the same wafer (A09-W14). Sensor S3, which was good at wafer level, remains so after bump bonding, with very similar leakage current and breakdown voltage values showing that no or little stress was applied to the wafer. Sensor S4 was bad on wafer and remains so after bump bonding, showing early breakdown. In general, the leakage current is smaller on the assembly, a fact that can be ascribed to the additional surface current contribution (MOS effect) caused by the temporary metal.

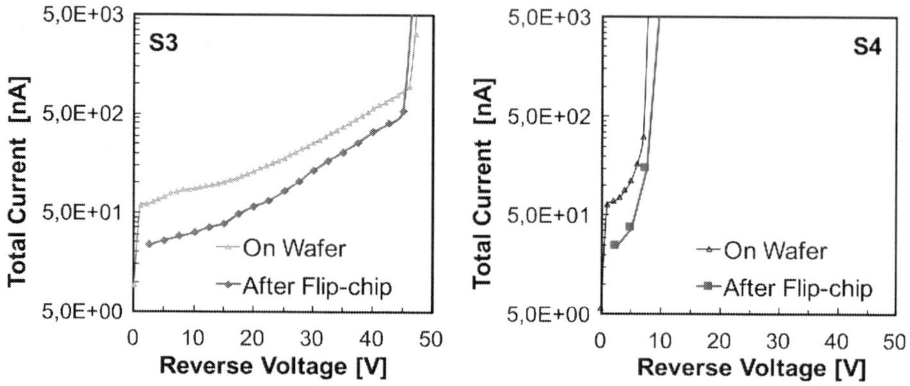

FIGURE 7.7 Comparison of total current versus voltage curves from two sensors belonging to one FBK wafer (A09-W14) measured on wafer and after bump bonding.

An example of the electrical characteristics measured on CNM sensors is shown in Fig. 7.8, where the guard ring currents versus voltages measured on wafer are compared to the total I-V curves measured after bump bonding. The leakage current measured on the guard ring is, as expected, much smaller than the total one measured after bump bonding, since the depleted volume contributing to the leakage is greater. In Fig. 7.8, the breakdown voltage is improved after bump bonding, sometimes by a large extent. This was thought to be due to the beneficial release of mechanical stress after wafer dicing, but, as will be shown later, it might not be completely the case.

Figure 7.9 shows a beautiful example of I-V characteristics measured on two "perfect" wafers from CNM (guard ring current) and FBK (total current), having all sensors meeting the specifications.

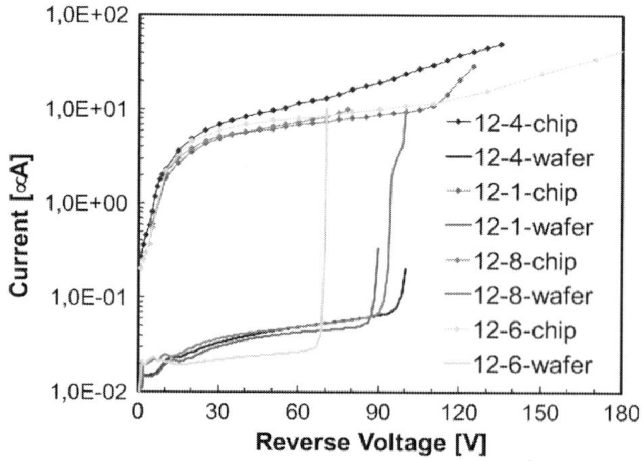

FIGURE 7.8 Guard ring current versus voltage curves measured on wafer (labeled "wafer") and total current versus voltage curves measured after bump bonding (labeled "chip") on some sensors from wafer W12 of the CNM qualification batch.

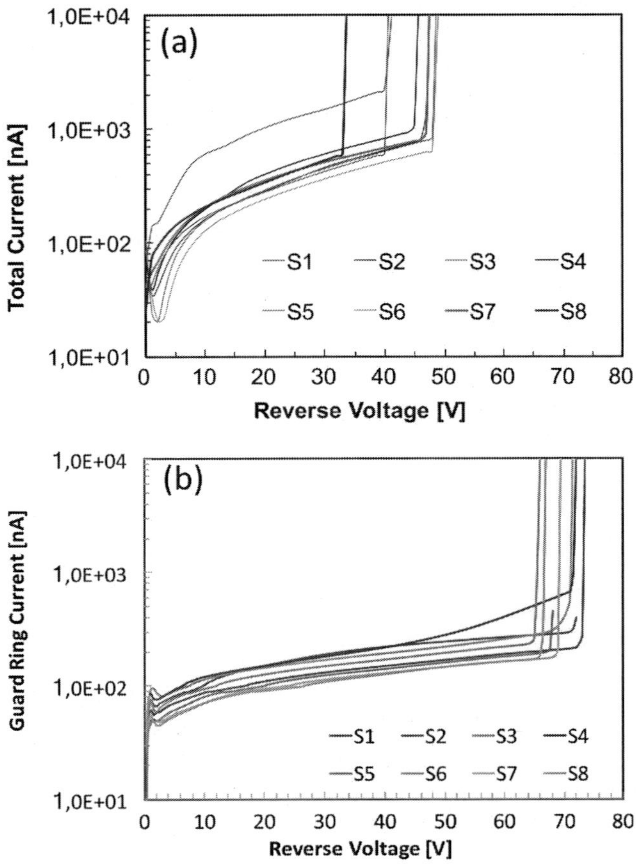

FIGURE 7.9 Examples of I-V curves measured on two "perfect wafers" from the IBL 3D sensor production at (a) FBK and (b) CNM.

Finally, Fig. 7.10 summarizes the results of the sensor production. The overall quality of the selected sensors was very good for both foundries, considering that this was the first medium-scale production for 3D sensors. Figure 7.10 (a,b,c) shows the distributions of the relevant parameters for 374 sensors (187 each for CNM and FBK), including all those selected for the IBL. The leakage current distributions shown in Fig. 7.10a are pretty uniform. Less than 10% of the considered sensors have leakage currents slightly beyond specification, but they are still useful for tests. The breakdown distribution (Fig. 7.10b) of FBK sensors peaks between 35 and 50 V, while for CNM the distribution is broader and extends to larger values, up to 100 V. This difference is due to the different surface insulating p-spray layers for FBK and p-stop ones for CNM [13]. The current slope visible in Fig. 7.10c shows that both distributions are very uniform and peak at values much lower than the required specification. In particular, CNM sensors show very small values of about 1.1. For FBK samples, slopes are slightly higher and correlate to the earlier breakdown voltages (see also the I-V curves in Fig. 7.9a): This effect is mainly due to a moderate avalanche multiplication of surface-generated charges [13].

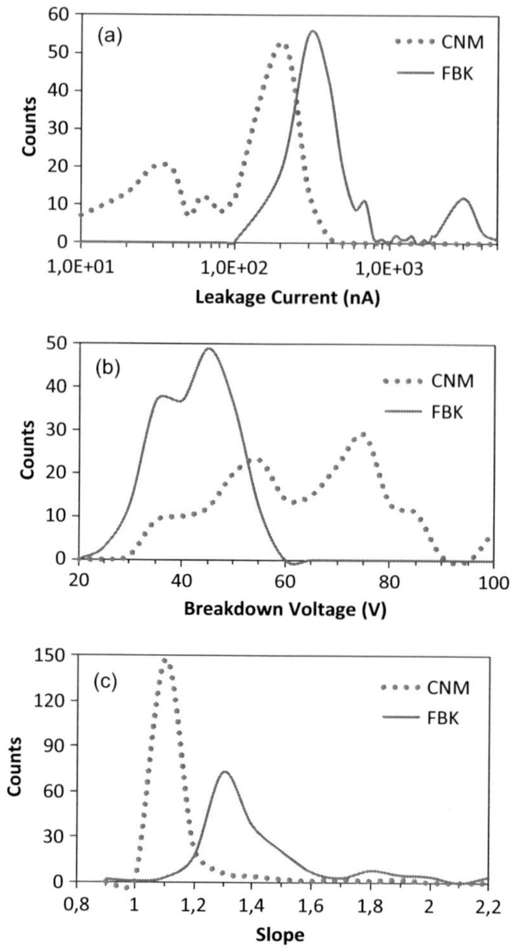

FIGURE 7.10 Distributions for all selected sensors of (a) leakage current at operating voltage, (b) breakdown voltage, and (c) slope of the I-V curves at operating voltage.

7.6 LESSONS LEARNED

The IBL production gave a significant boost to the development of 3D sensors. As can be seen from Fig. 7.10, the electrical characteristics of sensors from different batches are reproducible, evidence that good control of the fabrication process is in place. However, regardless of the specific approach and in spite of the significant advancement in the past few years, 3D sensor technology still remains quite complex when compared to more traditional sensors. Critical processing steps like deep reactive ion etching (DRIE) can produce major, irreversible damage, and the related defects are not easy to spot using standard techniques such as optical inspection, since they can also be hidden deep in the bulk. Therefore, an accurate electrical characterization at wafer level is necessary to identify bad pixel sensors at an early stage, before bump bonding.

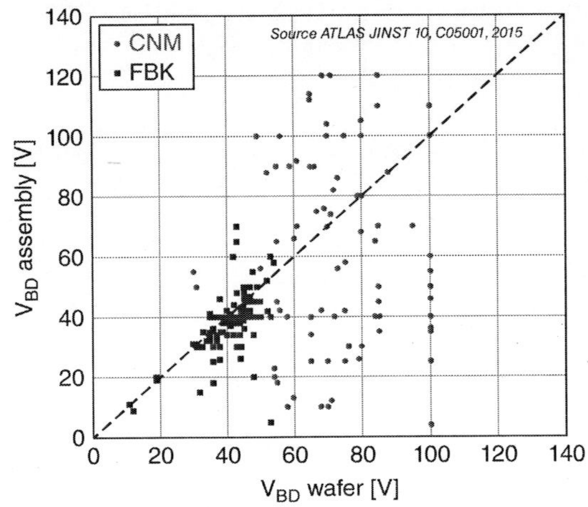

FIGURE 7.11 Measurement of sensor breakdown voltage (Vbd) after this is assembled into a detector module (referred to as "Assembly") versus on-wafer (referred to as "Wafer"). Measurements on wafer are done by contacting the fence guard ring for CMN and by temporary metal connecting pixels into strips for FBK. Only a subsample of CNM tiles is shown in the plot, for modules where I–V curves were available with a current compliance of 100 μA. Data are elaborated from ATLAS data published in [14].

The yield calculated according to on-wafer measurements, reported in Table 7.2, was significantly better for CNM than for FBK. However, it should be noted that while the I-V measurement technique based on a temporary metal layer used at FBK, SNF and SINTEF proved to be effective, the guard ring current method used at CNM was not so reliable, and caused several sensors to be rejected after bump bonding. This is evident from Fig. 7.11, which shows correlation plots of the breakdown voltage measured on wafer and bump-bonded assemblies for CNM and FBK sensors [14].

REFERENCES

1. C. Da Vià, S. I. Parker, G. Darbo, "Development, testing and industrialization of tull-3D active-edge and modified-3D silicon radiation pixel sensors with extreme radiation hardness: Results, plans." ATLAS Upgrade Document, 2007 (http://cern.ch/atlas-highlumi-3dsensor)
2. M. Capeans et al., "ATLAS insertable B-layer technical design report," CERNLHCC-2010-013; ATLAS-TDR-019. G. Darbo, M. Nessi (The ATLAS Collaboration), ATLAS Insertable B-Layer Technical Design Report Addendum, CERN-LHCC-2012-009; ATLAS-TDR-019-ADD-1.
3. C. Da Vià, M. Boscardin, G.-F. Dalla Betta, G. Darbo, C. Fleta, C. Gemme, P. Grenier, S. Grinstein, T.-E. Hansen, J. Hasi, C. Kenney, A. Kok, S. Parker, G. Pellegrini, E. Vianello, N. Zorzi, "3D silicon sensors: Design, large area production and quality assurance for the ATLAS IBL pixel detector upgrade," Nuclear Instruments and Methods A, vol. 694, pp. 321–330, 2012.
4. M. Garcia-Sciveres, D. Arutinov, M. Barbero, R. Beccherle, S. Dube, D. Elledge, J. Fleury, D. Fougeron, F. Gensolen, D. Gnani, V. Gromov, T. Hemperek, M. Karagounis, R. Kluit, A. Kruth, A. Mekkaoui, M. Menouni, J.-D. Schipper, "The FE-I4 pixel readout integrated circuit," Nuclear Instruments and Methods A, vol. 636, pp. S155–S159, 2011.

5. C. Da Vià, E. Bolle, K. Einsweiler, M. Garcia-Sciveres, J. Hasi, C. Kenney, V. Linhart, S. Parker, S. Pospisil, O. Rohne, T. Slavicek, S. Watts, N. Wermes, "3D active edge silicon sensors with different electrode configurations: Radiation hardness and noise performance," Nuclear Instruments and Methods A, vol. 604, pp. 505–511, 2009.
6. M. Povoli, A. Bagolini, M. Boscardin, G.-F. Dalla Betta, G. Giacomini, E. Vianello, N. Zorzi, "Slim edges in double-sided silicon 3D detectors," Journal of Instrumentation, JINST 7, C01015, 2012.
7. C. Gemme, "3D sensors for tracking detectors: Present and future applications," Proceedings of Science—22nd Workshop on Vertex Detectors (Vertex 2013), Paper 28, 2014.
8. I. Perić, L. Blanquart, G. Comes, P. Denes, K. Einsweiler, P. Fischer, E. Mandelli, G. Meddeler, "The FEI3 readout chip for the ATLAS pixel detector," Nuclear Instruments and Methods A, vol. 565, pp. 178–187, 2006.
9. H. Chr. Kästli et al., "Design and performance of the CMS pixel detector readout chip," Nuclear Instruments and Methods A, vol. 565, pp. 188–194, 2006.
10. S. D'Auria, T. Rohe, S. Seidel, R. Wunstorf, The ATLAS Pixel Sensor Group, "Technical specifications of the ATLAS pixel prototype sensors," 1999.
11. TOPSIL Semiconductor Materials A/S, Linderupvej 4, DK 3600 Frederikssund, Denmark. /http://www.topsil.com.
12. A. Kok, M. Boscardin, G.-F. Dalla Betta, C. Da Vià, G. Darbo, C. Fleta, P. Grenier, S. Grinstein, T.-E. Hansen, J. Hasi, C. J. Kenney, S. I. Parker, G. Pellegrini, E. Vianello, N. Zorzi, "3D silicon sensors—Large Area Production, QA and Development for the CERN ATLAS Experiment Pixel Sensor Upgrade," 2012 IEEE Nuclear Science Symposium and Medical Imaging Conference Record, Paper N14-154.
13. G.-F. Dalla Betta, C. Da Vià, M. Povoli, S. Parker, M. Boscardin, G. Darbo, S. Grinstein, P. Grenier, J. Hasi, C. Kenney, A. Kok, C.-H. Lai, G. Pellegrini, S. Watts, "Recent developments and future perspectives in 3D silicon radiation sensors," Journal of Instrumentation, JINST 7, C10006, 2012.
14. G. Darbo on behalf of the ATLAS Collaboration, "Experience on 3D silicon sensors for ATLAS IBL," Journal of Instrumentation, JINST 10 (2015) C05001.

CHAPTER 8

Planar Active-Edge Sensors

8.1 INTRODUCTION

Edge termination is one of the most critical aspects of radiation sensor design (for additional discussion, see Chapter 4). Regardless of the specific sensor type (pixel, strip, pad, etc.), the active area is normally surrounded by at least a main guard ring biased at the same voltage as the readout electrodes. This is necessary to collect the leakage current generated by imperfections in the peripheral region and to ensure a uniform electric field distribution in the innermost regions of the active area. Multiple floating guard rings, with progressive bias voltages, are also routinely used around the main guard ring to obtain higher breakdown voltage and better long-term stability [1]. Additionally, between the guard ring(s) and the cut line, a safety distance is normally left, so that the depletion region spreading laterally from the active area cannot reach the physical edge of the sensor, thus minimizing the risk of leakage current generation from defects introduced by the process of detector separation by dicing with standard diamond saws.

The geometrical details of these terminations depend on a number of parameters specific to the considered application; including substrate thickness, doping polarity and concentration, oxide charge density, environmental conditions (in particular humidity [2]), the irradiation scenario and the required bias conditions. Nevertheless, in all cases, a relatively wide insensitive region is introduced at the sensor edge, the size of which can span from a few hundreds of micrometers to more than 1 millimeter. As a consequence, within, for example, a particle tracking system, it is common practice to tilt sensors so that they slightly overlap at the edge, within the same tracking layer. This is a fact that complicates the detector assembly and increases the material budget.

In the past few years, there has been an increasing interest in "edgeless" sensors. This topic has become very popular since several applications of silicon pixel sensors in HEP experiments (like the HL LHC or the linear colliders) and in X-ray imaging are setting increasing demands on minimization of the dead area at the edge, calling for improved design/technological solutions for edge termination. As an example, new sensors for the inner pixel layers at the high luminosity LHC upgrade detectors, require reduced geometrical inefficiency because of material budget restrictions and tight mechanical constraints.

As noted in Chapter 7, pioneering work in this direction was already successfully implemented in the ATLAS Insertable B-Layer (IBL), with maximum dead regions at the edge of single-chip sensors of 200 μm along the direction parallel to the beam [3]. Moreover, experiments in the forward region with sensors inserted in the beam pipe to detect diffracted protons like the TOTEM experiment used one-side edgeless strip sensors [4]. More recent forward physics experiments, such as the AFP [5] in ATLAS and the CT-PPS in CMS [6], use one-side edgeless pixel sensors to maximize their physics potential. As previously mentioned, edgeless sensors are also appealing for X-ray imaging applications: Four-side edgeless sensors allow construction of seamless large-area imagers by tiling several detector modules into matrices, whereas one-side edgeless sensors are sufficient to largely improve the detection efficiency for X-ray imaging in the so-called "edge-on" configurations where the beam is pointed to the edge of the sensors rather than its surface [7].

When aiming at an edgeless solution, significant advantages are offered by the adoption of 3D fabrication technologies, which allow more design flexibility by exploiting the third dimension within the sensor substrate. In this chapter, after recalling some approaches that have so far been proposed to reduce the sensor dead area at the edge, the most important developments in active-edge and slim-edge planar sensors based on microfabrication will be reviewed.

8.2 DIFFERENT APPROACHES TO EDGELESS SENSORS

8.2.1 Early Attempts

Several technological and design solutions have been reported to minimize the insensitive area at the sensor edge. One of the first methods proposed was laser dicing through the sensor active region, in combination with chemical etching of the edge and/or aging treatments to reduce the cut damage by dangling bond passivation [8]. This approach, however, while providing a truly one-side edgeless device, as demonstrated with beam test experiments [9], did not provide optimal results in terms of leakage current and stability.

Another interesting solution was the so-called current terminating ring (CTR) used in combination with a normal bias ring [4]. In this case, the CTR was able to sink most of the high-leakage current caused by saw dicing, while only a small fraction of this current (~0.1%), was drawn by the bias ring, with values of 100s of nA, similar to standard sensors with full guard rings in place. By using this technique, the TOTEM experiment was able to obtain devices with a single slim edge as small as ~50 μm [10]. Further reduction of the dead area to ~25 μm could be obtained by replacing the saw dicing with a dry etching closer to the CTR, as proposed in [11].

8.2.2 The Scribe-Cleave-Passivate Technique

The Scribe-Cleave-Passivate (SCP) technique represents an interesting alternative for slim edges [12]. Since all treatments can be applied postprocessing and at low temperature, SCP is compatible with any kind of sensors, regardless of the specific processing technology, with the only requirement being to have <100> crystal orientation wafers with reasonably good alignment between sensor and lattice in order to obtain good-quality rectangular side cleaving. Wafer scribing and cleaving are used to minimize the number of defects with respect to a standard dicing saw [13], whereas surface passivation of the edge aims at allowing a bias voltage gradient along the sidewall surface.

The SCP technique actually includes a variety of methods:

- Classic diamond-stylus scribing steps can be replaced with laser dicing or by etching (e.g., by XeF_2 or DRIE).
- Cleaving steps can be enhanced by sidewall damage removal by a reactive agent.
- Passivation can be performed with different dielectric layers (e.g., native oxide, oxide or nitride from plasma-enhanced chemical vapor deposition), to be tailored to different substrate doping types in terms of interface charge sign.
- For p-type substrates, in particular, negative charge is required, and it can be obtained with alumina (Al_2O_3) by atomic layer deposition [14].

In the past few years, different SCP options have been explored with promising results, also in terms of charge collection uniformity in the edge region compared to the core of the sensors [15]. More recent work focused mainly on (i) wafer-level processing, to automatize the manual methods so far adopted for volume productions; and (ii) radiation hardness, which is a possible concern particularly for Al_2O_3 passivated devices, since high edge currents have been observed after irradiation at low fluences [15].

8.3 ACTIVE-EDGE TECHNOLOGIES

As already mentioned in Chapters 4 and 5, active edges were first introduced at the Stanford nanofabrication facility (SNF) as an extension of 3D sensor technology [16]. However, they were later successfully implemented also in planar sensors, in the so-called planar active-edge (or edgeless) configuration. These sensors have collection electrodes arranged by standard planar design, but with scribe lines replaced by trenches that are etched by deep reactive ion etching (DRIE) around the sensor active area and doped to act as electrodes [17].

Figure 8.1 shows four possible implementations of planar active-edge sensors, differing by the substrate type (n-type vs. p-type) and by the role played by the trench (+backside) electrode, which can act as either a junction or an ohmic contact. While a junction trench allows for an easier depletion of the corner region at the bottom [17], in practice ohmic trenches have so far been more frequently used.

Sensor dicing and related problems are avoided with active edges; moreover, arbitrary edge shapes can be achieved. As an example, torus shape structures are reported in [18]. These advantages come at the expense of an increased process complexity, owing to several DRIE steps and other critical features, as already addressed in Chapter 5 with reference to 3D sensors. In particular, according to the original process sequence developed at SNF, a support wafer was initially fusion bonded to the sensor wafer [19] to hold different

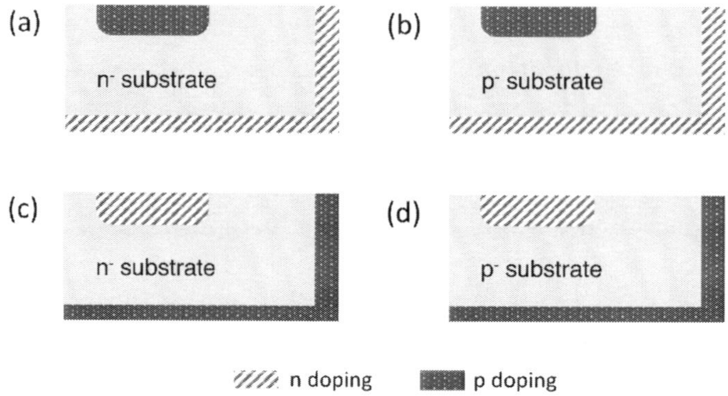

FIGURE 8.1 Schematic cross sections of different configurations of planar active-edge sensors: (a) p-on-n, (b) p-on-p, (c) n-on-n, and (d) n-on-p. In (b) and (c), the trench and backside act as a junction electrode, whereas in (a) and (d) they act as an ohmic electrode.

sensors together once the trenches have been etched, which should finally be removed. While wafer bonding is a reliable technology, support wafer removal is still quite challenging in the presence of etched trenches. Furthermore, in order to ease detector assembly within a system, the substrate bias should preferably be applied from the sensor backside. To this purpose, after support wafer removal, the residual insulating layers on the backside have to be etched and a metal layer has to be deposited. Extensive R&D is still ongoing (e.g., by using temporary wafer carriers) to engineer these back-end steps. In particular, in the case of pixel sensors, a proper combination of support wafer removal with sensor bump bonding is required.

Besides the technologies originally proposed at SNF, active-edge technologies are now available at other processing facilities. As an example, a similar fabrication process was developed at FBK (Trento, Italy) [20]. A schematic cross section of a p-on-n sensor made on an oxide-bonded substrate (referred to as SOI) is shown in Fig. 8.2. The only significant difference with respect to the SNF process is the use of sidewall phosphorus thermal diffusion for the active-edge doping, whereas undoped poly-Si is used to fill the trenches to cope with the severe topography on the etched side for subsequent lithographical steps [20].

Other details, also common to SNF technology, can be appreciated from Fig. 8.2. In particular, the n^- substrate (active layer) is implanted on the backside before fusion bonding, so that the bias voltage can be uniformly distributed to the entire sensor volume. Moreover, the trench doping also extends laterally at its opening, so that the sensor can be biased from the front surface.

To apply the sensor bias more easily from the backside, FBK has also developed active-edge sensors on epitaxial wafers [21] and Si-Si direct-wafer-bonded (DWB) substrates [22, 23]. In the active-edge sensors, a high-resistivity epitaxial layer is grown on top of a low-resistivity substrate; this approach is suited to relatively thin active layers (up to ~100 μm) and is not optimal for thicker layers due to an increasing number of defects observed. To overcome this problem, a high-resistivity float zone wafer can be directly bonded

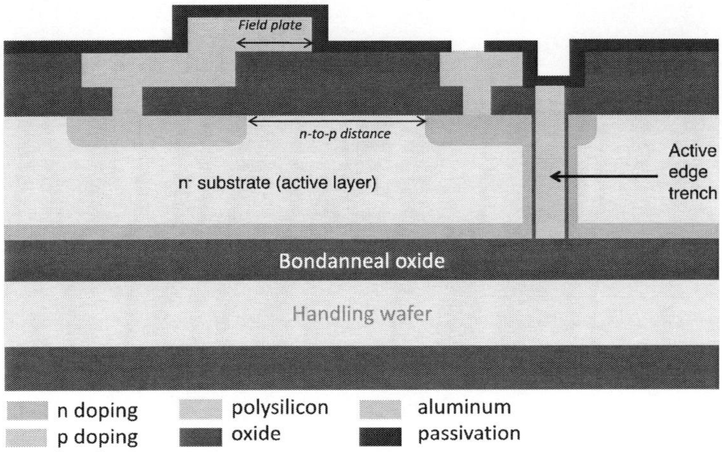

FIGURE 8.2 Schematic cross section of a planar p-on-n sensor with active edge fabricated at FBK on SOI wafers (Trento, Italy).

(i.e., without a bonding oxide) to a low-resistivity handle substrate. This approach has been shown to be successful regardless of the active layer thickness. A schematic cross section of these types of sensors is shown in Fig. 8.3. Common to both approaches is the presence of a highly doped substrate directly in contact with the high-resistivity active layer, so that the bias can propagate from the backside contact to the active layer and the trench. The lateral doping of the trench on the front surface and related contact with a metal pad can be used for test purposes or removed depending on the case.

Also the Technical Research Centre of Finland (VTT) developed planar active-edge sensors, using both a process similar to the one at SNF [24] and a simplified, original process on 6-inch wafers, which makes use of four-quadrant edge implantation for the trench doping [25]. A schematic cross section of a sensor made according to the latter technology is

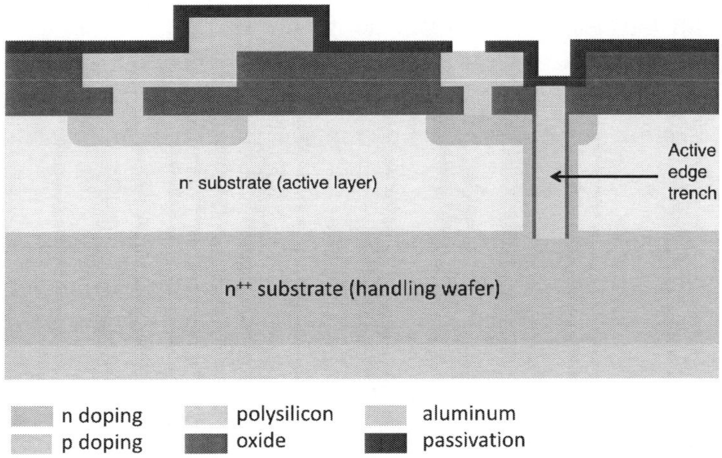

FIGURE 8.3 Schematic cross-section of a planar p-on-n sensor with active edge fabricated at FBK (Trento, Italy) on epitaxial and Si-Si direct wafer-bonded substrates.

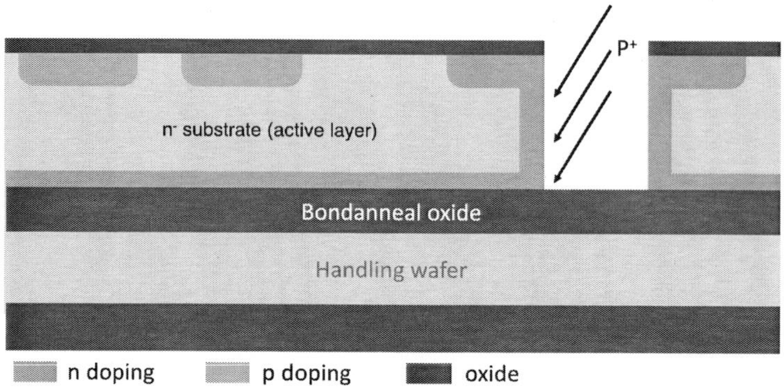

FIGURE 8.4 Schematic cross section of a planar p-on-n sensor with active edge fabricated at VTT (Helsinki, Finland) on SOI wafers using four-quadrant edge implantation (Trento, Italy).

shown in Fig. 8.4. Despite the need for multiple-edge implantations, such an approach has the advantage of using only one DRIE step and not using poly-Si for either doping or for trench filling, thus avoiding the related steps of poly-Si deposition and poly-Si etching from the surface.

One potentially critical aspect with unfilled trenches, however, is the severe topography of the wafers, which is strongly nonplanar. This can affect the quality of the lithography steps required to define the remaining features of the sensors (e.g., contacts, metal, etc.). VTT demonstrated successful mastery of this problem with very good yields in the case of 1-mm large trenches by using a specially tuned lithography process.

8.4 RESULTS

Excellent edge-sensitivity results were obtained from the first prototypes of planar active-edge-strip sensors made at SNF using a 12.5 keV X-ray microbeam [17]. The sensitive region was found to extend within 5 μm from the physical edge of the device (see Fig. 8.5). This outstanding performance, along with the great flexibility in defining the structure shapes, paved the way for use of planar active-edge sensors in several applications requiring measurement of the intensity, profile, and position of X-ray beams [18]. A large-area detection system for synchrotron science based on arrays of pixel sensors with active edges was also reported, specially tailored for protein molecular X-ray crystallography [26].

It should be mentioned that, depending on the application, planar detectors might pose some operational challenges for active edges in addition to the technological complications. In 3D detectors, the active edges represent just another vertical electrode, and their different geometry only requires a minor adjustment of the spacing to neighboring electrodes to ensure a uniform electric field distribution. On the contrary, in planar detectors, depending on the substrate thickness and radiation environment, relatively large-bias voltages could be required to fully deplete the active volume. These large-bias voltages, however, are difficult to withstand if the outermost charge-collecting electrode is very close to the active edge, requiring larger distances and dedicated design solutions, like, for

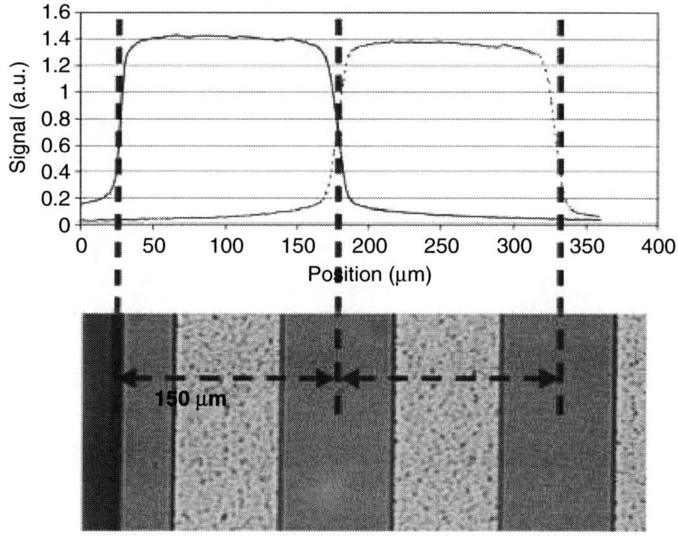

FIGURE 8.5 (Bottom) Part of an active-edge planar strip detector and (top) plot of the output signals from the two segments near the physical edge of the device. (Kenney et al., Nucl. Instrum. Methods A, 565, 2006. Copyright Elsevier. With permission). [17]

example, floating guard rings. As a result, while the peripheral region could still be made slim enough (~100 μm), sensitivity might not extend to the physical edge.

An additional drawback stemming from the close proximity of the trench to the readout electrodes can result in partial distortion of the electric field distribution at the edge, possibly leading to charge collection inefficiencies, especially in segmented detectors, as will be shown later in this chapter.

The first batch of planar sensors with active edges fabricated at FBK in 2010 was aimed mainly at evaluating the impact of the edge region design on the electrical and functional characteristics of the devices, besides assessing the critical steps of the technology [20]. Several devices were processed with p-on-n junctions and an active layer thickness of 200 μm. On test diodes, the trade-off between breakdown voltage and edge sensitivity was studied as a function of the distance (gap) between the outermost junction and the ohmic active edge as can be seen in Fig. 8.3, also exploring different field-limiting options like field plates and floating rings.

In nonirradiated samples, featuring a very low oxide charge density (~10^{10} cm^{-2}), the breakdown voltage was found to increase with the gap: with ~100 V at the minimum gap explored of 10 μm and increased to ~450 V at 125 μm with a saturated value at larger gaps [27].

This trend can be clearly observed in Fig. 8.6. Field plates were found to yield an increased breakdown voltage for gaps larger than 20 μm, in good agreement with numerical device simulations [20]. Selected diodes were functionally characterized using a position-resolved laser system and with X-ray beam scans, and proved to work properly with high efficiency up to a few micrometers from their physical edge [28].

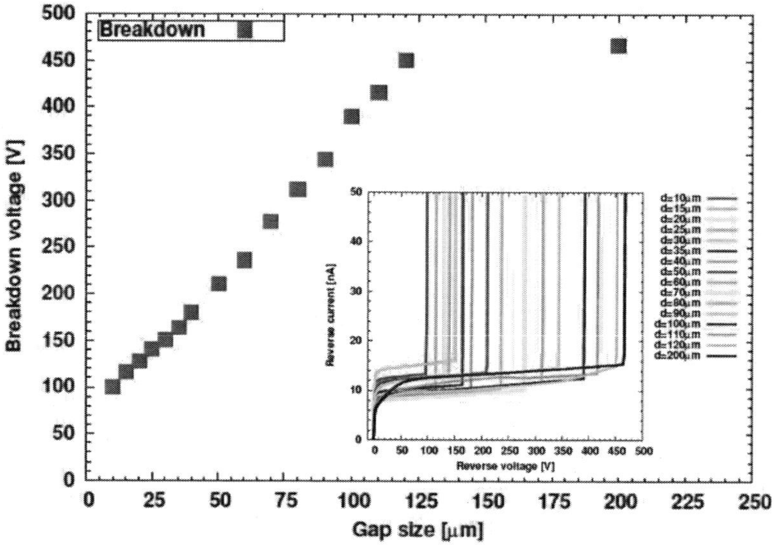

FIGURE 8.6 Breakdown voltage measured in p-on-n planar active-edge diodes with different n-to-p distance (gap size). The I-V curves of all measured structures are shown in the inset.

Further characterization of the edge region was performed on strip sensors from the same batch, featuring a readout strip pitch of 50 μm [29]. Figure 8.7 shows a micrograph of one of these devices, highlighting some of the layout details as well as a schematic cross section of the sensing areas. The nominal distance between the center of the outermost strip and the active edge (trench) is 50 μm. For simplicity, all strips were shorted together and read out in one single channel using a fast amplifier. The scanning direction is also indicated in Fig. 8.7, with the laser beam impinging perpendicularly to the surface.

Two sensors were characterized with this method: one denoted with S3 was not irradiated, while the other one (S13) was irradiated with reactor neutrons at JSI (Ljubljana, Slovenja)

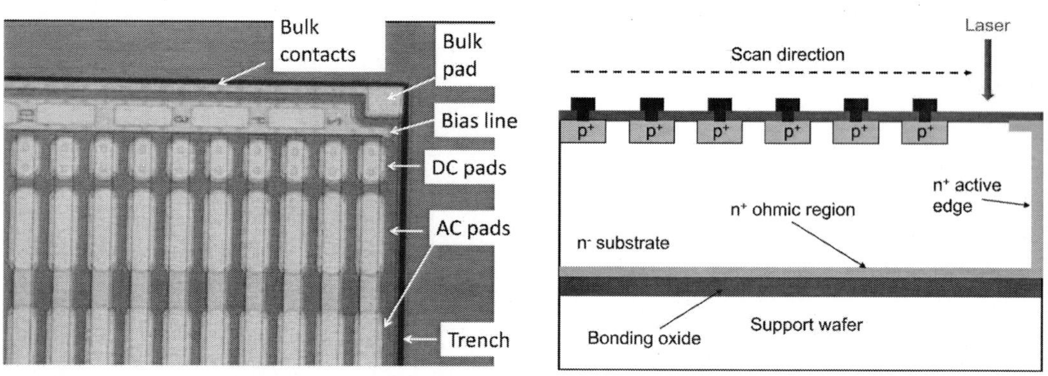

FIGURE 8.7 Micrograph (left) and schematic cross section (right) of a planar strip sensor with active edge tested with a position-resolved laser system.

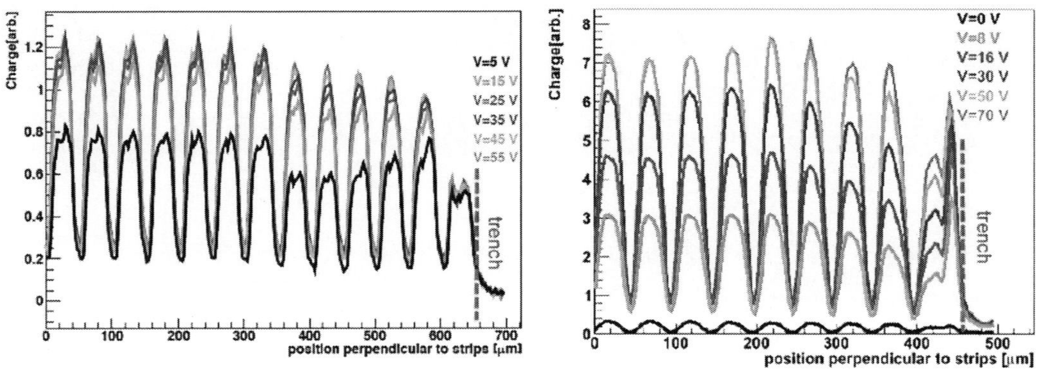

FIGURE 8.8 Collected charge as a function of the position for the nonirradiated strip sensor S3 tested with scanning laser setup at different bias voltages: red laser (left) and IR laser (right). (From Dalla Betta et al., Vertex 2013, Paper 42, 2014. Copyright owned by the authors under the terms of the Creative Commons Attribution-NonCommercial-ShareAlike Licence. http://creativecommons.org/licenses/by-nc-sa/3.0/.) [29]

at a fluence of 2×10^{15} n_{eq}/cm^2 and annealed for 80 minutes at 60°C before testing. Two different laser wavelengths were used for the tests: (i) 658 nm (red), with a penetration depth in Si of about 3 μm and a beam full width at half maximum (FWHM) of 6 μm, and (ii) 1064 nm (IR), with a penetration depth in Si larger than 1 mm (thus emulating an MIP in a 200 μm thick sensor) and a beam FWHM of 9 μm. The current pulses were integrated over 60 ns, yielding the collected charge signals.

Figure 8.8 shows the signals as a function of the impinging laser beam position for the nonirradiated strip sensor S3 at different bias voltages. The nominal position of the trench is also shown. It should be noted that signals can be observed when the trench is hit and also slightly beyond it due to the finite spatial resolution of the measurement, mainly on account of the width of the laser beam. Light reflection from metal layers covering the strips is the reason for the alternation of valleys in the data curves with peaks. The active-edge region is efficient at very low voltage, since lateral depletion is established soon. For the red laser, increasing the voltage has a negligible impact on charge collection near the edge, whereas the effect is much more pronounced for the IR laser. This is due to the much deeper absorption of IR light, which requires the depletion region to extend deep in the substrate at higher voltages before the maximum signal amplitude is reached. The presence of the active edge and its impact on the electric field distribution affect the response of the adjacent strips to a large extent: The transition region is as wide as ~300 μm in case of the red laser, and ~200 μm for the IR laser. After that, the charge collection becomes uniform as expected for a simple pad detector.

Figure 8.9 shows results relevant to the irradiated sensor S13. As explained in Chapter 3, the considered fluence would be high enough to cause substrate-type inversion, from n-type to effective-type—hence, a low field at the p-type strips and non-negligible charge-trapping effects, which would explain the relatively small collected charge with respect to the pre-irradiation case.

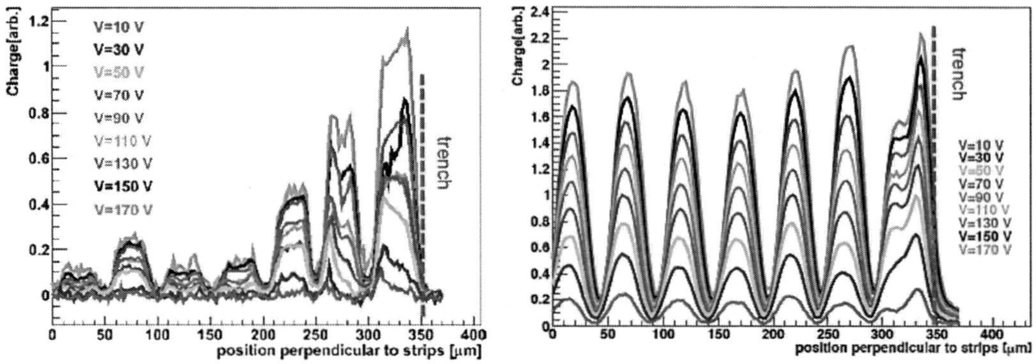

FIGURE 8.9 Collected charge as a function of the position for the irradiated strip sensor S13 tested with scanning laser setup at different bias voltages: red laser (left) and IR laser (right). (From Dalla Betta et al., Vertex 2013, Paper 42, 2014. Copyright owned by the authors under the terms of the Creative Commons Attribution-NonCommercial-ShareAlike Licence. http://creativecommons.org/licenses/by-nc-sa/3.0/.) [29]

Also after irradiation, the presence of the active edge affects the response a few strips away (~200 μm) from it but with an opposite trend; that is, the collected charge decreases as the distance from the edge increases: the reason for this behavior is the strong lateral electric field induced by the n-type trench in the type-inverted substrate. This effect is particularly evident for the red laser, since the electric field close to the surface (where 658 nm photons are absorbed) is very strong. This can also explain the large increase of charge from 150 V to 170 V at the edge, but not in the center of the detector, owing to impact ionization effects causing charge multiplication. In fact, the electric field is high enough to cause breakdown at ~170 V, as observed from the current-voltage curves (not shown).

For the infrared (IR) laser, on the one hand, the increase in the collected charge with bias voltage is more pronounced at all positions, since the electric field grows in a larger part of the detector where charge is generated by the IR photons. On the other hand, the increase in the charge at high voltage (150–170 V) in the edge region, though still visible, is not as big as that for the red laser because only a small fraction of the charge is generated close to the surface where the very high electric field region is confined.

The impact of the edge region design was also studied in p-on-n sensors made at FBK on epitaxial wafers (epilayer thickness of 100 μm) [21]. Similarly to previously reported results, the breakdown voltage was found to increase with the gap size and with the number of floating guard rings. The two quantities are of course correlated because a larger number of guard rings require a larger edge region. The study also involved irradiation with reactor neutrons at JSI (Ljubljana, Slovenia) up to a fluence of 2.5×10^{15} n_{eq}/cm^2. After irradiation, the breakdown voltage was found to be remarkably high, reaching ~400 V in sensors without a guard ring and ~600 V in sensors with three guard rings.

Planar active-edge pixel sensors (n-on-p), aimed at the ATLAS upgrade, were also developed at FBK in collaboration with Laboratoire de physique nucléaire et des hautes énergies

(LPNHE) (Paris, France). Sensors were initially studied using TCAD simulations [30, 31], which pointed at the dependence of the electrical characteristics and the edge efficiency on the design of the termination. As far as the breakdown voltage is concerned, similar qualitative conclusions hold as those previously reported for p-on-n sensors, with simulated values increasing with the gap size and the number of floating guard rings. However, an additional variable plays a major role in this case, namely, the presence of the p-spray surface isolation layer. As a result, the breakdown voltage values are smaller than those of p-on-n sensors and generally do not exceed 250 V before irradiation. Experimental results were in good agreement with simulations. After irradiation with neutrons to a fluence of 2.5×10^{15} n_{eq}/cm^2, the breakdown voltage was found to largely increase, reaching ~500 V in the best case [32–34]. The functional performance was evaluated in beam tests with pixel sensors compatible with the FEI4 readout chip [35, 36]: The hit efficiency was 98.5% at the center of the sensor and still as high as 97% 70 μm away from the outermost pixel, demonstrating very good edge efficiency. It was also demonstrated that the presence of floating guard rings within the edge region is not harmful in terms of hit efficiency, since the guard rings actually help in the lateral spreading of the depletion region [36].

Planar p-on-n active-edge pixel sensors were also developed on Si-SI DWB wafers with 450 μm thick active region. These devices are intended for applications at free electron lasers (FELs) for X-ray imaging applications [37]. The idea of combining fully active sensors and readout chips by vertical integration to make four-sided buttable tiles to be assembled into large-area arrays with minimal dead area was also reported in [38] for applications at next-generation trackers.

In the case of FELs, the challenging requirements in terms of high detection efficiency, very wide dynamic range, plasma effect, and tolerance to high ionizing radiation doses call for high-operation voltages. A comprehensive TCAD simulation study was performed aiming at the best trade-offs between the minimization of the edge region size and the sensor breakdown voltage [39, 40]. An edge termination structure with four floating guard-rings and a total size of ~150 μm was designed, withstanding more than 400 V in all oxide charge conditions. Experimental results from the electrical characterization are in good agreement with the simulation predictions [23, 41].

As previously mentioned, planar detectors with active edges were successfully developed at VTT/Advacam, Finland, several years ago [42–53]. Devices were made on SOI substrates of different active thicknesses, ranging from 50 μm to 675 μm. Most studies have been performed using pixel sensors compatible with readout chips of the Medipix/Timepix family. Besides investigating the electrical properties for different edge designs, which were generally very good with low leakage currents and high enough breakdown voltages, the response of the edge pixels has been thoroughly studied in order to quantify the impact of the trench electrode on the distortion of the electric field configuration. To this purpose, different experimental setups have been used, such as position-resolved laser systems, synchrotron beam scans, and charged particle beam tests. Depending on the detector structure (e.g., n-on-p, p-on-n, n-on-n, p-on-p), the electric field distortion was found to lead to either enhancement or reduction of the charge collected by the edge pixels. However, for imaging applications, this effect can be quantified and corrected during data acquisition or in the postprocessing phase.

Among results obtained from Advacam edgeless pixel sensors, particularly impressive are those relevant to a large-area X-ray imaging system, made of a 10 × 10 arrays of Timepix-based tiles for a total of 6.5 megapixels [48].

A TCAD simulation study of the charge collection behavior of edge pixels has been performed at Deutsches Elektronen-SYnchrotron (DESY) in view of FEL applications [42, 43]. A device model, also incorporating the functionality of possible readout chips (either photon counting or charge integration based), was also developed and provided results that were in very good agreement with experimental data of [46]. It is proposed that the distance between the outermost pixel and the active edge maintain at least 50% of the sensor thickness in order to ensure full sensitivity of the edge region.

Edgeless pixel sensors from VTT, compatible with the ATLAS pixel readout chips (FEI3 and FEI4), have also been characterized in view of the ATLAS pixel system upgrade, showing good results in terms of electrical characteristics and charge collection properties, both before and after irradiation up to 5×10^{15} n_{eq}/cm^2 [56–60]. As an example, Fig. 8.10 shows the hit efficiency measured in a beam test on a 150 μm thick sensor with active edges after irradiation at a fluence of 10^{15} n_{eq}/cm^2 [60]. It can be seen that the edge region beyond the last pixel becomes increasingly sensitive as the bias voltage is increased, so that full efficiency can be obtained in almost all edge regions, that is, 50 μm wide, when the bias voltage is raised to 250 V.

Thin planar active-edge pixel sensors have also been considered for application in the high-precision vertex detector of the future CLIC linear e^+e^- collider [61, 62]. Different designs of the edge region involving various guard ring layouts have been investigated by using TCAD simulations. In comparison with experimental data obtained from assemblies based on the TIMEPIX readout chip, the overall agreement is good. Results confirm that full efficiency up the physical edge of the sensor can be achieved in 50 μm thin sensors [61, 62].

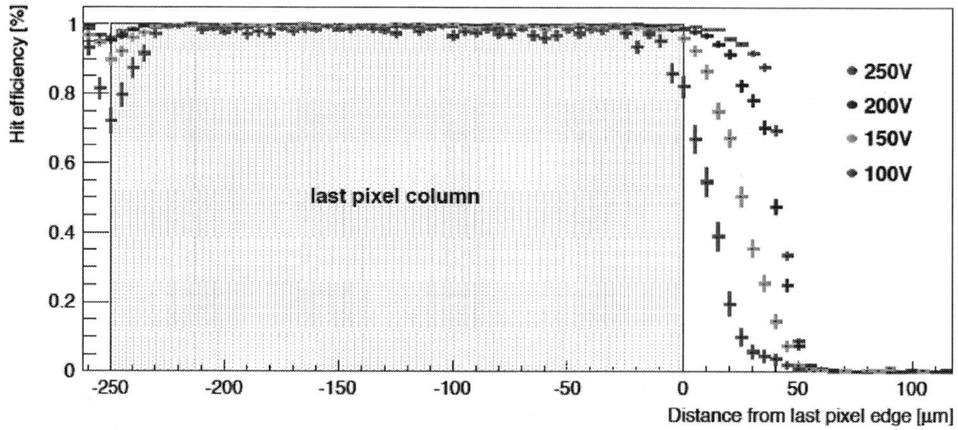

FIGURE 8.10 Hit efficiency measured with a 150 μm thick sensor with active edges after irradiation at a fluence of 10^{15} n_{eq}/cm^2. (From Macchiolo et al., JINST, 12, C01024, 2017. Copyright IOP. With permission.) [60]

8.5 ALTERNATIVE SOLUTIONS FOR SLIM EDGES

All previously described active-edge sensors require a fabrication process with a support wafer. Other structures fabricated with dedicated technologies without the use of a support wafer, are worth mentioning.

For X-ray sensors to be used in the edge-on configuration, the technique used to improve the detection efficiency for low-energy photons, a one-side active edge is sufficient. Such a device was processed at SINTEF, and data reported in [63] show that a device with a full trench is realized on one side, with a good mechanical stability ensured by the three non-etched sides.

Among other solutions, trenches partially etched and doped from one side, combined with different types of guard rings and a final dicing on the opposite side, were reported in [64] and proved to yield an effective slim edge of about 50 μm.

A similar slim-edge postprocessing technique, inspired by SCP, was proposed by Hamamatsu for p-type sensors oriented to detector upgrades at the high luminosity LHC, which require withstanding very high bias voltage. This design is based on trench etching by DRIE on three sides and sidewall passivation by Al_2O_3 [65]. Results demonstrated that with a 250 μm wide edge region, sensors can be reliably operated up to 1000 V bias voltage.

Finally, another design developed in Trento should be mentioned, which is based on dashed ohmic trenches alternating with nonetched silicon regions aimed at ensuring the mechanical integrity of the structures (see Fig. 8.11). The dashed trenches, typically disposed in two offset rows, are able to effectively stop the depletion region spreading from the active area, so that it does not reach the scribe line region.

First successfully applied to 3D sensors [27, 66], as described in Chapter 4, these terminations are fully compatible with planar sensors, provided the design is adapted for higher bias voltages. Their performance was proved by fabricating devices on Si-Si DWB wafers of different active thicknesses: 100 μm and 130 μm thick for the ATLAS upgrade [36] and 450 μm thick for FEL applications [23, 41]. In the latter case, the electrical performance was found to be very similar to true active-edge sensors made on the same wafers. The only penalty is represented by the need for wafer dicing slightly outside the dashed trench, which makes the dead area at the edge slightly wider (a few tens of micrometers) [23]. These terminations can also be made without using a support wafer, which represents a very important advantage for X-ray imaging applications. Without a support wafer, a thin X-ray entrance window could be optimized on the sensor backside, which would allow extending the range of X-ray energies to very low values (~250 eV).

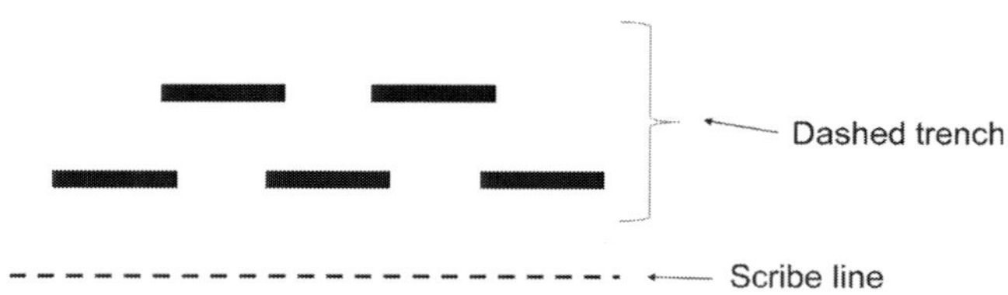

FIGURE 8.11 Schematic drawing of a slim edge based on dashed trenches.

A single-row of dashed trenches could be turned into a true active-edge termination by a long step of thermal diffusion of the dopant, until it fills the lateral gaps between adjacent trenches, forming a continuous wall. This solution, proposed in [67] and described in Chapter 4, was recently demonstrated experimentally at SINTEF.

REFERENCES

1. M. Da Rold, N. Bacchetta, D. Bisello, A. Paccagnella, G. F. Dalla Betta, G. Verzellesi, O. Militaru, R. Wheadon, P. G. Fuochi, C. Bozzi, R. Dell'Orso, A. Messineo, G. Tonelli, P. G. Verdini, "Study of breakdown effects in silicon multiguard structures," IEEE Transactions on Nuclear Science, vol. 46(4), pp. 1215–1223, 1999.
2. A. Longoni, M. Sampietro, L. Strüder, "Instability of the behavior of high resistivity silicon detectors due to the presence of oxide charge," Nuclear Instruments and Methods, vol. 288, pp. 35–43, 1990.
3. J. Albert et al. (The ATLAS IBL Collaboration), "Prototype ATLAS IBL modules using the FE-I4A front-end readout chip," Journal of Instrumentation, JINST 7, P11010, 2012.
4. G. Ruggiero, E. Alagoz, V. Avati, V. Bassetti, V. Berardi, V. Bergholm, V. Boccone, M. Bozzo, A. Buzzo, M. G. Catanesi, R. Cereseto, S. Cuneo, M. Deile, R. De Oliveira, K. Eggert, N. Egorov, I. Eremin, F. Ferro, J. Hasi, F. Haug, R. Herzog, P. Jarron, J. Kalliopuska, A. Kiiskinen, K. Kurvinen, A. Kok, W. Kundrát, R. Lauhakangas, M. Lokajíccek, D. Macina, M. Macrí, T. Mäki, S. Minutoli, L. Mirabito, A. Morelli, P. Musico, M. Negri, H. Niewiadomski, E. Noschis, F. Oljemark, R. Orava, M. Oriunno, K. Österberg, V. G. Palmieri, R. Puppo, E. Radicioni, R. Rudischer, H. Saarikko, G. Sanguinetti, A. Santroni, P. Siegrist, A. Sidorov, G. Sette, J. Smotlacha, W. Snoeys, S. Tapprogge, A. Toppinen, A. Verdier, S. Watts, E. Wobst, "Planar edgeless silicon detectors for the TOTEM experiment," IEEE Transactions on Nuclear Science, vol. 52(5), pp. 1899–1902, 2005.
5. L. Adamczyk et al., "Technical design report for the ATLAS forward proton detector," CERN-LHCC-2015-009; ATLAS-TDR-024, May 2015.
6. M. Albrow et al., "CMS-TOTEM precision proton spectrometer," Technical Report CERN LHCC-2014-021. TOTEM-TDR-003. CMS-TDR-13, September 2014.
7. F. Arfelli, V. Bonvicini, A. Bravin, G. Cantatore, E. Castelli, M. Fabrizioli, R. Longo, A. Olivo, S. Pani, D. Pontoni, P. Poropat, M. Prest, A. Rashevsky, L. Rigon, G. Tromba, A. Vacchi, E. Vallazza, "A multilayer edge-on silicon microstrip single photon counting detector for digital mammography," Nuclear Physics B (Proc. Suppl.), vol. 78(1), pp. 592–597, 1999.
8. Z. Li, M. Abreu, V. Eremin, V. Granata, J. Mariano, P. R. Mendes, T. O. Niinikoski, P. Sousa, E. Verbitskaya, W. Zhang, "Electrical and transient current characterization of edgeless Si detectors diced with different methods," IEEE Transactions on Nuclear Science, vol. 49(3), pp. 1040–1046, 2002.
9. B. Perea Solanoa, M. C. Abreu, V. Avati, T. Boccali, V. Boccone, M. Bozzo, R. Capra, L. Casagrande, W. Chen, K. Eggert, E. Heijne, S. Klauke, Z. Li, T. Mäki, L. Mirabito, A. Morelli, T. O. Niinikoski, F. Oljemark, V. G. Palmieri, P. Rato Mendes, S. Rodrigues, P. Siegrist, L. Silvestris, P. Sousa, S. Tapprogge, B. Trocmé, "Edge sensitivity of 'edgeless' silicon pad detectors measured in a high-energy beam," Nuclear Instruments and Methods A, vol. 550, pp. 567–580, 2005.
10. E. Alagoz et al. (The TOTEM Collaboration), "Performance of almost edgeless silicon detectors in CTS and 3D-planar technologies," Journal of Instrumentation, JINST 8, P06009, 2013.
11. G. Pellegrini, M. Lozano, F. Campabadal, C. Fleta, J. M. Rafí, M. Ullán, "Edgeless detectors fabricated by dry etching process," Nuclear Instruments and Methods A, vol. 563, pp. 70–73, 2006.
12. V. Fadeyev, H. F.-W. Sadrozinski, S. Ely, J. G. Wright, M. Christophersen, B. F. Phlips, G. Pellegrini, S. Grinstein, G.-F. DallaBetta, M. Boscardin, R. Klingenberg, T. Wittig,

A. Macchiolo, P. Weigell, D. Creanza, R. Bates, A. Blue, L. Eklund, D. Maneuski, G. Stewart, G. Casse, I. Gorelov, M. Hoeferkamp, J. Metcalfe, S. Seidel, G. Kramberger, "Scribe–Cleave–Passivate (SCP) slim edge technology for silicon sensors," Nuclear Instruments and Methods A, vol. 731, pp. 260–265, 2013.

13. M. Christophersen, V. Fadeyev, S. Ely, B. F. Phlips, H. F.-W. Sadrozinski, "The effect of different dicing methods on the leakage currents of n-type silicon diodes and strip sensors," Solid-State Electronics, vol. 81, pp. 8–12, 2013.
14. M. Christophersen, V. Fadeyev, B. F. Phlips, H. F.-W. Sadrozinski, C. Parker, S. Ely, J. G. Wright, "Alumina and silicon oxide/nitride sidewall passivation for P- and N-type sensors," Nuclear Instruments and Methods A, vol. 699, pp. 14–17, 2013.
15. V. Fadeyev, S. Ely, Z. Galloway, J. Ngo, C. Parker, H. F.-W. Sadrozinski, M. Christophersen, B. F. Phlips, G. Pellegrini, J. M. Rafi, D. Quirion, G. F. Dalla Betta, M. Boscardin, G. Casse, I. Gorelov, M. Hoeferkamp, J. Metcalfe, S. Seidel, E. Gaubas, T. Ceponis, J. V. Vaitkus, "Update on Scribe-Cleave-Passivate (SCP) slim edge technology for silicon sensors: Automated processing and radiation resistance," Nuclear Instruments and Methods A, vol. 765, pp. 59–63, 2014.
16. C. J. Kenney, S. I. Parker, E. Walckiers, "Results from 3-D silicon sensors with wall electrodes: near-cell-edge sensitivity measurements as a preview of active-edge sensors," IEEE Transactions on Nuclear Science, vol. 48(6), pp. 2405–2410, 2001.
17. C. J. Kenney, J. D. Segal, E. Weestbrook, S. Parker, J. Hasi, C. Da Vià, S. Watts, J. Morse, "Active-edge planar radiation sensors," Nuclear Instruments and Methods A, vol. 565, pp. 272–277, 2006.
18. C. Kenney, J. Hasi, S. Parker, A. C. Thompson, E. Westbrook, "Use of active-edge silicon detectors as X-ray beam monitors," Nuclear Instruments and Methods A, vol. 582, pp. 178–181, 2007.
19. S. H. Christiansen, R. Singh, U. Gösele, "Wafer direct bonding: From advanced substrate engineering to future applications in micro/nanoelectronics," Proceedings of the IEEE, vol. 94(12), pp. 2060–2106, 2006.
20. M. Povoli, A. Bagolini, M. Boscardin, G.-F. Dalla Betta, G. Giacomini, E. Vianello, N. Zorzi, "Development of planar detectors with active edge," Nuclear Instruments and Methods A, vol. 658, pp. 103–107, 2011.
21. M. Boscardin, L. Bosisio, G. Contin, G. Giacomini, V. Manzari, G. Orzan, I. Rashevskaya, S. Ronchin, N. Zorzi, "Development of thin edgeless silicon pixel sensors on epitaxial wafers," Journal of Instrumentation, JINST 9, P09013, 2014.
22. G. F. Dalla Betta, M. Boscardin, M. Bomben, M. Brianzi, G. Calderini, G. Darbo, R. Mendicino, M. Meschini, A. Messineo, S. Ronchin, D. M. S. Sultan, N. Zorzi, "The INFN-FBK 'Phase-2' R&D Program," Nuclear Instruments and Methods A, vol. 824, pp. 388–391, 2016.
23. M. A. Benkechkache, S. Latreche, S. Ronchin, M. Boscardin, L. Pancheri, G.-F. Dalla Betta, "Design and first characterization of active and slim-edge planar detectors for FEL applications," IEEE Transactions on Nuclear Science, vol. 64(4), pp. 1062–1070, 2017.
24. S. Eranen, J. Kalliopuska, R. Orava, N. van Remortel, T. Virolainen, "3D processing on 6 in. high resistive SOI wafers: Fabrication of edgeless strip and pixel detectors," Nuclear Instruments and Methods A, vol. 607, pp. 85–88, 2009.
25. J. Kalliopuska, S. Eranen, T. Virolainen, "Alternative fabrication process for edgeless detectors on 6 in. wafers," Nuclear Instruments and Methods A, vol. 633, pp. S50–S54, 2011.
26. S. I. Parker, C. J. Kenney, D. Gnani, A. C. Thompson, E. Mandelli, G. Meddeler, J. Hasi, J. Morse, E. M. Westbrook, "3DX: An X-ray pixel array detector with active edges," IEEE Transactions on Nuclear Science, vol. 53(3), pp. 1676–1688, 2006.
27. G.-F. Dalla Betta, A. Bagolini, M. Boscardin, G. Giacomini, M. Povoli, E. Vianello, N. Zorzi, "Development of active and slim edge terminations for 3D and planar detectors," 2011 IEEE Nuclear Science Symposium, Valencia (Spain), October 23–29, 2011, Conference Record, Paper N25-4.

28. M. Povoli, A. Bagolini, M. Boscardin, G.-F. Dalla Betta, G. Giacomini, J. Hasi, A. Oh, N. Zorzi, "Functional characterization of planar sensors with active edges using laser, X-ray beam scans," Nuclear Instruments and Methods A, vol. 718, pp. 350–352, 2013.
29. G. F. Dalla Betta, M. Povoli, M. Boscardin, G. Kramberger, "Edgeless and slim-edge solutions for silicon pixel sensors," Proceedings of Science—22nd Workshop on Vertex Detectors (Vertex 2013), Paper 042, 2014.
30. M. Bomben, A. Bagolini, M. Boscardin, L. Bosisio, G. Calderini, J. Chauveau, G. Giacomini, A. La Rosa, G. Marchiori, N. Zorzi, "Development of edgeless n-on-p planar pixel sensors for future ATLAS upgrades," Nuclear Instruments and Methods A, vol. 712, pp. 41–47, 2013.
31. M. Bomben, A. Bagolini, M. Boscardin, L. Bosisio, G. Calderini, J. Chauveau, G. Giacomini, A. La Rosa, G. Marchiori, N. Zorzi, "Novel silicon n-on-p edgeless planar pixel sensors for the ATLAS upgrade," Nuclear Instruments and Methods A, vol.730, pp. 215–219, 2013.
32. M. Bomben, A. Bagolini, M. Boscardin, L. Bosisio, G. Calderini, J. Chauveau, G. Giacomini, A. La Rosa, G. Marchiori, N. Zorzi, "Performance of irradiated thin edgeless N-on-P planar pixel sensors for ATLAS upgrades," 2013 IEEE Nuclear Science Symposium, Seoul (Republic of Korea), October 27–November 2, 2013, Conference Record, paper N13-2.
33. M. Bomben, A. Bagolini, M. Boscardin, L. Bosisio, G. Calderini, J. Chauveau, G. Giacomini, A. La Rosa, G. Marchiori, N. Zorzi, "Electrical characterization of thin edgeless N-on-p planar pixel sensors for ATLAS upgrades," Journal of Instrumentation, JINST 9, C05020, 2014.
34. G. Giacomini, A. Bagolini, M. Bomben, M. Boscardin, L. Bosisio, G. Calderini, J. Chauveau, A. La Rosa, G. Marchiori, N. Zorzi, "Selected results from the static characterization of edgeless n-on-p planar pixel sensors for ATLAS upgrades," Journal of Instrumentation, JINST 9, C01063, 2014.
35. M. Bomben, A. Ducourthial, A. Bagolini, M. Boscardin, L. Bosisio, G. Calderini, L. D'Eramo, G. Giacomini, G. Marchiori, N. Zorzi, A. Rummler, J. Weingarten, "Performance of active edge pixel sensors," Journal of Instrumentation, JINST 12, P05006, 2017.
36. A. Ducourthial, M. Bomben, G. Calderini, L. D'Eramo, G. Marchiori, I. Luise, A. Bagolini, M. Boscardin, L. Bosisio, G. Darbo, G.-F. Dalla Betta, G. Giacomini, M. Meschini, A. Messineo, S. Ronchin, N. Zorzi, "Thin and edgeless sensors for ATLAS pixel detector upgrade," Journal of Instrumentation, JINST 12, C12038, 2017.
37. G. Rizzo, D. Comotti, L. Fabris, M. Grassi, L. Lodola, P. Malcovati, M. Manghisoni, L. Ratti, V. Re, G. Traversi, C. Vacchi, G. Batignani, S. Bettarini, G. Casarosa, F. Forti, F. Morsani, A. Paladino, E. Paoloni, G. F. Dalla Betta, L. Pancheri, G. Verzellesi, H. Xu, R. Mendicino, M.A. Benkechkache, "The PixFEL project: Development of advanced X-ray pixel detectors for application at future FEL facilities," Journal of Instrumentation, JINST 10, C02024, 2015.
38. R. Lipton, G. Deptuch, U. Heintz, M. Johnson, C. Kenney, M. Narian, S. Parker, I. Planell-Mendez, E. Sawyer, A. Shenai, L. Spiegel, J. Thom, Z. Ye, "Combining the two 3Ds," Journal of Instrumentation, JINST 7, C12010, 2012.
39. G. F. Dalla Betta, G. Batignani, M. A. Benkechkache, S. Bettarini, G. Casarosa, D. Comotti, L. Fabris, F. Forti, M. Grassi, S. Latreche-Lassoued, L. Lodola, P. Malcovati, M. Manghisoni, R. Mendicino, F. Morsani, A. Paladino, L. Pancheri, E. Paoloni, L. Ratti, V. Re, G. Rizzo, G. Traversi, C. Vacchi, G. Verzellesi, H. Xu, "Design and TCAD simulations of planar active-edge pixel sensors for future XFEL applications," 2014 IEEE Nuclear Science Symposium, Seattle (USA), November 9–15, 2014, Conference Record, paper N08-7.
40. G.-F. Dalla Betta, G. Batignani, M. A. Benkechkache, S. Bettarini, G. Casarosa, D. Comotti, L. Fabris, F. Forti, M. Grassi, S. Latreche, L. Lodola, P. Malcovati, M. Manghisoni, R. Mendicino, F. Morsani, A. Paladino, L. Pancheri, E. Paoloni, L. Ratti, V. Re, G. Rizzo, G. Traversi, C. Vacchi, G. Verzellesi, H. Xu, "Design and TCAD simulation of planar p-on-n active-edge pixel sensors for the next generation of FELs," Nuclear Instruments and Methods A, vol. 824, pp. 384–385, 2016.

41. L. Pancheri, M. A. Benkechkache, G.-F. Dalla Betta, H. Xu, G. Verzellesi, S. Ronchin, M. Boscardin, D. Comotti, M. Grassi, L. Ratti, L. Lodola, P. Malcovati, C. Vacchi, L. Fabris, M. Manghisoni, V. Re, G. Traversi, G. Batignani, S. Bettarini, G. Casarosa, M. Giorgi, F. Forti, A. Paladino, E. Paoloni, G. Rizzo, F. Morsani, "First experimental results on active-edge and slim-edge silicon sensors for XFEL," Journal of Instrumentation, JINST 11, C12018, 2016.
42. M. J. Bosma, E. Heijne, J. Kalliopuska, J. Visser, E. N. Koffeman, "Edgeless planar semiconductor sensors for a Medipix3-based radiography detector," Journal of Instrumentation, JINST 6, C11019, 2011.
43. J. Kalliopuska, L. Tlustos, S. Eranen, T. Virolainen, "Characterization of edgeless pixel detectors coupled to Medipix2 readout chip," Nuclear Instruments and Methods A, vol. 648, pp. S32–S36, 2011.
44. X. Wu, J. Kalliopuska, S. Eranen, T. Virolainen, "Recent advances in processing and characterization of edgeless detector," Journal of Instrumentation, JINST 7, C02001, 2012.
45. J. Kalliopuska, X. Wu, J. Jakubek, S. Eranen, T. Virolainen, "Processing and characterization of edgeless radiation detectors for large area detection," Nuclear Instruments and Methods A, vol. 731, pp. 205–209, 2013.
46. R. Bates, A. Blue, M. Christophersen, L. Eklund, S. Ely, V. Fadeyev, E. Gimenez, V. Kachkanov, J. Kalliopuska, A. Macchiolo, D. Maneuski, B. F. Phlips, H. F.-W. Sadrozinski, G. Stewart, N. Tartoni, R. M. Zain, "Characterisation of edgeless technologies for pixellated and strip silicon detectors with a micro-focused X-ray beam," Journal of Instrumentation, JINST 8, P01018, 2013.
47. X. Wu, J. Kalliopuska, M. Jakubek, J. Jakubek, A. Gadda, S. Eranen, "Study of edgeless radiation detector with 3D spatial mapping technique," Journal of Instrumentation, JINST 9, C04004, 2014.
48. J. Jakubek, M. Jakubek, M. Platkevic, P. Soukup, D. Turecek, V. Sykora, D. Vavrik, "Large area pixel detector with full area sensitivity composed of 100 timepix assemblies with edgeless Sensors," Journal of Instrumentation, JINST 9, C04018, 2014.
49. C. Ponchut, M. Ruat, J. Kalliopuska, "X-ray imaging characterization of active edge silicon pixel sensors," Journal of Instrumentation, JINST 9, C05017, 2014.
50. D. Maneuski, R. Bates, A. Blue, C. Buttar, K. Doonan, L. Eklund, E. N. Gimenez, D. Hynds, S. Kachkanov, J. Kalliopuska, T. McMullen, V. O'Shea, N. Tartoni, R. Plackett, S. Vahanen, K. Wraight, "Edge pixel response studies of edgeless silicon sensor technology for pixellated imaging detectors," Journal of Instrumentation, JINST 10, P03018, 2015.
51. K. Akiba, M. Artuso, V. van Beveren, M. van Beuzekom, H. Boterenbrood, J. Buytaert, P. Collins, R. Dumps, B. van der Heijden, C. Hombach, D. Hynds, D. Hsu, M. John, E. Koffeman, A. Leflat, Y. Li, I. Longstaff, A. Morton, E. Perez Trigo, R. Plackett, M. M. Reid, P. Rodriguez Perez, H. Schindler, P. Tsopelas, C. Vazquez Sierra, M. Wysokinski, "Probing active-edge silicon sensors using a high precision telescope," Nuclear Instruments and Methods A, vol. 777, pp. 110–117, 2015.
52. T. Peltola, X. Wu, J. Kalliopuska, C. Granja, J. Jakubek, M. Jakubek, J. Harkonen, A. Gadda, "Characterization of thin p-on-p radiation detectors with active edges," Nuclear Instruments and Methods A, vol. 813, pp. 139–146, 2016.
53. R. Plackett, K. Arndt, D. Bortoletto, I. Horswell, G. Lockwood, I. Shipsey, N. Tartoni, S. Williams, "X-ray metrology of an array of active edge pixel sensors for use at synchrotron light sources," Nuclear Instruments and Methods A, vol. 879, pp. 106–111, 2018.
54. J. Zhang, D. Tartarotti Maimone, D. Pennicard, M. Sarajlic, H. Graafsma, "Optimization of radiation hardness and charge collection of edgeless silicon pixel sensors for photon science," Journal of Instrumentation, JINST 9, C12025, 2014.
55. M. Sarajlic, J. Zhang, D. Pennicard, S. Smoljanin, T. Fritzsch, M. Wilke, K. Zoschke, H. Graafsma, "Development of edgeless TSV X-ray detectors," Journal of Instrumentation, JINST 11, C02043, 2016.

56. A. Macchiolo, L. Andricek, H.-G. Moser, R. Nisius, R. H. Richter, S. Terzo, P. Weigell, "Development of active edge pixel sensors and four-side buttable modules using vertical integration technologies," Nuclear Instruments and Methods A, vol. 765, pp. 53–58, 2014.
57. S. Terzo, A. Macchiolo, R. Nisius, B. Paschen, "Thin n-in-p planar pixel sensors and active edge sensors for the ATLAS upgrade at HL-LHC," Journal of Instrumentation, JINST 9, C12029, 2014.
58. S. Terzo, L. Andricek, A. Macchiolo, H. G. Moser, R. Nisius, R. H. Richter, P. Weigell, "Heavily irradiated N-in-p thin planar pixel sensors with and without active edges," Journal of Instrumentation, JINST 9, C05023, 2014.
59. N. Savic, J. Beyer, A. La Rosa, A. Macchiolo, R. Nisius, "Investigation of thin n-in-p planar pixel modules for the ATLAS upgrade," Journal of Instrumentation, JINST 11, C12008, 2016.
60. A. Macchiolo, J. Beyer, A. La Rosa, R. Nisius, N. Savic "Optimization of thin n-in-p planar pixel modules for the ATLAS upgrade at HL-LHC," Journal of Instrumentation, JINST 12, C01024, 2017.
61. M. Munker (on behalf of the CLICdp collaboration), "Silicon pixel R&D for CLIC," Journal of Instrumentation, JINST 12, C01096, 2017.
62. S. Spannagel (on behalf of the CLICdp collaboration), "Silicon technologies for the CLIC vertex detector," Journal of Instrumentation, JINST 12, C06006, 2017.
63. T. E. Hansen, N. Ahmed, A. Ferber, G. Bouquet, "Edge-on detectors with active edge for X-ray photon counting," 2011 IEEE Nuclear Science Symposium, Valencia (Spain), October 23–29, 2011, Conference Record, Paper N25-6.
64. M. J. Bosma, J. Visser, O. Evrard, P. De Moor, K. De Munck, D. Sabuncuoglu Tezcan, E. N. Koffeman, "Edgeless silicon sensors for Medipix-based large-area X-ray imaging detectors," Journal of Instrumentation, JINST 6, C01035, 2011.
65. S. Kamada, K. Yamamura, Y. Unno, Y. Ikegami, "Development of N+ in P pixel sensors for a high-luminosity large hadron collider," Nuclear Instruments and Methods A, vol. 765, pp. 118–124, 2014.
66. M. Povoli, A. Bagolini, M. Boscardin, G. F. Dalla Betta, G. Giacomini, F. Mattedi, R. Mendicino, N. Zorzi, "Design and testing of an innovative slim-edge termination for silicon radiation detectors," Journal of Instrumentation 8, C11022, 2013.
67. G. F. Dalla Betta, C. Da Vià, M. Povoli, S. Parker, M. Boscardin, G. Darbo, P. Grenier, S. Grinstein, J. Hasi, C. Kenney, A. Kok, C.-H. Lai, G. Pellegrini, S. J. Watts, "Recent developments and future perspectives in 3D silicon radiation sensors," Journal of Instrumentation, JINST 7, C10006, 2012.

CHAPTER 9

Applications

The configuration of 3D sensors with p^+ and n^+ electrodes, as well as active edges, penetrating the semiconductor's substrate either partially or fully, allows tailoring radiation sensors to fit special requirements in many fields. As the previous chapters have described in detail, among the key properties of such devices are extreme radiation tolerance, low bias voltage and low power dissipation, well-defined detection volume as well as high speed and four-side tiling capability. Being able to access the bulk of the sensors with trenches, holes, or other geometries could open other possibilities such as the use of converters inside the apertures for enhanced neutron or gamma detection or the fabrication of very small enclosed volumes mimicking single cells for biologically compatible microdosimetry.

It took about 20 years of hard work to obtain a reproducible and reliable process, but finally today several commercial companies are able to provide 3D technology both in its original design and with variations, which better adapt to targeted applications apart from ATLAS pixels, which was the original target for these sensors. In high-energy physics, 3D pixels are now being used in the forward region in the so-called ATLAS–Forward Proton project (AFP); see Chapter 8, in CMS [1], where tests with 3D sensors from different foundries have been performed with promising results [2–6], and in LHCb, another experiment at the Large Hadron Collider specialized in the study of CP violation where the speed properties of 3D sensors could be useful [7].

As will be shown shortly, this innovative technique can also be used in medical and biological applications to overcome challenges still unresolved by traditional semiconductor imagers. This chapter explores which novelties 3D technology could bring with respect to traditional planar semiconductors and investigates its potential for different disciplines, such as future high-energy physics detectors and systems, medicine, and neutron detection.

9.1 HIGH-ENERGY PHYSICS

So far, 3D silicon has been developed mainly for high-energy physics vertex detectors, taking advantage of the superior radiation tolerance of such geometry, as described in Chapter 6. The particle tracking systems employed at CERN-LHC are based primarily on silicon detectors of microstrip, pixel and drift type. The radiation levels expected in the vertex detectors

FIGURE 9.1 $Z \rightarrow \mu\mu$ event (run number 201289, event number 24151616) from 15 April 2012, $L = 4 \times 10^{33}$ cm^{-2} s^{-1}, 25 vertices (ATLAS Experiment © 2014 CERN).

regions are unprecedented in high-energy physics experiments and are the consequence of the very high multiplicity of tracks expected when several proton–proton collisions occur at the same time. A visual example of such multiple vertex events from the ATLAS experiment can be seen in Fig. 9.1 where a $Z \rightarrow \mu\mu$ event is visible from the April 15, 2012 run with a luminosity of 4×10^{33} cm^{-2} s^{-1} generating 25 primary vertices [8].

As pointed out in Chapter 6, what makes 3D sensors one of the radiation-hardest detector design ever made is the fact that the distance between the p+ and n+ electrodes can be tailored to match the best signal efficiency, the best signal amplitude, and the best signal-to-noise ratio to the expected radiation fluence.

Since the radiation tolerance of 3D sensors is a purely geometrical effect, the signal, which can be calculated theoretically, shows a direct dependence on the ratio of the carrier trapping time and the distance between the two electrodes [9]. Therefore, if the signal is inversely proportional to the p-n electrode distance, the smallest this distance is the larger is the left-over signal when compared with its original pre-irradiation signal. This is also possible in planar sensors, but since the electrode distance almost equals the substrate thickness, which is needed to provide the signal in the first place for MIPs, it will quickly exhaust the available signal charge. In 3D sensors, the generation thickness can therefore be as large as needed without compromising the radiation hardness properties of the sensor; the exception is a large noise due to the increased capacitance, which one will need to take into account.

Since June 2014, 112 of the above-mentioned sensors have been installed in the ATLAS IBL to cover 25% of the surface of a pixel detector, which literally surrounds the beam pipe at just 3.4 cm from the beam. A sketch of such a pixel layer is visible in Fig. 9.2a, together

FIGURE 9.2 (a) A sketch of the ATLAS Insertable B-Layer and (b) a picture of a stave composed by planar and 3D modules (ATLAS Experiment © 2014 CERN).

with the previously installed pixels and strip detectors, while Fig. 9.2b shows one of the 16 "staves" containing 32 modules, of which the 8th most external ones are 3D while the remaining are planar sensors [10].

The performance of the IBL 3D and planar sensors has been monitored since their installation to control their degradation after irradiation. The parameters used to define the radiation levels were the variation in time of the leakage currents and the bias voltages. Figure. 9.3 shows IBL leakage currents measured as a function of integrated luminosity and temperature collected during the 2015 and 2016 runs. Since the fluence is proportional to the integrated luminosity and the leakage current is linearly proportional to the fluence, the leakage current increases almost linearly with the luminosity. As discussed in Chapter 3, one part of the irradiation process in silicon is beneficial annealing in which defects can restore with time and faster at elevated temperatures. This effect is indicated at the bottom of Fig. 9.3 in correspondence with the transition between the 2015–2016 runs where the detector was warmed up. If all IBL modules were exposed to the same fluence, the slope of the leakage current curves in Fig. 9.3 would be the same. The differences observed among module groups are an indication of a possible position-dependent irradiation profile, with lower irradiation affecting the modules further away from the interaction point [11].

Other experiments that are presently using 3D sensors are the Atlas Forward Physics (AFP) [12] and the CMS TOTEM precision proton spectrometer (CT-PPS) [13]. These detectors measure the diffractive proton–proton scattering at the ATLAS and CMS experiments, respectively. Very simply: in diffractive scattering, two protons will not collide to produce

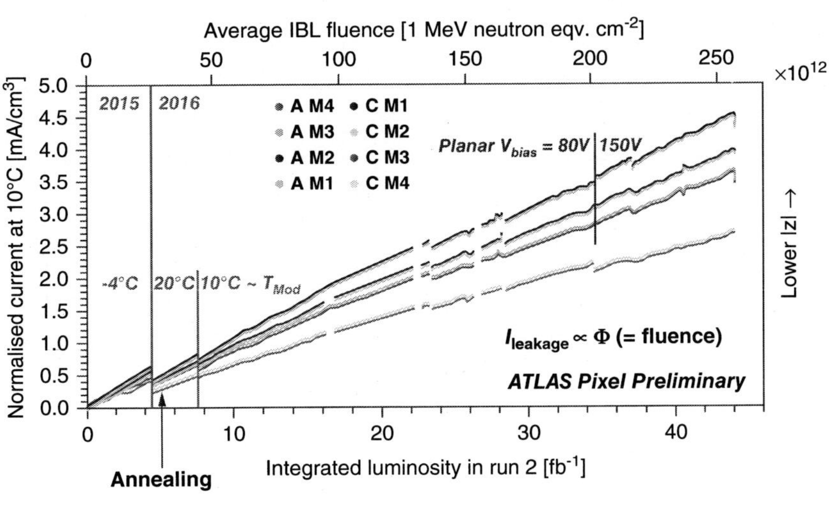

FIGURE 9.3 Leakage currents measured on the IBL as a function of integrated luminosity collected during the 2015 and 2016 runs. The IBL staves are divided into two main sides, A and C, depending on their direction from the interaction point. There are four groupings of modules on each side, with M1 closest to the detector center (small |z|) and M4 furthest away (large |z|). Module groups M1–M3 use planar sensors while M4 is composed of modules with 3D sensors. "(ATLAS Collaboration, Evolution of LV Currents and sensor leakage currents in the IBL, https://atlas.web.cern.ch/Atlas/GROUPS/PHYSICS/PLOTS/PIX-2016-006/. With permission.) [11]

other particles but will pass close enough to one another to exchange a so-called pomeron and then continue their trajectory intact in the beam pipe. This pomeron exchange, however, would have modified the momentum, and hence the trajectory of each of the two emerging protons that can now be detected if a sensor is placed as close as possible to the beam and at reasonable distance from the point where the two protons met. The radial distance from the beam at which the protons will appear is a direct measurement of their new momentum. The radiation tolerance requirements for the sensors used to measure such momentum are equivalent to the one experienced by the innermost layers of the central detectors, with the difference, however, that they are distributed asymmetrically over the sensing volume. Most of the damage caused by the protons will concentrate in the region near the beam. Ideally, a detector in this region would be divided into separate sections, with separate bias voltages to respond to this asymmetric irradiation [14]. However, 3D sensors have been shown to be robust enough to withstand the increase of bias voltage required after irradiation, even in the moderately irradiated regions, making them ideal for the task. Furthermore, as shown in detail previously, 3D sensors offer the possibility of reducing the dead area at the module's edges by using "active edges" or "slim edges" to record as many events close to the beam as possible. Results from 3D sensors with the same geometry as those installed in the IBL, but with a modified slim edge, are encouraging. As a further example of the requirements of detectors for such experiments, AFP plans to place such detectors symmetrically at 220 m from the ATLAS experiment beam interaction point to precisely measure such events and extract the physics information related to the optical properties of the protons.

Each good event will have to correlate the presence of two protons appearing simultaneously in the two stations at the opposite side of their central experiment. Since there might be several interesting interactions when the beams collide, as can be seen, for example, in Fig. 9.1, it is mandatory to identify correctly the trajectory of the two protons to make sure they are coming from the same collision point or vertex. For this purpose, an unprecedented time resolution is needed. Fast time response is possible with the recent development of fast readout electronics chips that can resolve signals as rapidly as 10–20 ps [15].

The limitation, however, could come from the sensor's response, which needs to fight with the finite intrinsic drift velocities of the carriers and their strict dependence on the strength and homogeneity of the internal electric field. To resolve this issue, thin low-gain avalanche detectors (LGADs) have been successfully developed to enhance the signal height, with extremely promising results [16, 17]. 3D sensors could also contribute to generate fast response signals with a specifically designed geometry [18], as will be shown shortly.

9.2 3D SPEED PROPERTIES

There are several reasons why 3D sensor signals could be intrinsically fast:

a. The 3D electrode spacing can be less than the wafer thickness, so carrier drift distances are shorter.

b. Depletion voltages are low, facilitating the use of overdepletion voltages, so that collection fields can be more uniform and uniformly high.

c. In 3D, the field lines end on surfaces, in this case cylinders, which normally have more area than corresponding planar electrodes, so the ratio of the average drift fields to the peak field can be larger for 3D sensors. This advantage comes with a price, however: an increased electrode capacitance.

d. Common, low-energy delta rays, with a short range compared to the gap, are collected together with the adjacent track ionization during a few picoseconds. If the track is parallel to the electrode, everything is collected during those few picoseconds at the end of the current pulse. Before this time, they are drifting close to the track and give the same induced signals. Except for noise, the pulse will be very smooth, unlike those pulses in planar sensors where the delta ray ionization is absorbed at random times during the pulse. Higher energy delta rays, with longer ranges, are rare and recognizable. Diffusion of about 2.6 μm for electrons and 1.5 μm for holes in the first ns and growing as the square root of the diffusion time modifies these signal shapes only slightly.

e. Both types of 3D electrodes can be contacted on a single surface, so if capacitive readout for at least one type of electrode can be used, pulses from both the n-type and p-type electrodes may be recorded. This was reported in Chapter 4 as "dual-readout" [19, 20], and it may improve both time and spatial resolution. The same magnitude of charge goes to spatially separated electrodes, providing additional track information, and can improve the signal-to-noise ratio, allowing better timing.

f. Finally, the signal height for minimum ionizing particles is determined by the sensor substrate, which in 3D can theoretically be chosen without disturbing the charge collection mechanism and the electric field. Again, the price to pay is higher capacitance, which can be tolerated with high signals.

To prove these points, timing measurements were performed on a hexagonal-shaped detector similar to the one shown in Fig. 9.4 connected to a fast amplifier with a 1.5 ns risetime [21]. The operating temperature was 20°C. Figure 9.5 shows a scatterplot of the timing error (dt) versus pulse height and, on the top and right-side axes, the projected distributions of each. The median and average values are 129 and 155 ps, and there is clear pulse-height dependence. More recently, 3D sensors were connected to faster amplifiers where average responses below 100 ps are expected. More optimized designs for 3D sensors, readout electronics, and fast tracking algorithms promise even better resolutions with pixel systems [22]. As already pointed out, further applications include possible upgrades of the CERN LHC LHCb experiment, which plans to unravel the matter–antimatter asymmetry in the universe among the rest.

Beyond high-energy physics, high-speed detectors would be useful in medicine and biological applications where fast-changing events, such as protein folding, could be monitored as the result of a chemical reaction [23], or the imaging precision could be enhanced in time of flight positron emission tomography (ToF PET) [24]. Furthermore, ultrafast detection could help quantum imaging involving photon entanglement or speckle fluctuation in "ghost" imaging or quantum computing [25].

FIGURE 9.4 Picture of a 3D hexagon-cell active-edge sensor fabricated at SNF. The one used for the measurements reported in the text was tiled with 16 columns, each with 20 hexagons with sides of 50 μm connected to 16 pads for readout.

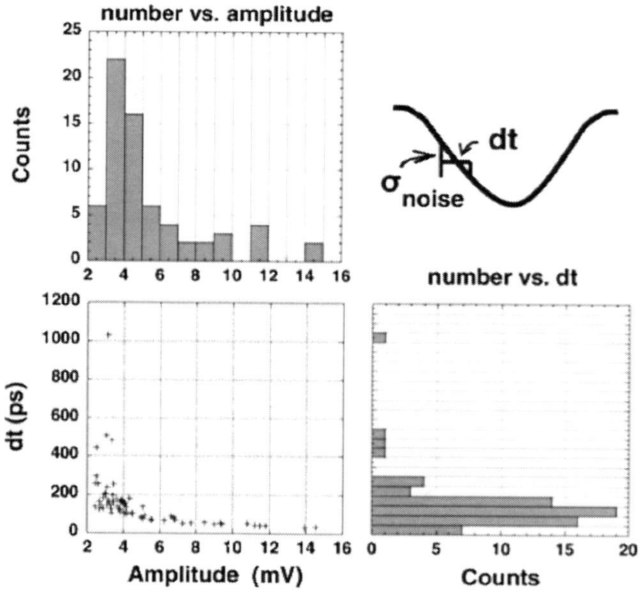

FIGURE 9.5 Scatterplot of expected noise-induced timing errors (dt) versus pulse amplitude (V), for 67 pulses and the projections of the dt (right) and V (top) distributions. The projections above and to the right have the same scales as the corresponding axes of the scatterplot. The relationship between dt, the pulse slope, and σ noise is shown in the upper right diagram. The signal-to-noise ratio is 3 times the pulse height in mV. (From Parker et al., IEEE Trans. Nucl. Sci., 58, 2, 2011. With permission.) [21]

FIGURE 9.6 Sketch of curved 3D detectors where Graytone lithography is used to remove silicon from the top and bottom of the wafer to leave a curved structure capable of conforming to a curved, shaped structure like beam-pipes.

Another interesting idea intended for high-energy physics applications extended the concept of application-driven "adaptive" geometry to the next level and is reported in [26]. With regard to this concept, the authors used a combination of microfabrication and so-called Gray-tone lithography to fabricate structures with 3D sensors with well-defined, curved surface contours. These curved structures, a sketch of which is shown in Fig. 9.6, were originally meant for radiation-hard vertex detectors for LHC upgrades, in particular as single-layer surroundings of beam-pipes.

Grayscale lithography is a specialized lithography process that results in three-dimensional resist profiles. A grayscale optical mask is used to transmit only a portion of the incident intensity of light, partially exposing sections of a positive photoresist to a certain depth. This exposure renders the top portion of the photoresist layer more soluble in a developer solution, while the bottom portion of the photoresist layer remains unchanged [27–30]. Unfortunately, after the first prototypes were developed, this idea was not pursued any further.

A further attempt to improve the 3D silicon performance in terms of radiation tolerance and speed for high-energy physics applications was made by the proponents of "core–shell" silicon structures where 200 μm deep and 2 μm channels with high aspect ratio were etched into silicon [31]. This excellent aspect ratio is possible by using a metal-assisted wet chemical etching by immersion of the sample in a solution containing HF (5M):H(30%) by a volume ratio of 20:1. After the metal-assisted wet chemical etching, the metal has to be completely removed, first by extended immersion of the sample in aqua regia (HNO6%):HCl (35%) = 1:3 in volume) to remove the Au at the bottom of the channel, and second, by treating the sample with a cleaning procedure developed specifically for high-aspect ratio structures [32, 33]. After this procedure, the residual metal is completely removed and suitable for further device processing. The boron-doped CZ-Si wafers used in the reported experiment had <100> crystal orientation, with a doping level of $\sim 3 \times 10^{15}$ cm^{-3} (~1.5 Ω·cm) and a thickness of 525 μm. The wafers were patterned by electron beam (e-beam) lithography with various pitch sizes, and a 50 nm Au layer was deposited through the photomask on the channel. Then the photoresist was lifted off, allowing the gold meshes to appear. Prototypes were successfully made, promising interesting results in future tests.

9.3 MEDICAL IMAGING

Several medical diagnostic applications look at the contrast that X-rays make with elements of different density, like bone and fat, calcifications, and so on. For this reason, the thickness of the X-ray sensor material is important since it determines its so-called detection efficiency. The thicker (and the heavier in terms of atomic number Z) the material, the higher the probability that an X-ray or photon of a defined energy will deposit its energy inside the sensor. It is for this reason that in radiographic examinations the X-ray machines should be tuned to the best energy, depending on the type of image needed. The aim is to distinguish structures of different density inside the human body; therefore, X-rays, which stop where the material density is higher (e.g., bones, calcifications) and pass through elsewhere (fat, water, etc.), are used. As an example, the X-ray energy used for bone radiography is about 60 keV because the contrast, or the visual difference, between bones and fat is the highest. For many years, the most common detector used for X-ray imaging in medicine was simply an X-ray film exposed for a certain time to as many X-rays as possible to obtain the right contrast. Since films have a relatively small stopping power for X-rays, a lot of X-rays were needed so that details, in particular soft tissue details, could be seen. More recently with the development of semiconductor detectors, more efficient digital imagers with higher X-ray stopping power have been developed.

As was just pointed out, the X-ray absorption process, which relates to the detection efficiency of a material like silicon, is strongly dependent on photon energy. As seen in Chapter 2, a photon is a massless and neutral particle for which the interaction mechanisms with matter, including semiconductors, are photoelectric, Compton, or, if their energy is high enough, pair production [34]. In practice, depending on the energy and on the material atomic number Z, there is a given probability that a photon will interact somewhere in the sensing material and release, at one particular point, all its energy. For a material with known Z with a given thickness, its detection efficiency will therefore decrease when the photon or X-ray energy E increases. Thus, silicon, with its $Z = 14$ and thicknesses as high as 1–2 mm can be used efficiently only for few applications, where the X-ray energy is at most slightly above 20 keV. In this case, "efficiently" means that photons are mostly absorbed inside the sensor's volume and therefore low photon doses can be used. If this is not the case, high doses will be needed to compensate for the low probability of an interaction of the photon to occur. This could result in biological damage, with unwanted consequences to the patient. Recalling the previous example of bone–fat contrast radiology, which requires X-ray energies around 60 keV to maximize the image contrast, this is not optimized for silicon, even silicon as thick as 2 mm, since its efficiency is still only a low percentage. Therefore, only a few X-rays would stop in the detector, and several photons (hence high dose) would be required to form an image with enough details to be useful.

The solution to X-ray detection above 20 keV was found by using either an additional converting stage in front of the silicon detector with higher absorption efficiency, usually scintillators, or another semiconducting material with higher efficiency and therefore higher atomic number Z. These materials include the well-established germanium, which unfortunately does not work at room temperature: it is too noisy because its band gap is too small, and so it requires a sophisticated cooling system or other compound semiconductors such as gallium arsenide (GaAs) or cadmium telluride (CdTe) and cadmium zinc telluride (CdZnTe). These semiconductors are currently being investigated with excellent results, but they are not yet as well established as silicon, in particular for fabrication of homogeneous large-area imagers.

Unfortunately, the use of a thick slab of silicon, which is well known, abundant, and cheap and can make a detector that can absorb high-energy X-rays, is not feasible using a traditional planar process. Apart from the difficulty of using the processing tools that are normally tuned for thinner substrates and the need for high-bias voltages in order to reach full depletion, the main reason is charge diffusion. This phenomenon occurs when electrons and holes, generated by an impinging photon, lie in a region where the electric field lines are not strong enough to capture them in their stream toward the collecting electrode, so they wander until they fall into a strong electric field line. In other words, one can imagine making many, very small electrodes, such as those in modern cameras, so that even the smallest details can be appreciated. However, if carriers can diffuse during their path to the collecting electrode to generate a signal and therefore be seen, they can move long enough to fall into the stream of the neighboring electric field and therefore be collected from an electrode that is not the closest to where the photon impinged the sensor. This effect is also known as charge sharing and needs to be avoided if sharp information is the goal. A useful image is sharp, so small details can be seen, while charge sharing would result in a "blurred" image, with less resolution. This is also true when a converting material is used in front of the pixelated sensor since its emitted signal, usually light, is isotropic over 360 degrees and can be detected by pixels that are distant from where the X-ray impinged.

3D sensors have been shown to be a good way to overcome charge sharing in thick substrates since electrodes go all the way through so that the same field will have equivalent strength independently from the substrate's depth. Therefore, even if diffusion happens in the same way as planar sensors, the probability of sharing charge among neighboring pixels is considerably reduced, even for thick substrates, so the image stays sharp [35]. The price to pay in imaging, however, is loss of information in the electrode region. This loss could be overcome by optimizing the electrode filling with doped polysilicon (which proved to be sensitive to radiation up to 60%, see Chapter 5) [36] or by piling thin 3D sensors, with smaller electrode diameter, one on the top of another, using bump bonding, like "active" interposers [37].

Some innovative detectors for imaging applications with columnar electrodes filled with scintillators have been proposed. In particular, the confinement of scintillator materials into the columns can provide an intrinsically better spatial resolution with respect to the standard coupling with planar detectors [38, 39]. Moreover, owing to the peculiar properties of charge collection dynamics in 3D structures, detectors for fast imaging of hard X-rays or gamma rays would be feasible. With 3D sensors, the operating voltage can be kept low, solving system problems such as increased noise due to power dissipation.

As an additional example, thick (up to ~cm) trenched sensors have been proposed for gamma-ray detection, with good initial results from 2 mm thick prototypes in terms of low leakage current and energy resolution in gamma-ray detection from ^{241}Am and ^{57}Co sources [40].

The previously mentioned Graytone lithography has also been proposed to create thick (~5 mm) 3D silicon drift detectors, where arrays of modulated-depth trenches act as 3D drift cathodes, with demonstration of the operation principle by TCAD simulations [41]. A similar concept was also presented in [42], but because of the technological complexity involved, no prototype has yet been fabricated.

9.4 PROTEIN CRYSTALLOGRAPHY AND MICRODOSIMETRY

Active-edge 3D sensors have also been proposed for fast imaging in protein crystallography [43]. In this particular case, active edges were key in defining the geometry of a fully sensitive concave detector plane capable of always receiving diffracted X-rays in the crystallized protein perpendicularly. The Bragg spots obtained would therefore be very sharp at all diffractive angles, with better image resolutions of the reconstructed protein structure. Furthermore, a fast detector response would allow dynamic imaging of the protein folding, which could lead to better understanding of the cause of diseases like Alzheimer's and Parkinson's.

Active-edge 3D sensors could also be used to define very precisely molecular size-sensitive volumes for microdosimetric tests. Microdosimetry, which measures stochastic (or probabilistic) energy deposition events at the cellular level, was developed to provide a comprehensive description of the spatial and temporal distribution of absorbed energy in irradiated matter at the micrometric level. For this purpose, a detector with well-defined sensitive volume (SV) equivalent to the cell size is needed. This can be obtained by using 3D silicon technology in arrays of coaxial structures, each the size of a cell as illustrated in Fig. 9.7.

Applications of these devices include dosimetry during particle beam radiotherapy with protons or carbon ions for cancer treatment. This type of microdosimetry would allow more precise monitoring of the dose delivered to tumor cells as well as to cells in the regions surrounding the tumors, thus ensuring better control of damage done to healthy tissues of the areas surrounding the tumor(s). More precise knowledge of the dose received at the cellular level could also help professionals, most notably airplane pilots and cabin crews, who work in radiation-rich environments.

Pioneering work on these devices was performed by Anatoly Rozenfeld and coworkers at the University of Wollongong in Australia [43]. More recently, the same group collaborated on other projects in which dosimeters were based on arrays of 3D sensors; each element was connected to a metal network that provided the bias and carried the measured signal to the readout electronics to determine the absorbed energy and thereafter the deposited amount of radiation (dose), using a similar layout as the one shown in Fig. 9.8 [45–48]. The results

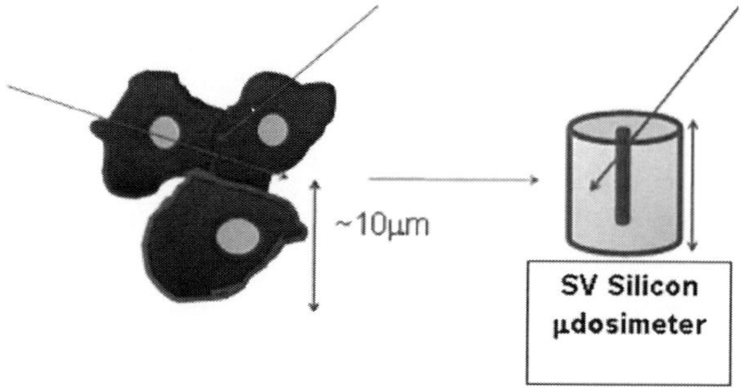

FIGURE 9.7 Determination of radiation deposited at the cellular level during radiation treatments require a microdetector with molecular size and well-defined sensitive volume (SV).

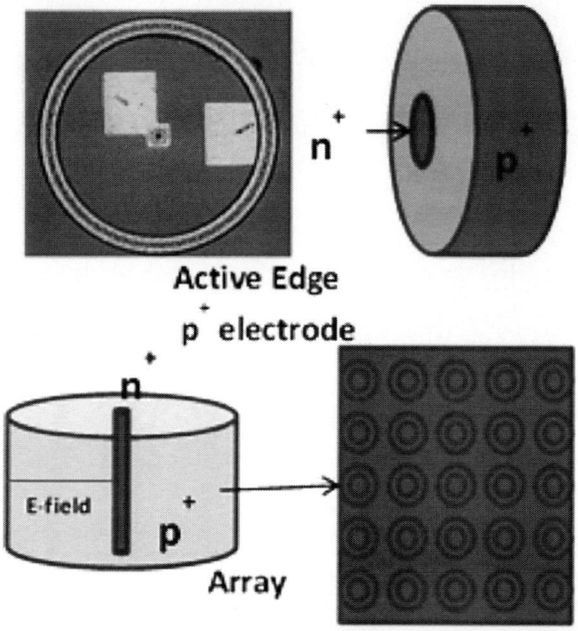

FIGURE 9.8 Coaxial 3D silicon fabricated at the Stanford Nanofabrication Facility, Stanford, USA, together with side-view and array sketches. On the top left of the figure one can see a picture of a coaxial 3D silicon sensor. Details of the central electrode and trench with the respective electrical connections pads are also visible as large yellow rectangles. In the figure also visible a 3D sketch of the coaxial device (top left) and at the bottom how a single element (on the left) would be positioned inside a two dimensional array (on the bottom right).

obtained confirmed that this design could reproduce the radiobiological effectiveness that previously was possible only with tissue-equivalent proportional counters (TEPC) which, however are not suitable for "in vivo" dosimetry because of their bulky structure [49].

9.5 NEUTRON DETECTORS

Neutrons can be complementary to X-rays for many applications. Since X-ray attenuation increases with the element atomic number, there is very little contrast between light materials and the background. This is the case for a large variety of materials rich in water or hydrogen, where it is almost impossible to obtain a sufficiently high contrast of their details to visualize with traditional imaging techniques. For this reason, neutron tomography finds application in a broad spectrum of fields, including archaeological dating, analysis of historical artworks, materials science, fuel cells research, biological and medical applications and homeland security. Two-dimensional and three-dimensional neutron imaging are both nondestructive and noninvasive techniques, properties that are essential if information about the material structure, texture, phase contrast, and composition of objects is required. Neutron radiography can be obtained from attenuation of a neutron beam through the object in question. Since the attenuation coefficient depends on the material characteristics as well as on the neutron wavelength, the transmitted beam intensity decreases through beam absorption and scattering in the object itself. This technique is useful, for example, in the analysis of rare artifacts because the information about the material composition can be extracted since each element has a different cross section for neutrons. This quality allows detailed analysis of materials such as bronze where X-rays cannot penetrate deeply. Different bronze alloy compositions were used in different historical periods and can therefore be used to establish precise dating. Similarly, imaging of biological samples can be obtained by contrast, provided mainly by neutron scattering with hydrogen atoms, with precisions of the order of 75 microns [50].

Silicon is not sensitive to neutrons. For an incident neutron to be detected in silicon, a scattering or capture reaction event must occur leading to the generation of e–h pairs, which are separated and collected with a signal distinctly above the noise level. This is why in planar silicon sensors layers of neutron-converting materials with high cross section, and therefore high conversion probability like boron or lithium, are used. Since the products of scattering or capture reactions (usually short-range alpha particles and gammas) are emitted in all directions, only a few (< 5%) of such particles are reaching the active silicon junction generating a signal [51, 52]. This is why etching was originally used in silicon neutron detectors to carve its surface in chevron structures. In this way, the secondary emission of neutron-converting material deposited in them could reach the active silicon deplete volume with higher efficiencies [53].

After this first attempt, more traditional 3D sensors with and without active electrodes shaped as cylinders or trenches were used to host neutron-converting materials. By doing so, a further extended interaction surface could be achieved, as well as higher probabilities for reaction products to reach the semiconductor's active area. By tailoring the shapes and geometries of the apertures, it was shown that the neutron detection efficiency could be significantly increased, up to ~50%.

The most significant results in this direction have been obtained by using three-dimensional silicon design for thermal neutron detection at Kansas State University (USA) [54–57]. The greatest efficiency obtained by perforated neutron detectors filled by ^{6}Li was 29%, but it could be increased to 42% with back-to-back stacking.

Using a different 3D technology, scientists at the Lawrence Livermore National Laboratory (USA) developed a geometry composed of few micron silicon pillars surrounded by an enriched boron converter [58–61]. Efficiencies ranged from 22% for 26 μm deep pillars to 48.5% for 50 μm pillars, with discrimination set to 30 keV. Honeycomb-shaped, perforated structures filled with boron converter were developed by the Rensselaer Polytechnic Institute (USA), with an efficiency of 26% [62–65]. All these efforts were mostly motivated by the need to replace ^{3}He gas filled proportional neutron counters in USA homeland security applications.

Similar motivations urged European research groups to be active in this field. Arrays of micromachined pixels with encapsulated crystals of neutron-sensitive scintillating materials were fabricated at Delft University (Holland). The emitted light at 600 nm was used to detect the presence of neutrons [66]. As mentioned previously, the Czech

Technical University of Prague also reported thermal neutron test results from detectors fabricated with pyramidal dips on the surface and covered with ^{6}LiF. The reported efficiency was 6.3% for a single layer compared to 4.9% obtained with a reference planar detector [53]. Among other interesting devices based on microfabrication are the ultra-thin 3D sensors with planar converting coating ($^{10}B_4C$) proposed by CNM Barcelona (Spain) to minimize γ-ray signal contamination [67–70]. Furthermore, another version of superficial pyramidal groove sensors filled with neutron-sensitive TiB_2 and ^{6}LiF were fabricated at SINTEF (Oslo, Norway) with a relative efficiency improvement of 38% compared to planar structures [71].

Finally, the University of Trento, in collaboration with FBK, developed pixelated 3D hybrid neutron detectors, compatible with the MEDIPIX family readout chips, within the HYDE (HYbrid DEtectors for neutrons) project [72]. The first HYDE batch used relatively large cavities that could be easily filled with different converter materials: polysiloxane (for fast neutrons), $^{10}B_4C$, ^{6}LiF, and $_{10}B$ (for thermal neutrons). Despite the largely non-optimized geometries, results from the electrical and functional characterization were encouraging [73, 74] and led to the design of a new web-shaped structure. The currently observed limit in neutron detection efficiency of ~7%, much lower than the theoretical and simulated values of ~12%, was understood, and ways to improve it have been defined. The first prototypes of pixel sensors modules are currently being assembled. An extensive, up-to-date review of three-dimensional structures for thermal neutron detection and imaging can be found in [75].

9.6 VERTICALLY INTEGRATED SYSTEMS WITH MICROCHANNEL COOLING

Microfabrication also offers alternative options to minimize the full system's material budget while responding to key system requirements such as temperature management. In high-energy physics pixel vertex detectors, for example, reducing the sources of fake tracks produced by multiple scattering is one of the goals when high-particle multiplicity is present. Deep reactive ion etching can in fact be used to fabricate "microchannels" directly in contact with the electronic/detector modules, where coolant could be circulated to effectively remove heat in enclosed volumes.

Several studies to embed cooling within electronics wafers have been reported in the microprocessor industry [76]. For high-energy physics applications, two main lines of research have been pursued: (1) microchannels etched on the silicon surface cupped with wafer-bonded silicon and (2) superficial microtrenches with overetched bottom. Examples of these methods are reported in [77–79]. Both approaches have shown very promising cooling performances after tests in realistic conditions and the formers have been or will be implemented in experiments [80–82].

Furthermore, a 3D sensor module with 100 μm thin FE-I4 electronics was successfully integrated vertically with surface silicon–silicon microchannels for CO_2 cooling. The performance forecast for such a system is a reduction in radiation length of a factor four when compared with traditional CO_2 cooling systems like, for example, the one used in the

ATLAS IBL [83]. A sketch (on the top) and a picture (bottom) of such a prototype is visible in Figure 9.9, mounted on an FE-I4 readout board [84]. The 3D and FE-I4 front-end electronics module is visible in the center of the board on the top of the microchannel cooling module. The FE-I4 readout electronics chip is the CMOS 130 nm technology, 4 cm^2 chip with 26880, 250 × 50 μm square pixels, designed for the ATLAS-IBL pixel detector. The input–output microfluidics ports are also visible on the left of the module.

The microchannels were fabricated using an LHCb design [85] on a 0.38 mm thick silicon and 0.45 mm thick silicon wafers, connected after etching using wafer bonding. The

FIGURE 9.9 (Top) Sketch of the vertically integrated microfabricated system composed from the top of a 3D sensor bump-bonded to an ATLAS FEI4 electronic chip glued to Si-Si microchannels for thermal management for a total thickness of about 1mm. (Bottom) picture of the system mounted on a measurement board. Visible are the metal input and output inlets for the CO_2.

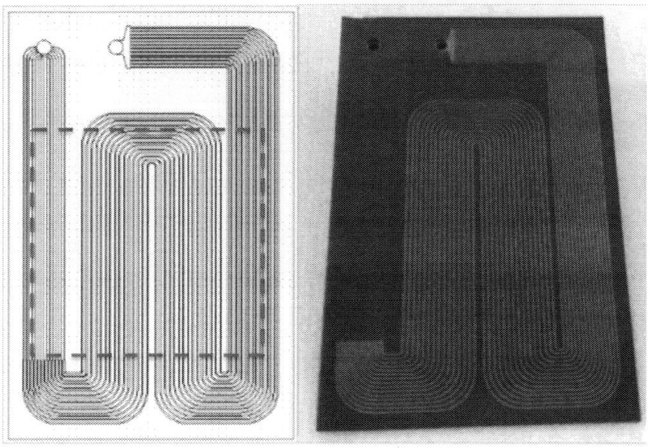

FIGURE 9.10 Drawing and photograph of the "snake" design etched onto the silicon substrate. The red square on the left side drawing indicate the approximated position of the FEI4-3D module [85].

microchannels' rectangular shape measuring 40 mm by 60 mm can be seen in the design (on the left) and picture (on the right) of Fig. 9.10. The 22 × 20 mm^2 FE-I4-3D module was positioned in the center of it, as indicated by the dotted square on the left.

The photograph in Fig. 9.10 clearly shows the microchannel network through a cupping glass [85]. The design is conceived with the inlet and outlet to be positioned on the same side of the sample, while roughly equalizing the length of all channels. All etching is performed on one of the two silicon wafers only. In this design, the microchannels are 70 μm, deep while the inlet and outlet have a 2 mm diameter through holes separated by 10 mm. Kovar connectors, bent at 90 degrees and visible in Fig. 9.9 are laser soldered to the outside surface of the silicon wafer corresponding to the inlet and outlet holes, visible at the top left of the drawing and picture of Fig. 9.10, for easy reach of the cooling system.

In normal operation, the detector modules in experiments must be cooled well below zero degrees to control the overall temperature of the detector-electronics system, in particular after exposure to radiation. However it is crucial for the system to be able to tolerate a warm-up to room temperature or to be operated at room temperature for certain test or emergency situations. In this condition, the microchannels as well as the connectors must sustain a CO_2 overpressure, which could reach levels as high as 65 bars. Relatively large areas of the inlet and outlet would therefore undergo considerable forces when pressurized, making the input–output manifolds the weakest points to withstand the high pressure. The manifold area therefore needs to be as small as possible to reduce the force and to guarantee operational reliability. The current designs use 15 microchannels starting from the inlet with 30 μm width small capillaries. These inlet capillaries dominate the channel-flow impedance and, therefore, ensure a uniform flow distribution across all channels [86]. Without this impedance, power fluctuations in a parallel, two-phase system would lead to varying flow rates causing dry-out in the channels with the highest heat load.

Tests with the above system were performed at different temperatures with CO_2 to test the cooling reliability of the microchannels. It was shown that the system could support the high pressure at room temperature and managed to obtain a thermal figure of merit (TFoM), corresponding to the thermal resistance of the medium and defined as:

$$TFoM = \frac{\Delta T \times A}{Power} = \frac{4K\ cm^2}{W}$$

with ΔT being the difference of the temperatures measured on the heating source in Kelvins with the cooling system on and off and A the surface of the heat source, in this case the size of the FEI4 chip 4 cm^2. To get an idea of the quality of this method, the measured value of 4K cm^2/W obtained with the microchannels should be compared to 13K cm^2/W measured in the laboratory on an IBL module using CO_2 cooling [83].

9.7 MULTIBAND SPECTROSCOPY

Etching conductive holes at different depths in electrically separated silicon volumes could also be very useful for multiband spectroscopy since photons with different energies have different penetration depths in silicon. Applications in astronomy have been reported in [87–88] where a charge-coupled device (CCD) architecture containing multiple vertically stacked active layers independently connected to the external readout through etched conductive "vias" was shown to be capable of generating color images without external filters. This innovative technology would allow, for example, collecting multiple-band astronomical imaging contemporarily with the same device, considerably speeding up the overall survey time of a telescope with high quantum efficiency.

9.8 3D SENSORS WITH OTHER SUBSTRATES

Versatility in 3D sensor microfabrication has surpassed that of silicon for several years now and has moved to other materials like diamond, cadmium telluride, and gallium arsenide for specific motivations and applications. All these materials are characterized by an intrinsic high resistivity mainly due to whide bandgap or compensation and a consequent low leakage current. The first proposal to process 3D structures in materials different than silicon can be found in Parker's original paper [89] recognizing the importance of having a high Z material in 3D geometry for medical applications. All these materials in fact could highly benefit from a 3D structure since wafers tend to be rather nonhomogeneous, making large-area device processing difficult, while 3D electrodes would make charge generation and collection within each crystalline boundary more probable throughout the sensor's volume.

Single-crystal diamond is an exception to the above considerations since the grain boundaries causing a nonhomogeneous response are no longer present. This comes, however, at the price of being rather costly as a detector material. Nevertheless, successful attempts have been made to process 3D sensors on single-crystal diamond with other advantages such as the low-bias voltage needed to fully collect the charge generated by a particle for a 450 micron thick sample, which was reduced from 500 V for a planar sensor to 25 V for a 3D one [90].

The technique of choice to process diamond as well as compounds like CdTe and GaAs 3D sensors is infrared laser drilling. So far diamond, in both its single and polycrystalline forms, has been shown to be the most interesting in terms of electrode fabrication since femtosecond laser drilling transforms the diamond structure in graphite, which, being conductive, acts as an ideal electrode without requiring further processing. This is why electrode fabrication of diamond 3D with femptosecond laser drilling does not require elegant cleanrooms with traditional semiconductor processing to be made, but just photolithography and gold sputtering [91, 92]. At present, several groups have processed devices that have yielded extremely positive results and various dimensions that are even compatible with large-area readout electronic chips [93].

Electrode processing after femtosecond laser drilling proved to be more challenging for CdTe and GaAs due to the complexity of the materials, which, contrary to diamond, would still require stacked metals to form barriers and ohmic contacts to work properly. The attempts that have been made, however, look promising for future use in nuclear medicine and other imaging applications requiring high Z materials [94–96].

REFERENCES

1. Compact Muon Solenoid experiment at CERN, http://cms.web.cern.ch.
2. O. Koybasi, E. Alagoz, A. Krzywda, K. Arndt, G. Bolla, D. Bortoletto, T. -E. Hansen, T. A. Hansen, G. U. Jensen, A. Kok, S. Kwan, N. Lietaer, R. Rivera, I. Shipsey, L. Uplegger, C. Da Vià, "Electrical characterization and preliminary beam test results of 3D silicon CMS pixel detectors," IEEE Transactions on Nuclear Science, vol. 58(3), pp. 1315–1323, 2011.
3. E. Alagoz, M. Bubna, A. Krzywda, G.-F. Dalla Betta, A. Solano, M. M. Obertino, M. Povoli, K. Arndt, A. Vilela Pereira, G. Bolla, D. Bortoletto, M. Boscardin, S. Kwan, R. Rivera, I. Shipsey, L. Uplegger, "Simulation and laboratory test results of 3D CMS pixel detectors for HL-LHC," Journal of Instrumentation, JINST 7, P08023, 2012.
4. A. Kryzda, E. Alagoz, M. Bubna, M. M. Obertino, A. Solano, K. Arndt, L. Uplegger, G.-F. Dalla Betta, M. Boscardin, J. Ngadiuba, R. Rivera, D. Menasce, L. Moroni, S. Terzo, D. Bortoletto, A. Prosser, J. Adreson, S. Kwan, I. Osipenkov, G. Bolla, C. M. Lei, I. Shipsey, P. Tan, N. Tran, J. Chramowicz, J. Cumalat, L. Perera, M. Povoli, R. Mendicino, A. Vilela Pereira, R. Brosius, A. Kumar, S. Wagner, F. Jensen, S. Bose, S. Tentindo, "Pre- and post-irradiation performance of FBK 3D silicon pixel detectors for CMS," Nuclear Instruments and Methods A, vol. 763, pp. 404–411, 2014.
5. M. Bubna, E. Alagoz, M. Cervantes, A. Krzywda, K. Arndt, M. M. Obertino, A. Solano, G. F. Dalla Betta, D. Menace, L. Moroni, L. Uplegger, R. Rivera, I. Osipenkov, J. Andresen, G. Bolla, D. Bortoletto, M. Boscardin, J. M. Brom, R. Brosius, J. Chramowicz, J. Cumalat, M. Dinardo, P. Dini, F. Jensen, A. Kumar, S. Kwan, C. M. Lei, M. Povoli, A. Prosser, J. Ngadiuba, L. Perera, I. Shipsey, P. Tan, S. Tentindo, S. Terzo, N. Tran, S. R. Wagner, "Testbeam and laboratory characterization of CMS 3D pixel detectors," Journal of Instrumentation, JINST 9, C07019, 2014.
6. M. Fernandez, R. Jaramillo, M. Lozano, F. J. Munoz, G. Pellegrini, D. Quirion, T. Rohe, I. Vila, "Radiation resistance of double-type double-sided 3D pixel sensors," Nuclear Instruments and Methods A, vol. 732, pp. 137–140, 2013.
7. A. Lai, "TIMESPOT: TIME & SPace real-time Operating Tracker," presented at New Dimensions in Silicon Pixel Detectors, Manchester (UK), October 19, 2017 (http://indico.hep.manchester.ac.uk/getFile.py/access?contribId=9&sessionId=4&resId=0&materialId=slides&confId=5238).
8. ATLAS Event: ATLAS experiment @2014 CERN, http://atlas.web.cern.ch/Atlas/Collaboration.
9. C. Da Vià, S. Watts, "The geometrical dependence of radiation hardness in planar and 3D silicon detectors," Nuclear Instruments and Methods A, vol. 603, pp. 319–324, 2009.

10. B. Abbott et al., "Production and Integration of the ATLAS Insertable B-Layer," Journal of Instrumentation, JINST 13, T05008, 2018.
11. ATLAS Collaboration, "Evolution of LV Currents and sensor leakage currents in the IBL", https://atlas.web.cern.ch/Atlas/GROUPS/PHYSICS/PLOTS/PIX-2016-006/.
12. S. Grinstein (on behalf of the AFP Collaboration), "The ATLAS forward proton detector (AFP)," Nuclear and Particle Physics Proceedings, vol. 273–275, pp. 1180–1184, 2016.
13. Ksenia Shchelina, "Measurement of high-mass dilepton production with the CMS-TOTEM precision proton spectrometer," Proceedings of Science, The European Physical Society Conference on High Energy Physics (EPS-HEP 2017), Paper 405, 2018.
14. S. Parker, N. V. Mokhov, I. L. Rakhno, I. S. Tropin, C. Da Vià, S. Seidel, M. Hoeferkamp, J. Metcalfe, R. Wang, C. Kenney, J. Hasi, P. Grenier, "Proposed triple-wall, voltage isolating electrodes for multiple-bias-voltage 3D sensors," Nuclear Instruments and Methods A, vol. 685, pp. 98–103, 2012.
15. A. Rivetti, "ASIC design for precision timing," 11th Workshop on Pico-Second Timing Detectors for Physics and Medical Applications, Turin (Italy), May 16–18, 2018.
16. N. Cartiglia, A. Staiano, V. Sola, R. Arcidiacono, R. Cirio, F. Cenna, M. Ferrero, V. Monaco, R. Mulargia, M. Obertino, F. Ravera, R. Sacchi, A. Bellor, S. Durando, M. Mandurrino, N. Minafra, V. Fadeyev, P. Freeman, Z. Galloway, E. Gkougkousis, H. Grabas, B. Gruey, C. A. Labitan, R. Losakul, Z. Luce, F. McKinney-Martinez, H. F. W. Sadrozinski, A. Seiden, E. Spencer, M. Wilder, N. Woods, A. Zatserklyaniy, G. Pellegrini, S. Hidalgo, M. Carulla, D. Flores, A. Merlos, D. Quirion, V. Cindro, G. Kramberger, I. Mandić, M. Mikuž, M. Zavrtanik, "Beam test results of a 16 ps timing system based on ultra-fast silicon detectors," Nuclear Instruments and Methods A, vol. 850, pp. 83–88, 2017.
17. G. Kramberger, M. Carulla, E. Cavallaro, V. Cindro, D. Flores, Z. Galloway, S. Grinstein, S. Hidalgo, V. Fadeyev, J. Lange, I. Mandić, G. Medin, A. Merlos, F. McKinney-Martinez, M. Mikuž, D. Quirion, G. Pellegrini, M. Petek, H. F.-W.Sadrozinski, A. Seiden, M. Zavrtanik, "Radiation hardness of thin low gain avalanche detectors," Nuclear Instruments and Methods A, vol. 891, pp. 68–77, 2018.
18. A. Loi, "3D detectors with high space and time space resolution," Journal of Physics: Conference Series, vol. 956, 012012, 2018.
19. S. Parker, C. Da Vià, M. Deile, T.-E. Hansen, J. Hasi, C. Kenney, A. Kok, S. Watts, "Dual readout: 3D direct/induced-signals pixel systems," Nuclear Instruments and Methods A, vol. 594, pp. 332–338, 2008.
20. C. Da Vià, S. Parker, M. Deile, T.-E. Hansen, J. Hasi, C. Kenney, A. Kok, S. Watts, "Dual readout— strip/pixel systems," Nuclear Instruments and Methods A, vol. 594, pp. 7–12, 2008.
21. S. I. Parker, A. Kok, C. Kenney, P. Jarron, J. Hasi, M. Despeisse, C. Da Vià, G. Anelli, "Increased speed: 3D silicon sensors; fast current amplifiers," IEEE Transactions on Nuclear Science, vol. 58(2), pp. 404–417, 2011.
22. L. Piccolo, "The Time Spot Project: Fast timing with 3D detectors," 11th Workshop on Pico-Second Timing Detectors for Physics and Medical Applications, Turin (Italy), May 16–18, 2018.
23. G. Rhodes, Crystallography made crystal clear: A guide for users of macromolecular models, 2nd Edition, San Diego, CA, Academic, 2000.
24. Suleman Surti, "Update on time of flight PET imaging," Journal of Nuclear Medicine, vol. 56(1), 98–105, January 2015.
25. R. E. Meyers et al. "Space-yime quantum imaging," Entropy 2015, 17, 1508–1534.
26. B. F. Phlips, M. Christophersen, "Curved radiation detector," 2008 IEEE Nuclear Science Symposium, Dresden (Germany), October 19–25, 2008, Conference Record, N30-132.
27. G. Gal, "Method for fabricating microlenses," Patent no. 5,310,623, May 1994.
28. W. Henke, H. Quenzer, P. Staudt-Fischbach, B. Wagner, "Simulation assisted design of processes for gray-tone lithography," Microelectronic Engineering, vol. 27(1–4), pp. 267–270, 1995.

29. C.-K. Wu, "High energy beam sensitive glasses," Patent no. 5,285,517, February 1994.
30. M. Christophersen, B. F. Phlips, "Gray-tone lithography using an optical diffuser and a contact aligner," Applied Physics Letters, vol. 92, pp. 194102–194102-3, May 2008.
31. G. Jia, U. Hübner, J. Dellith, A. Dellith, R. Stolz and G. Andrä, "Core-shell diode array for high performance particle detectors and imaging sensors: status of the development," Journal of Instrumentation, JINST, vol. 12, C02044, 2017.
32. G. Jia, M. Steglich, I. Sill, F. Falk, "Core-shell heterojunction solar cells on silicon nanowires arrays," Solar Energy Materials and Solar Cells, vol. 96, pp. 226–230, 2012.
33. G. Jia, B. Eisenhawer, J. Dellith, F. Falk, A. Thøgersen, A. Ulyashin, "Multiple core-shell silicon nanowire-based heterojunction solar cells," Journal of Physical Chemistry C, vol. 117(2), pp. 1091–1096, 2013.
34. G. F. Knoll, "Radiation detection and measurement," 3rd Edition, John Wiley & Sons, New York, 2000.
35. L. Tlustos, M. Campbell, E. Heijne, X. Llopart, "Signal variations in high-granularity Si pixel detectors," IEEE Transactions on Nuclear Science, vol. 51(6), pp. 3006–3012, 2004.
36. J. Hasi, "3D—The next step to silicon particle detection," PhD Thesis, Brunel University, London, 2004.
37. C. Da Vià, "3D sensors and micro-fabricated detector systems," Nuclear Instruments and Methods A, vol. 765, pp. 151–154, 2014.
38. P. Kleimann, J. Linnros, C. Fröjdh, C. S. Petersson, "An X-ray imaging pixel detector based on scintillator filled pores in a silicon matrix," Nuclear Instruments and Methods A, vol. 460, pp. 15–19, 2001.
39. X. Badel, J. Linnros, M. S. Janson, J. Österman, "Formation of pn junctions in deep silicon pores for X-ray imaging detector applications," Nuclear Instruments and Methods A, vol. 509, pp. 96–101, 2003.
40. M. Christophersen, B. F. Phlips, F. J. Kub, "Trenched gamma-ray detector," 2008 IEEE Nuclear Science Symposium, Dresden (Germany), October 19–25, 2008, Conference Record, Paper N68-5.
41. M. Christophersen, B. F. Phlips, "Thick silicon drift detectors," 2008 IEEE Nuclear Science Symposium, Dresden (Germany), October 19–25, 2008, Conference Record, Paper N34-5.
42. T. Knežević, L. K. Nanver, T. Suligo, "Silicon drift detectors with the drift field induced by PureB-coated trenches," Photonics, vol. 3(4), 54, 2016.
43. S. I. Parker, C. J. Kenney, D. Gnani, A. C. Thomson, E. Mandelli, G. Meddeler, J. Hasi, J. Morse, E. M. Westbrook, "3DX: an X-ray pixel array detector with active edges," IEEE Transactions on Nuclear Science, vol. 53(3), pp. 1676–1688, 2006.
44. A. B. Rosenfeld, G. I. Kaplan, T. Kron, B. J. Allen, A. Dilmanian, I. Orion, B. Ren, M. L. F. Lerch, A. Holmes-Siedle, "MOSFET dosimetry of an X-ray microbeam," IEEE Transactions on Nuclear Science, vol. 46(6), pp. 1774–1780, 1999.
45. L. T. Tran, D. A. Prokopovich, M. Petasecca, M. L. F. Lerch, A. Kok, A. Summanwar, T. E. Hansen, C. D. Vià, M. I. Reinhard, A. B. Rosenfeld, "3D radiation detectors: Charge collection characterization and applicability of technology for microdosimetry," IEEE Transactions on Nuclear Science, vol. 61(4), pp. 1537–1543, 2014.
46. L. T. Tran, D. A. Prokopovich, M. Petasecca, M. L. F. Lerch, C. Fleta, G. Pellegrini, C. Guardiola,M. I. Reinhard, A. B. Rosenfeld, "Ultra-thin 3-D detector: Charge collection characterization and application for microdosimetry," IEEE Transactions on Nuclear Science, vol. 61(6), pp. 3472–3478, 2014.
47. L. T. Tran, L. Chartier, D. A. Prokopovich, D. Bolst, M. Povoli, A. Summanwar, A. Kok, A. Pogossov, M. Petasecca, S. Guatelli, M. I. Reinhard, M. Lerch, M. Nancarrow, N. Matsufuji, M. Jackson, A. B. Rosenfeld, "Thin silicon microdosimeter utilizing 3-D MEMS fabrication technology: Charge collection study and its application in mixed radiation fields," IEEE Transactions on Nuclear Science, vol. 65(1), pp. 467–472, 2018.

48. L. T. Tran, L. Chartier, D. Bolst, J. Davis, D. A. Prokopovich, A. Pogossov, S. Guatelli, M. I. Reinhard, M. Petasecca, M.L.F. Lerch, N. Matsufuji, M. Povoli, A. Summanwar, A Kok, M. Jackson, A. B. Rosenfeld, "In-field and out-of-file application in 12C ion therapy using fully 3D silicon microdosimeters," Radiation Measurements, vol. 115, pp. 55–59, 2018.
49. L. De Nardo, V. Cesari, G. Donà, G. Magrin, P. Colautti, V. Conte, G. Tornielli, "Mini-TEPCs for radiation therapy," Radiation Protection Dosimetry, vol. 108(4), pp. 345–352, 2004.
50. H. Z. Bilheux, J.-Ch. Bilheux, W. B. Bailey, W. S. Keener, L. E. Davis, K. W. Herwig, M. Cekanova, "Neutron imaging at the Oak Ridge National Laboratory: Application to biological research," IEEE Biomedical Science and Engineering Center Conference (BSEC 2014), Oak Ridge National Laboratory (USA), May 6–8, 2014, pp. 1–4.
51. D. S. McGregor, M. D. Hammig, Y.-H. Yang, H. K. Gersch, R. T. Klann, "Design considerations for thin film coated semiconductor thermal neutron detectors—i: basics regarding alpha particle emitting neutron reactive films," Nuclear Instruments and Methods A, vol. 500, pp. 272–308, 2003.
52. J. Jakubek, T. Holy, E. Lehmann, S. Pospíšil, J. Uher, J. Vacik, D. Vavrik, "Neutron imaging with Medipix-2 chip and a coated sensor," Nuclear Instruments and Methods A, vol. 560, pp. 143–147, 2006.
53. J. Uher, C. Fröjdh, J. Jakubek, C. Kenney, Z. Kohout, V. Linhart, S. Parker, S. Petterson, S. Pospíšil, G. Thungström, "Characterization of 3D thermal neutron semiconductor detectors," Nuclear Instruments and Methods A, vol. 576, pp. 32–37, 2007.
54. S. L. Bellinger, R. G.Fronk, W. J. McNeil, T. J. Sobering, D. S. McGregor, "Enhanced variant designs and characteristics of the microstructured solid-state neutron detector," Nuclear Instruments and Methods A, vol. 652, pp. 387–391, 2011.
55. S. L. Bellinger, R. G. Fronk, W. J. McNeil, T. J. Sobering, D. S. McGregor, "Improved high efficiency stacked microstructured neutron detectors back filled with nanoparticle LiF," IEEE Transactions on Nuclear Science, vol. 59, pp. 167–173, 2012.
56. S. L. Bellinger, R. G. Fronk, T. J. Sobering, D. S. McGregor, "High-efficiency microstructured semiconductor neutron detectors that are arrayed, dual-integrated, and stacked," Applied Radiation and Isotopes, vol. 70, pp. 1121–1124, 2012.
57. D. S. McGregor, S. L. Bellinger, J. K. Shultis, "Present status of microstructured semiconductor neutron detectors," Journal of Crystal Growth, vol. 379, pp. 99–110, 2013.
58. R. Nikolic, C. Li Cheung, C. E. Reinhardt, T. F. Wang, "Roadmap for high efficiency solid-state neutron detectors, Proceedings of SPIE, vol. 6013, 601305, 2005.
59. R. J. Nikolic, A. M. Conway, C. E. Reinhardt, R. T. Graff, T. F. Wang, N. Deo, C. L. Cheung, "6:1 aspect ratio pillar based thermal neutron detector filled with 10B," Applied Physics Letters, Vol. 93(13), 133502, 2008.
60. A. M. Conway, T. F. Wang, N. Deo, C. L. Cheung, R. J. Nikolic, "Numerical simulations of pillar structured solid state thermal neutron detector: Efficiency and gamma discrimination," IEEE Transactions on Nuclear Science, vol. 56, pp. 2802–2807, 2009.
61. Q. Shao, L. F. Voss, A. M. Conway, R. J. Nikolic, M. A. Dar, C. L. Cheung, "High aspect ratio composite structures with 48.5% thermal neutron detection efficiency," Applied Physics Letters, vol. 102, 063505, 2013.
62. R. Dahal, K. C. Huang, J. Clinton, N. LiCausi, J.-Q. Lu, Y. Danon, I. Bhat, "Self-powered micro-structured solid state neutron detector with very low leakage current and high efficiency," Applied Physics Letters, vol. 100, 243507, 2012.
63. Y. Danon, J. Clinton, K. C. Huang, N. LiCausi, R. Dahal, J. J. Q. Lu, I. Bhat, "Towards high efficiency solid-state thermal and fast neutron detectors," Journal of Instrumentation, JINST 7, C03014, 2012.
64. K. C. Huang, R. Dahal, N. LiCausi, J. J.-Q. Lu, Y. Danon, I. B. Bhat, "Boron filling of high aspect ratio holes by chemical vapor deposition for solid-state neutron detector applications," Journal of Vacuum Science and Technology B, vol. 30, 051204, 2012.

65. K.-C. Huang, R. Dahal, J. J.-Q. Lu, A. Weltz, Y. Danon, I. B. Bhat, "Scalable large-area solid-state neutron detector with continuous p-n junction and extremely low leakage current," Nuclear Instruments and Methods A, vol. 763, pp. 260–265, 2014.
66. C. P. Allier, R. W. Hollander, C. W. E. van Eijk, P. M. Sarro, M. de Boer, J. B. Czirr, J. P. Chaminade, C. Fouassier "Thin photodiodes for a neutron scintillator silicon-well detector," IEEE Transactions on Nuclear Science, vol. 48, pp. 1154–1157, 2001.
67. C. Guardiola, C. Fleta, G. Pellegrini, F. García, D. Quirion, J. Rodríguez, M. Lozano, "Ultra-thin 3D silicon sensors for neutron detection," Journal of Instrumentation, JINST 7, P03006, 2012.
68. C. Guardiol, F. Gómez, F. C. Fleta, J. Rodríguez, D. Quirion, G. Pellegrini, A. Lousa, L. Martínez-de-Olcoz, M. Pombar, M, Lozano, "Neutron measurements with ultra-thin 3D silicon sensors in a radiotherapy treatment room using a Siemens PRIMUS linac," Physics in Medicine and Biology, vol. 58, pp. 3227–3242, 2013.
69. C. Fleta, C. Guardiola, S. Esteban, C. Jumilla, G. Pellegrini, D. Quirion, J. Rodríguez, A. Lousa, L. Martínez-de-Olcoz, M. Lozano, "Fabrication and nuclear reactor tests of ultra-thin 3D silicon neutron detectors with a boron carbide converter," Journal of Instrumentation, JINST 9, P04010, 2014.
70. S. Esteban, C. Fleta, C. Guardiola, C. Jumilla, G. Pellegrini, D. Quirion, J. Rodríguez, M. Lozano, "Microstructured silicon neutron detectors for security applications," Journal of Instrumentation, JINST 9, C12006, 2014.
71. A. Kok et al, Silicon sensors with pyramidal structures for neutron imaging, JINST 9 (2014) C04011.
72. http://medipix.web.cern.ch/medipix
73. R. Mendicino, M. Boscardin, S. Carturan, M. Cinausero, G. Collazuol, G.-F. Dalla Betta, M. Dalla Palma, F. Gramegna, T. Marchi, E. Perillo, M. Povoli, A. Quaranta, S. Ronchin, N. Zorzi, "Novel 3D silicon sensors for neutron detection," Journal of Instrumentation, JINST 9, C05001, 2014.
74. R. Mendicino, M. Boscardin, S. Carturan, G.-F.Dalla Betta, M. Dalla Palma, G. Maggioni, A. Quaranta, S. Ronchin, "Characterization of 3D and planar Si diodes with different neutron converter materials," Nuclear Instruments and Methods A, vol. 796, pp. 23–28, 2015.
75. R. Mendicino, G.-F. Dalla Betta, "Three-dimensional detectors for neutron imaging," Nuclear Instruments and Methods A., vol. 878, pp. 129–140, 2018.
76. P. Parida, M. Schultz, T. Chainer, "Beat the heat in 3D chip stacks with embedded cooling," Electronics Cooling, March 9, 2018.
77. A. Mapelli, A. Catinaccio, J. Daguin, H. van Lintel, G. Nuessle, P. Petagna, P. Renaud, "Low material budget microfabricated cooling devices for particle detectors and front-end electronics," Nuclear Physics B (Proc. Suppl.), vol. 215 pp. 349–352, 2011.
78. F. Bosi, G. Balestri, M. Ceccanti, P. Mammini, M. Massa, G. Petragnani, A. Ragonesi, A. Soldani, "Light prototype support using micro-channel technology as high efficiency system for silicon pixel detector cooling," Nuclear Instruments and Methods A, vol. 650, pp. 213–217, 2011.
79. M. Boscardin, P. Conci, M. Crivellari, S. Ronchin, S. Bettarini, F. Bosi, "Silicon buried channels for pixel detector cooling," Nuclear Instruments and Methods A, vol. 718, pp. 297–298, 2013.
80. A. Francescon et al., "Application of micro-channel cooling to the local thermal management of detectors electronics for particle physics," Microelectronics Journal, vol. 44(7), pp. 612–618, July 2013.
81. A. Francescon et al., Thermal management of the ALICE ITS detector at CERN with ultra-thin silicon micro-channel devices, Proceedings of ExHFT-8, 8th World Congress on Experimental Heat Transfer, Fluid Mechanics and Thermodynamics, Lisbon, 2013.
82. R. Dumps et al., Micro-channel CO2 cooling for the LHCb VELO upgrade, Proceedings of PIXEL2012, International workshop on Semiconductor Pixel Detectors for Particles and Imaging, Inawashiro, Japan, 2012.

83. B. Verlaat, M. Ostrega, L. Zwalinski, C. Bortolin, S. Vogt, J. Godlewski, O. Crespo-Lopez, M. Van Overbeek, T. Blaszcyk, "The ATLAS IBL CO2 cooling system," Journal of Instrumentation, JINST 12, C02064, 2017.
84. The FE-I4 A testing board was designed and produced by the colleagues at the Physikalisches Institut, University of Bonn, Bonn, Germany.
85. A. Nomerotski, J. Buytart, P. Collins, R. Dumps, E. Greening, M. John, A. Mapelli, A. Leflat, Y. Li, G. Romagnoli, B. Verlaat, "Evaporative CO2 cooling using microchannels etched in silicon for the future LHCb vertex detector," Journal of Instrumentation, JINST 8, P04004, 2013.
86. Junemo Koo and Clement Kleinstreuer, Liquid flow in microchannels: Experimental observations and computational analyses of microfluidics effects, Journal of Micromechanics and Microengineering, vol. 13, 568, 2003.
87. C.-E. Chang, J. Segal, A. J. Roodman, R. T. Howe, C. J. Kenney, "Multiband charge-coupled device," 2012 IEEE Nuclear Science Symposium, Anaheim (USA), October 27–November 3, 2012, Conference Record, paper N12-1.
88. C.-E. Chang, J. Segal, A. J. Roodman, C. J. Kenney, R. T. Howe, "Experimental demonstration of a stacked SOI multiband charge-coupled device," 2014 IEEE International Electron Device Meeting, San Francisco (USA), paper 4.5, December 15–17, 2014.
89. S. I. Parker, C. J. Kenney, J. Segal, "3D—A proposed new architecture for solid-state silicon detectors," Nuclear Instruments and Methods A, vol. 395, pp. 328–343, 1997.
90. F. Bachmair, L. Bäni, P. Bergonzo, B. Caylar, G. Forcolin, I. Haughton, D. Hits, H. Kagan, R. Kass, L. Li, A. Oh, S. Phan, M. Pomorski, D. S. Smith, V. Tyzhnevyi, R. Wallny, D. Whitehead, "A 3D diamond detector for particle tracking," Nuclear Instruments and Methods A, vol. 786, pp. 97–104, 2015.
91. T. Kononenko, M. Meier, M. S. Komlenok, S. M. Pimenov, V. Romano, V. P. Pashinin, V.I. Konov, "Microstructuring of diamond bulk by IR femtosecond laser pulses," Applied Physics A, vol. 90, pp. 645–651, 2008.
92. T. V. Kononenko, M. S. Komlenok, V. P. Pashinin, S. M. Pimenov, V. I. Konov, M. Neff, V. Romano, W. Lüthy, "Femtosecond laser microstructuring in the bulk of diamond," Diamond and Related Materials, vol. pp. 196–199, 2009.
93. S. A. Murphy, M. Booth, L. Li, A. Oh, P. Salter, B. Sun, D. Whitehead, A. Zadoroshnyj, "Laser processing in 3D diamond detectors," Nuclear Instruments and Methods A, vol. 845, pp. 136–138, 2017.
94. M. -L. Avenel, D. Farcage, M. Ruat, L. Verger, E. Gros d'Aillon, "Development and characterization of a 3D CdTe:Cl semiconductor detector for medical imaging," Nuclear Instruments and Methods A, vol. 671, pp. 144–149, 2012.
95. E. Gros d'Aillon, M. -L. Avenel, D. Farcage, L. Verger, "Development and characterization of a 3D GaAs X-ray detector for medical imaging," Nuclear Instruments and Methods A, vol. 727, pp. 126–130, 2013.
96. M. Christophersen, B. Phlips, "Laser-micromachining for 3D silicon detectors," 2010 IEEE Nuclear Science Symposium, Knoxville (USA), October 30–November 6, 2010, Conference Record, paper N15-2.

Silicon Detectors: A Partial History

i. PARALLEL PATHS TO THE FIRST HIGH-RESOLUTION SILICON VERTEX DETECTOR

The work that led to development of the technology used in the world's first high-channel density vertex detector for tracking particles started with the development of the high-resolution silicon microstrip sensor and the Microplex chip, a 128-channel integrated circuit amplifier [1]. But many parallel paths led to that, in both Europe and the United States. The originators of many of the projects were not aware of the others and did not, at that time, have the goal of making high-resolution tracking detectors.

The first path started during World War II, in Nazi-occupied Netherlands by a graduate student—Pieter Jacobus Van Heerden, who made the first solid-state particle sensor—and was published not long after the end of the war [2]. The low-resolution detectors made on both sides of the Atlantic for nuclear physics purposes are another early example. One of the people working on that, Jack Walton, directly helped with the start of what turned out to be the Microplex project.

A later example was a project started by Luis Alvarez. Sometime before 1969, Luis wrote a far-sighted report on the coming need for high spatial resolution detectors in future high-energy experiments, due to the decrease in magnetic bending and the increase in the number of particles coming from interactions. He believed both of these requirements could be met by the use of liquid or solid materials for particle sensors since they have a higher density of ionization, shorter ion diffusion distances, and shorter delta ray ranges than the gas-based sensors then in use. (Scintillators, though usually made of solid materials, used light generated by the track rather than ionization and had to be thick to generate enough light. They were also wide, given the size of the light-measuring device, the photomultiplier, and so had even less resolution than gas detectors.) Luis and his group then went on to make what they thought would be an appropriate detector using liquid argon proportional chambers [3]. (At that time, the National Accelerator Laboratory [NAL] had the highest energy machine in the world. NAL was later renamed Fermilab.) Luis was certainly correct in believing that the detection elements would have a higher density than that of gas. But liquid argon must not allow radiation absorption from the avalanche region, at a place where it could restart a second pulse. We suspected Luis's group would have a difficult time. They did. Eventually the project was abandoned.

ii. SLAC, SHATTERED SILICON

1980–1982

In 1981, a series of meetings were held to plan for the coming Stanford Linear Collider (SLC) in which a narrow beam of electrons from the Linac at the Stanford Linear Accelerator Center (SLAC) was to be hit head-on, with a narrow positron beam, also from the Linac. The total energy was to be at and on either side of the mass of the Z^0 resonance at about 91 GeV (according to the electro-weak theory the Z^0, along with the $W^{+/-}$, are the particles that carry the weak interactions). Charge-coupled devices (CCDs) were considered as a possible front-end to be attached to a microstrip detector, but there was no way to get the charge on the microstrip to move to the much smaller capacitance of the CCD.

I had already been making trips to SLAC since I, with others, had searched for free quarks at the SLAC Positron-Electron Project (PEP) Collider. That experiment also had a thin front end to detect possible quarks with an interaction cross section of many times geometric [4]. We found no free quarks in either part of that experiment. However, there was one way we could have missed finding the quarks. With only one-third or two-thirds of the charge of the normal particles, they would have less than half the number of ion pairs in their tracks. Often, the quark pulse heights would be within the range of normal particle pulse heights, due to Landau fluctuations in the quark tracks. If the quarks traveled too close to that normal particle, they would not be discovered as separate tracks because of the low double-track resolution of all earlier detectors—other than emulsion, which was not used in the most sensitive quark searches.

The recently developed silicon detectors could possibly have far higher resolution and solve the problems mentioned by Luis. But first I wanted to see if I was going to be working with a difficult material.

Jack Walton designed and fabricated silicon detectors for X-rays, synchrotron radiation, and gamma-ray detection for the nuclear chemistry department at Lawrence Berkeley Lab (LBL). I got a sample of silicon a few centimeters long from him, placed it hanging over the edge of my desk, and weighed it down with a stapler. I then started hanging weights from a rubber band on the other end of the silicon, until at more than a 20-degree bend, I lost courage, gently took the weights off, and carefully placed the silicon in my desk drawer. Then I calculated its resistance to bending and, to my surprise, found that it was the same as stainless steel! I started to learn everything I could about silicon detectors, photocopying many articles. Several weeks later, I looked in that drawer and saw hundreds of pieces of shattered silicon, but by then it was too late. I had already done too much work, and so I accidentally avoided the first place where the entire project might have been abandoned.

Jack gave me the necessary fabrication steps for making silicon detectors. I started following him around his lab in building 70A. Every time I did, I found he was doing another step he had not mentioned. Eventually, I listened to Jack dictating all of the steps via telephone. The process consisted of 98 steps, far more than the number he had originally given me and far more than any article I had found on fabrication.

I later found that oxide passivated p-n junction detectors had been in use for over 17 years [5]. But their article did not describe in such detail how they were made. Later, Kemmer in his 1980 paper [6] described the process in two paragraphs, and in his 1984 paper [7] he listed 7 key fabrication steps. All had less detail than in the 98 steps.

iii. SILICON DETECTORS BEFORE CUSTOM VLSI READOUT CHIPS

1982

High-resolution silicon sensors were being developed for use in fixed-target beam lines at the European Organization for Nuclear Research (CERN) and Fermilab. The ACCMOR collaboration at NA32 in CERN was one such experiment. Had they read out their 2 cm × 3 cm silicon sensor at our initial 25 micron pitch, they would have required a 6-foot diameter fan outboard, even before the start of their non-VLSI electronics.

The photo shown in Figure A.1 is part of a similar experiment at Fermilab. Paul Karchin is shown in front of a silicon microstrip detector. If only we had been able to finish the Microplex work in time for Paul!

The beams entering the planned SLC were narrow. The crossing point was tiny. The stay-away distances were smaller than those of any other experiment. One could not imagine cramming all those fan outboards and discrete-component electronics around that tiny region.

FIGURE A.1 Part of the University of Santa Barbara's instrumentation for their Fermilab experiment on charmed particles (1985).

iv. MICROPLEX CHIP: THE FIRST VLSI READOUT

1982, Mid-July

One day, David Leith, a SLAC faculty member, suggested that I attend an evening talk at his home to be given by Bernard Hyams, who had built the first silicon microstrip detector. Bernard had come to Stanford to meet with Terry Walker, a Stanford electrical engineer, to see if Terry could design and have Stanford fabricate a custom very-large-scale integrated circuit (VLSI) readout chip.

1982, July 20

Bernard gave a beautiful talk in which he mentioned one unsolved problem: there was no way to connect the planned 128 dense microstrip channels with the equally dense VLSI inputs. By the end of his talk, when questions were being taken, I had figured it out and told him I knew how to make the connections. We discussed it over David's kitchen table and decided to work together. This is shown in Figure A.2.

1982, July 29

The Hawaii High Energy Physics Group leader was so enthusiastic that he flew to California to attend some of the initial meetings with Terry. The first two were in Terry's office in the Applied Electronics Lab. Terry decided it could possibly be built, even though their minimum element width was 5 microns and the circuit overall width had to be no more than 50 microns, with alternate strips going to readout amplifiers, one at each end. Terry had a design that used only seven elements, for a width of 35 microns. After the second meeting, we all went to Jacques Beaudouin's office. As the chief engineer of the integrated circuits lab (and, while a Berkeley student, he was a former employee of Shockley's original transistor company), he would make the final decision. My impression of his attitude was quite clear. I could see an imaginary dotted clock ticking above his head. Terry had, I think, 15 minutes to make our case. Just as the 15 minutes were up, I noticed several pictures of trains on

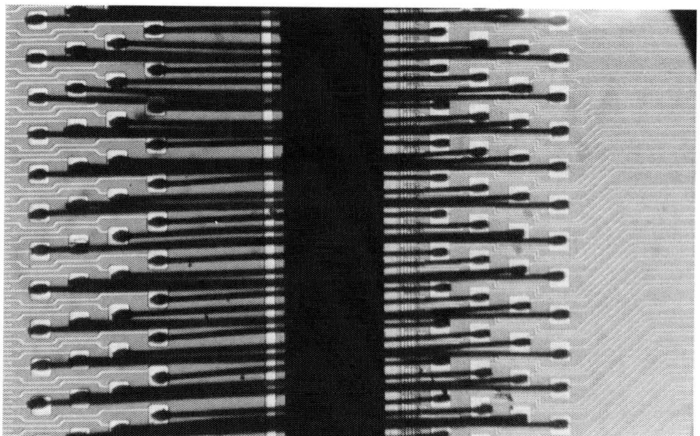

FIGURE A.2 The four-level interconnect I suggested to Bernard.

the wall, and as an avid, lifelong railroad fan, I commented on them. The clock evaporated. We got into a conversation about trains while everyone else waited impatiently. But he did decide "yes." After Bernard returned to CERN, my job would be to meet Terry every week to follow his work.

v. NURSERY SCHOOL / SPICE

1982, August–October

All went well for a few weeks, but then there was no apparent progress. And the reason was clear: A few minutes after each meeting started, there would be a knock on the door, a worried person would tell Terry of some machine or computer that wasn't working and Terry would go out and fix it. So to make him feel guilty for not working on our project as he had promised, I volunteered to help, knowing that he knew that I knew almost nothing about integrated circuit design, simulation, and testing, and would only volunteer out of desperation, with the real hope that he would get back to work. That plan blew up when he accepted, and I found myself in Integrated Circuit Nursery School, making daily trips to the Stanford campus to work in the Applied Electronics Lab.

Terry could not calculate how any new design would behave. There was, however, a recently developed program, the Simulation Program with Integrated Circuit Emphasis (SPICE) [8]. Circuits were described in SPICE by equations in which a matrix of admittances multiplied a vector of voltages and set equal to a vector of currents. This provided a set of equations to be solved. Internal circuit junctions were numbered. Components were represented by their admittance values. For instance, a resistor, R_{ij}, in parallel with a capacitor, C_{ij}, linking nodes i and j would be written as $(1/R_{ij}) + j\ \omega\ C_{ij}$. Where necessary, in the case of nonlinear elements, such as transistors and diodes, a detailed model of their behavior was included. Diagonal elements M_{ii} of the matrix would be represented by the sum of admittances attached to junction i. Nondiagonal elements M_{ij} would be the negative of the admittances between junction i and junction j. In most circuits, most elements would be zero. In a simple DC analysis, capacitors would be treated as open circuits, and inductors, if any, would be treated as short circuits. The voltages in the voltage vector and currents in the current vector either would have specific values or would be unknown and found by solving the equations.

SPICE could then directly do a DC and an AC analysis for small signals. It also could, for transient input signals, calculate a possibly large signal transient output by calculating the nodal values, their rate of change, and a time step size. The AC analysis and the transient analysis were the main calculations I did. SPICE, then in its early stages, was very difficult to use. Anytime we changed the properties of a layer of the integrated circuit, we had to recalculate the input values, one by one typing the old, incorrect input, followed by the new changed input. So I finally wrote a Fortran program, which, with a few simple inputs, calculated as its output all the many SPICE input instructions.

vi. 128 OSCILLATORS

1983, July

Alan Litke of the University of California Santa Cruz Physics Department had worked with me on the SLAC quark search experiments and was interested in joining the Microplex integrated circuit development. I suggested he drop by the Applied Electronics Lab on the Stanford campus.

Now in first grade, and intensely curious, I had just completed the SPICE statements for a transient analysis. It was to look at the output when a pulse was given to the input, while the circuit was in reset. Most sensible electrical engineers would have left it to the very last, if they would even do it all. There should have been almost no output. I ran the transient output calculations while Alan was looking, and started the pen recorder output. (Pen recorder: that's how primitive our computation was!) It gave a little jog, then a slightly larger one, then still larger, and larger yet again. Finally, it was shooting from the minimum possible, staying there a moment and then flying back to the maximum possible, then back to the minimum, and so on. Obviously, something very wrong was going on. I took the pen recorder output to Terry. The SPICE code that was used would have been happy to balance a pencil on its sharp point forever. Under this state of false equilibrium, noise simulation capabilities that were available in SPICE were not used.

Terry had never designed a circuit where the input had such a tiny capacitance. The reset circuit had a transistor that, when turned on, connected the output to the input, thus shorting out any charge left by a preceding signal on the integrating capacitor. Alan and I were looking at the consequences of a signal being put on the input during reset. That signal, not being loaded down by the large-input capacitance expected by Terry, went right back through the amplifier to the output and started another round trip. In real life, a noise signal would always be present at the input of the circuit in reset, and that noise would have been amplified on the next round trip. Each subsequent round trip would have resulted in further amplification until the output was as large as possible.

We had inadvertently designed a circuit that would work beautifully until it was reset and then would turn into a 128-channel oscillator. Had we built it, it would have presented us with a complete disaster.

Alan and I both learned that what we were doing was not straightforward. Terry needed several months to redesign the circuit so that it would work in DC, AC, and transient mode with the small source impedance, without turning into that oscillator on reset. When that was done and optimized with SPICE runs, we had a finished design, which was quickly put into fabrication in the Stanford Integrated Circuit Lab. Each fabricated wafer had many test chips, each with 128 channels as well as many process monitoring devices.

vii. MUNICH

1983–1984

During the time the Microplex chip was being fabricated, the annual conference on semiconductor detectors was to be held in Munich, Germany. Bernard told me that the Munich group was going to present a talk on their chip, which was being designed for a similar readout task. He was going to try to get my talk placed right next to theirs. This was the first time I had even known about another effort.

Gerhard Lutz, of the Munich Max Planck Institute, described what they were doing. They were having the first section of the circuit fabricated, measured, and refabricated, if necessary, until its performance was satisfactory. Then they went on to the next section, while we just simulated the whole circuit in a quick SPICE computer run to look at the simulated output and we then changed the circuit or the integrated circuit fabrication parameters until the output was satisfactory. We had the entire 128-channel chip in fabrication before the Munich conference [9].

After the fabrication was finished and a few months after the conference, Terry had another surprise for me. I was now a second-grade student and was going to be the person checking the circuits and measuring their performance. Terry designed and assembled the driving electronics and added an output amplifier for the Microplex chip. I then used a probe station which held the fabricated wafer and a probe card that supplied all the operating and driving voltages to its many contacts. The probe card had to be aligned and gently set down, contacting the voltage supply and monitoring points of one of the circuits on the wafer.

During the first four months of 1984, a number of small problems were found. These were temporarily corrected with a minor supply voltage shift. They were ultimately corrected with some minor design changes. After several months of testing, we found that a reasonable number of chips were satisfactory.

viii. BERNARD'S LAB AT CERN/GOLD AND BETTER THAN GOLD

1984–1985

At the same time as the electrical work at Stanford was underway, Bernard had a large group of postdocs, machinists, Alan Litke, and a visitor from Poland, Michal Turala (who, along with Bernard, was already an ACCMOR collaborator) at CERN combining the sensors with 128 channel Microplex chips. Here, two minor problems were found involving signal pickup from adjacent parts of the circuitry.

The connections were made with gold ball-bonds: the end of a heated gold wire was melted, and surface tension forced it into a ball, which was then smashed into the circuit board. A similar set of 128 wires was made at the other end of the sensor. Each of the four levels was separated from all of the other levels by a 0.2 millimeter quartz rod that was glued in place. The full 256 wire connections at the ends of the microstrip sensor, shown in Figure A.3, took one to two days, a real pain.

I hated the gold ball-bonding machines because I kept burning myself. And so when Leo Hubbeling, who did the gold ball bonding, wondered out loud whether an aluminum wedge bonding machine might be better, I jumped at the chance to get rid of hot gold. Gold was considered superior because you can take off in any direction from the smashed gold ball. I called my cousin who worked for Kulicke and Soffa in Pennsylvania. They manufactured aluminum wedge bonders in which a room temperature aluminum wire was smashed into a metal pad on the circuit board, but then

FIGURE A.3 On the left part of this picture is a view of a 128-wire connection. The top quartz rod is visible, and the Microplex chip is seen on the right.

the wire had to lead off in the direction of the wedge. They could rotate the wire-handling part, so the wedge could be in any desired direction, including the offset needed because the microstrip sensor and the Microplex chip had slightly different channel-to-channel spaces. My cousin had told me just what I had hoped to hear—they had just come out with an automated aluminum wedge bonder, which. before the smashing, could rotate in any direction.

He gave us a list of their nearby customers in Switzerland, where we discovered Switzerland's rough equivalent of Silicon Valley: the Swiss watch industry, which now used integrated circuits and connected them with Kulicke and Soffa aluminum wedge bonds. No more tiny, mechanical jewel movements. Roland Horisberger, a postdoc in Bernard's group, was from the German-speaking part of Switzerland and preferred speaking German, not the French of Geneva. He went to a watch company with a chip that already had all of the quartz rods glued in place, with the idea of seeing how they could bond the most difficult layer—the top one that had to go over all of the rods. I will never understand how a German-speaking-native in the German-speaking Swiss watch valley could be so misunderstood. When we got the bonds back, we found they had made a four-level bonding, with the lowest level floating above all the rods and with each level still comfortably separated from the other three. I gently poked a wire with the eraser of a pencil and was surprised to see it vibrating but not touching any other wire as the vibrations died down. The watch company was happy to make our initial bonds with their automated Kulicke and Soffa wedge bonder. The 256 wire bonds could now be made in less than five minutes. No more glued rods, no more burnt fingers, but most importantly we had a process that could rapidly connect many detectors.

Testing a Microplex chip from the first fabrication run completed just after the Munich conference, I found a noise level that would predict a marginally adequate signal-to-noise ratio (SNR). The people in Bernard's group, and I as well, would have liked a higher ratio. One Saturday morning in Bernard's lab, I realized we must be looking at reset noise: The noise from the reset transistor had been saved as a charge on the integrating capacitor. Clearly, it would benefit by having a circuit that could subtract out that noise.

I was just desperate enough, with no integrated circuit design apparatus available, but with a nearby Xerox machine, a scissors, and scotch tape, to design and lay out extra circuitry that would cancel out that source of noise, and to give up my treasured Saturday afternoon in downtown Geneva. Such circuitry was called "double correlated sample and hold" and had been included in the Munich chip circuit from the beginning. The circuit I designed had, in addition to the integrating capacitor, two storage capacitors. One stored the output voltage from reset noise, before the pulse came in. The other stored the sum of the reset noise and the signal from the pulse. The circuit was different from the Munich circuit, which had a single coupling capacitor linking its input and output sections with the negative of the reset noise placed on the coupling capacitor, therefore, subtracting the reset noise from the output after the pulse, which contained both the signal and the reset noise. My scheme provided an additional subtraction of possible pickup noise between the Microplex outputs and a differential input in the following circuit, some distance away.

Back in the United States again, I found out that Terry did not include a double correlated sample and hold because he could not fit it in the available space. After Terry made more improvements in my design, I redesigned the microstrip sensor to have lower capacitance, and we had the new Microplex chip fabricated at Stanford, incorporating the double correlated sample and hold, along with a few minor corrections. A new fabrication run at Stanford for the Microplex2 chip was completed.

A detector was assembled at CERN using the new Microplex2, still using gold ball bonds. At the time, no beams were available at CERN or SLAC, so Bernard set up a beam, using a ^{106}Ru beta source. The performance was adequate with an SNR of 12:1. When it went through a real beam test at SLAC, it showed an SNR of 17:1, 99.99+ −0.01% efficiency, and a double-track resolution smaller than 4 microns, with full accuracy for each track of a pair when their separation reached 100–150 microns [10]

ix. THE MARK II SILICON VERTEX DETECTOR AT THE SLC

1983–1992

We were now ready to propose a silicon detector to all of the groups competing for inclusion in the planned SLAC linear collider. All but two responded, asking us to give them a paragraph and they would include it in their proposal. If they were chosen, we would be able to make the vertex detector.

The Time Projection Chamber group did not respond, but Martin Perl of the SLAC-Lawrence Berkeley Labs' Mark II group said that they were not sure about the silicon detector, but they would like us to join their group anyway. We did. In Section 3.6 of "The Mark II at SLC" (pages 111–115) [11], three possible technologies are described; high-pressure drift proportional wire chambers, silicon microstrip detectors, and CCD arrays. Hoped-for resolutions were given, and it was noted that with an expected installation date of 1998, adequate development time would be available. No preference was indicated. The Mark II proposal was accepted.

On an earlier run of the Mark II detector, and even before, when we were at LBL making an efficient gamma detector using a wall of gamma-sensitive 318 lead glass Cherenkov, Martin and I had gotten along very well. We both liked the same things. Whoever got there first would clean up any mess left by the preceding shift. I would connect monitoring cables to my favorite oscilloscope, so I could monitor the operation of the detector.

On the SLC experiment there was one major problem: Martin's postdoc was coming up for tenure at SLAC. The postdoc had built what was then the world's most accurate vertex detector, a high-pressure multiwire proportional chamber—which would be made completely obsolete by our silicon detector. They both worried that if our detector did not work, the material in it would degrade the resolution of his detector through multiple Coulomb scattering. A committee was appointed and decided that a silicon vertex detector would be used. After that, we again all got along very well.

As this was going to be the first high-precision collider vertex detector ever made, we had a brand-new problem. The detector had to be made of thin, lightweight material in order to minimize Coulomb scattering, but it also had to be held accurately in position relative to the beam pipe and its colliding beams inside. So x,y,z and the two space angles of its pointing direction, θ and φ, all had to be constant despite possible temperature changes. The requirement of the support system was that it be highly accurate, but not overconstraining. Specifying more than one value of either x, y, z, θ, or φ might introduce buckling from unequal thermal expansion. The beams traveling in the high vacuum of the beam pipe introduced mirror-image charges on the inside of the beam pipe, traveling with the beam. Any place where there was a break in the continuous electrically conductive beam pipe gave the image charge the opportunity to escape to the outside, and from there induce similar currents on any outside conductor. Such electric fields caused by all this outside charge motion could be easily picked up and seen using a simple loop antenna connected to an oscilloscope—no amplifier needed. The silicon sensors and their amplifiers could be protected from the fields by keeping them inside a lightweight metal Faraday cage with insulating supports between it and the beam pipe.

I looked for a mechanical engineer expert in such systems, and from the staff of SLAC, LBL, and its sister lab in Livermore, I found one such person in Livermore, who declined to take the job. At this point, a SLAC collaboration member, Vera Luth, working with a SLAC designer, turned herself into such an engineer and did a beautiful job.

There was a delay of some years while the linear collider was being constructed and made to work properly. After it finally worked, the Mark II detector was moved in and data was taken. Then in December 1989, the two vertex detectors—silicon inside, high-pressure drift chamber outside—were moved inside the Mark II and just outside of the interaction point and its stay-clear region. With beam on, in 1990, the occupancy (the fraction of channels showing hits) was 10 to 15% for the drift chambers and about 1% for the silicon. When a sudden brief increase in intensity would jam up the gas detector, the silicon detector would display the many individual tracks. An additional difficulty with high-pressure drift chambers was that they could not be adapted to form a pixel detector.

For completeness, we should point out that the SLC came on late and did not generate many Z^0's. It did, however, prove that there were only three families of light neutrinos, which was a very important result. The SLC was rendered obsolete when the Large Electron-Positron Collider (LEP) came on at CERN.

The final result was that, in the early 1990s, we had successfully made the world's first VLSI microstrip readout—more than three years ahead of all others, in Germany, England, and at LBL—solved the difficult hybridization problems, and made and installed the first silicon vertex detector at a collider [12–17]. That technology is now standard for all main collider detectors worldwide [18].

It is Bernard's belief that, if left alone, the project would have died at the start, as Terry put out one fire after another. Had that happened, it would have delayed the advance of silicon detectors for high-energy physics by more than our lead of three plus years, as other groups could see from reading our intermediate papers that we were making progress, which I suspect, sped them up.

Beyond that, it is clear that the frequent trips to SLAC, Stanford, and companies in the rapidly developing Silicon Valley were vital. We saw what was being done, attended talks, purchased material and services from the companies, and worked in the semiconductor laboratories on the Stanford campus.

x. PIXEL DETECTORS: MONOLITHIC, BUMP BONDING

1989–1995

In 1989, I proposed a method for fabricating a monolithic pixel detector with a sensor and an amplifier on the same piece of silicon—the readout was made by column selectors and row selectors on the edges of the array. The design was greatly improved by Walter Snoeys, a Stanford electrical engineering graduate student. Walter fabricated it as part of his thesis work, and then Walter, Chris Kenney, and I tested it in a 300–600 GeV muon beam at Fermilab. It had a very high SNR, which gave it the most accurate readout ever achieved with silicon. (For many years its accuracy was listed in the Particle Data Book as being characteristic of the accuracy that could be achieved with such detectors.)

During this same period, a parallel path using the Hughes Aircraft bump-bonding process was pursued by physicists at SLAC and the University of California Berkeley's Space Sciences Lab. The monolithic approach, which did not need bump bonding, sent the charge directly to a transistor, with the substrate below forming a concentrating DC electric field. It also acted as a Faraday cage shielding the gate. Both technologies used a pixel organization; the monolithic detectors had the higher SNR of 80:1 and collected 99.995% of the charge from perpendicular beam traversals while detecting 5889 traversals with zero misses.

The spatial resolution of two microns was the best ever achieved [19] Nevertheless, in 1996 we ended up choosing the bump-bonding method for three reasons:

1. We were entering the era of billion dollar fabs: industrial fabrication facilities on which billions of dollars were spent for advanced machinery. Up until then, the size ratio of the smallest transistor between industry and university labs was approximately constant, with the advantages of monolithic devices more than making up for that fixed size ratio. The smaller size from the billion dollar fabs allowed for more transistors, each using lower power and producing higher speeds. The increasingly smaller size available from the billion dollar fabs was clearly going to open up a gap that would not be closed by the advantages of monolithic technology available at university labs. With the university labs no longer able to keep up, we decided to switch to the superior industry fabricated readout chips.
2. The chips provided by industry were limited in overall size, but many chips could be mounted on a single supporting structure, with one bump connecting each chip to a sensor pixel, therefore allowing the fabrication of detectors that were much larger than the size of a single amplifier chip.
3. Charge put on the gate electrode properly changes the behavior of a transistor. A problem with high-intensity colliders was severe radiation damage, which resulted in significant amounts of charge being generated directly in the insulating gate oxide, which causes an improper change in transistor operations. The thinner gate oxides of the smaller industrial transistors increased their radiation hardness because less charge is generated in the gate oxide. The induced charge layer, being closer to the gate, produces a lower voltage shift, and the lower mechanical strength of the thinner oxide layer results in less motion relative to the substrate, and so less damage, coming down from high-temperature steps. Furthermore, charge can actually escape from the thinnest gate oxide. Radiation hardening processes that did not require thinner gate oxide had been developed for military applications. However, in order to do so, sacrifices had to be made in the other device characteristics, and the amount of radiation hardness was less than that of the new commercial thin gate oxide, which required no sacrifices.

The high intensity of the colliders also produced the need for higher speeds, and the higher speeds of the industrial transistors helped here. An equally serious problem was damage to the sensor substrate, which limited the collection distance in the sensor of ionization charges. This problem would be greatly reduced by going to a new sensor electrode geometry: 3D, which will be described later.

xi. DEEP REACTIVE ION ETCHERS/HARDBALL

Although I did not realize it at the time, Stanford again played a key role in our next development, 3D-electrode silicon detectors. They installed new equipment loaned to them by Surface Technology Systems (STS) that could etch deep, narrow holes and trenches in silicon: a deep reactive ion etcher that used two plasmas: (1) an SF_6 plasma for etching, and (2) a C_4F_8 plasma, to protect the sidewalls and for a much shorter time, the bottom. Etchers without sidewall protection would have aspect ratios of much smaller values.

In the deep reactive ion etcher, the layer on the bottom was attacked by etchant molecules heading straight down, whereas the sidewalls were only struck by grazing incident etchant molecules on their way down and by bounce-back etchant molecules that did not react on the bottom. Those etchant molecules hit the sidewalls at possibly large angles, but at lower velocity, so the C_4F_8 protective layer on the sidewalls lasted longer than the layer on the bottom.

After a number of years, competing companies developed etchers with better aspect ratios of hole depth to hole width. When Stanford was ready to buy one, STS told Stanford that if they didn't buy their newest etcher and instead bought one of the competitors' etchers, they would pull out their loan. If that were to happen, they would end up with one etcher, instead of two, as was originally intended. A competitor told them if STS did that, they would sell them two machines for the price of one. Faced with the STS threat, the Stanford employee in charge of the decision decided to get a second machine from STS, keeping the original loaned etcher. I never asked him exactly why he made that choice, but the first machine had been tested, had many process steps worked out, and was immediately available for ongoing work—whereas the new machine would have to be characterized first.

1993

On a trip on August 23, 1993, I spoke at the Department of Energy (DOE) review about the next key development that would be affecting tracking detectors: pixels in which the formerly long strips would be broken up into short pixels, each with its own front-end readout electronics. It was clear from the pioneering work of Chris Damerell using a CCD as a pixel detector, that pixels would be the best way to handle dense track arrays. However, the slow readout of a CCD meant that we would need to have a fast silicon detector readout by a dedicated circuit. Working with Walter and Chris, we had already made a monolithic pixel detector.

xii. QUANTAM MAMMOGRAPHY

From the fall of 1993 to 1995, I was concentrating on monolithic detectors and on their use for high-energy physics and for mammography. In 1993, a member of our group told us that her mother's metastatic breast cancer had been missed by mammography. This turned out not to be an infrequent occurrence, and after some thought it seemed possible that the monolithic pixel detector could be used as an X-ray detector for mammography.

We formed a collaboration with the Stanford University Medical center, where we ran tests with the prototype pixel detector—which detected calcification in tissue samples with much greater accuracy than in clinical practice. We had received spin-off R&D funding for this from the Texas National Research Laboratory Commission (TNRLC) and had started working on a detector of X-rays from a mammography setup at the Stanford hospital.

October 21, 1993. Congress Canceled the Texas Superconducting Super Collider
After the cancellation of the Superconducting Super Collider, our funds from TNRLC continued for one more year, and then we applied for funding with the National Institutes of Health (NIH). Calculating the properties of a 10-layer detector from the measured properties of a single-layer detector, quantum efficiency of greater than 90% would be seen with higher spatial resolution than the less efficient film system then in use. The evaluation of the NIH study group (their term for a *proposal review panel*) showed a lack of understanding of the capabilities of high-energy physics instrumentation. For example, the NIH quoted an instrument speed we had exceeded decades before as being *very* fast, and so likely to be difficult to reach. We were not given funding.

Several years later, through a different route, I found myself serving as a member of the same NIH study group. There was another physicist on the panel, Ed Westbrook, who also had his own medical device company. It was here that I realized that though the members of the study group were intelligent, dedicated, and hardworking, they did not have firm knowledge of the capabilities of present detector instrumentation. They also lacked a similar understanding of electrical engineering and integrated circuits capabilities. For example, they all voted (over my objection) to fund a beautifully written but fatally flawed proposal that required the fabrication of an integrated circuit with more than 100 times the area of the then largest, available integrated circuit.

During this period, Ed and I were able to persuade Chris to rejoin our group, and later I was able to get Chris hired by SLAC. He is now the leader of a SLAC group that specializes in advanced detector design and fabrication.

xiii. 3D SILICON DETECTORS

1995
I had gone to the 1992 International Conference on Advanced Technology and Particle Physics at Villa Dell'Olmo, in Como, Italy. Cinzia Da Vià was right behind me in line, and seeing me write "University of Hawaii" on my registration form became interested. We met and eventually had dinner together. Then on my way back to the United States, I stopped at CERN and met her again.

On October 25 during the 1995 IEEE Nuclear Science Symposium and Medical Imaging Conference in California, David Nygren, the inventor of the time projection chamber, and I were having lunch when I saw someone walking toward me and smiling. It does not happen that a very attractive woman I do not know walks towards me, smiling. It turned out to be Cinzia, who had recognized me from the meetings in 1992. We invited her to join us for lunch. It was at the start of her planned work at Lawrence Berkeley Labs for her PhD thesis, which was to be on the use of gallium arsenide as a particle sensor. At that time, gallium arsenide transistors were faster than silicon transistors, and so her professor thought that gallium arsenide sensors would also be faster. During that time, we often met for dinner, and she would tell me about her work, which was to find out why the electrons released by ionization did not reach the surface where they could be collected by an electrode. This was not too surprising to me, as the distance they would have to travel was far larger than the dimensions of a typical transistor. The long collection distances we had were in float-zone silicon, where the impurity level was about 10^{12} per cubic centimeter, about one part impurity in 10^{10} silicon atoms. In gallium arsenide, at its operating temperature, if there was an excess of even 1 per million, then the impurity concentration in gallium arsenide would be ten thousand times higher than that in silicon. Cinzia's missing electrons could easily have been captured long before they reached the surface.

On November 6, waiting out the rush-hour, having dinner alone at Stickney's restaurant in Palo Alto—Cinzia was 50 miles away—a vision of the outline of two cylinders made with dashed lines came to my mind, and I realized that if we used the Stanford etching machine to make cylindrical holes in the gallium arsenide, and if we could somehow deposit material in them to make a back-biased diode, then the electrons from ionization would not need to go so far. It was two full days before I realized they could also improve the speed of silicon detectors and the collection fraction of radiation-damaged silicon detectors.

I found, from a book on polycrystalline silicon by T. Kamins, that in the correct temperature region, silane (SiH_4) would indeed make a conformal layer and fill the entire electrode volume [20]. I confirmed this with a telephone call to Kamins, who worked nearby at Hewlett Packard.

In that temperature range, the sticking probability of silane would be low, and so the typical molecule would enter, bounce around several times, and then exit the electrode volume. When it did stick, the four hydrogens were lost and the silicon atom became part of a small crystal. This sticking location was likely to be as deep in the hole as it was shallow. On November 11, I started calculating a number of properties for a sensor using silicon. With 3D, collection electrodes were no longer limited to the sensor surface but penetrated through the sensor substrate. The electrodes were made by etching narrow holes or trenches and filling them with dopant material; then a second set of holes or trenches were etched and filled with the opposite type of dopant material. Those short collection distances increased the sensor speeds. A sensor would be faster because the ion in an ion column would not have to travel the length of the column; the ion column simply needed to translate horizontally to the nearest electrode, which would be placed with separation distances less than the thickness of the silicon sensor. For the same reason, they would be more radiation hard with collection distances shorter than capture distances on radiation-induced defects.

Julie Segal, a Stanford electrical engineering student working with us, became so interested in it that she did accurate calculations and gave a seminar to her fellow graduate students. I held off on telling Cinzia to be sure it would work. On her last day, before she returned to Europe, I took her to see the fabrication lab at Stanford—by now in a new building—and then told her about 3D when we had lunch at the Sundeck, on the way to the San Francisco airport. It made a greater impression on her than I realized at the time.

1996–2001

I wrote a paper describing 3D and sent it to friends and colleagues, as well as the *IEEE Transactions on Nuclear Science.* I did not hear anything from the journal, but I was not concerned since copies had been sent to a large number of people. Finally, Erik Heijne asked me what was happening with my paper, and I said I had not heard back. He said he was organizing a conference in Bari, Italy, and offered for me to come and give a talk on 3D, promising that my article would be published rapidly in *Nuclear Instruments and Methods* with the other presentations from the conference. So I did. At the conference (24–27, 1996 *3rd International Workshop on Semiconductor Pixel Detectors for Particles and X-Rays*) another physicist told me he had been thinking about 3D, and here I was giving a talk with all the details worked out. The paper was published soon thereafter [21]. Somewhat later, a person who had been very helpful to my work apologized to me, saying he had been sent a copy of my paper for evaluation by the IEEE journal. He had put it in a pile on his desk and forgot about it until he saw the published article in *Nuclear Instruments and Methods.*

In the first fabrication run for 3D sensors, Chris Kenney laid out and fabricated sensors with internal trench electrodes in which, rather than a cylinder, the trenches were etched, the faces were doped, and the trenches were filled with polycrystalline silicon. Later, we made active-edge sensors in which the dope trenches formed the edges after a dicing edge was made through the middle of the trenches. They were fully efficient, right up to their physical edge, with no dead area.

In the following years, we published several papers describing both the interior trench and active-edge sensors [22, 23].

Preparing a pixel detector for tracking at the coming Large Hadron Collider at CERN was clearly one of the most important tasks. With the advantages of speed and radiation hardness, 3D sensors and developing fast pixel readouts would be of major importance. LBL had a competent group of electrical engineers and physicists working on readout chips, so we concentrated on 3D sensors. Cinzia finished her PhD in Glasgow and received an appointment to the faculty of Brunel University, just west of London.

On March 8, 2001, Cinzia sent her first graduate student, Jasmine Hasi, with salary support, to work with us at Stanford for her PhD on 3D detectors. In addition, a new DOE program, the Advanced Detector Research Program under Michael Procario, supporting generic developments on detectors, was started, and I received one of the first grants on April 19, 2001.

On my second trip to Japan to attend the 5th International Symposium on the Development and Application of Semiconductor Tracking Detectors in Hiroshima, Hitoshi asked me to visit Sendai to present a talk on 3D sensors at Tohoku University. A side observation typical of that trip: While walking the short distance from my hotel to the Sendai train station, I found it impossible, as I did in most Japanese cities, to walk more than a few steps without someone coming up and offering to carry my suitcase.

2002

Angela Kok, who grew up in Hong Kong and was sent by her parents to England because they thought she could get a better education there, became Cinzia's second graduate student at Brunel [24]. She came to California but did most of her graduate work at CERN, and so the work I did with her was done at CERN. One interesting example was what we saw while observing the speed of pulses from two different radioactive sources. One source was the high-energy electrons produced by ^{90}Sr beta source. The other one was the ^{133}Ba source, which captured a K electron producing a ^{133}Cs with a missing K-shell electron. When another electron fell into the K-shell vacancy, a 35 keV X-ray was produced. I would have expected the X-ray pulse to be shorter because of the short range of the electrons kicked out in the silicon center by the 35 keV X-ray. The secondary ionization electrons and holes came from a very small region in the silicon, due to the short range of the primary. The high-energy electrons produced by the ^{90}Sr source had long ranges, and its secondary ionization electrons and holes had greatly different collection distances. Instead of what I had expected, we were seeing the opposite: long X-ray pulses and shorter ^{90}Sr pulses, [25]. Both pulses were slowed down due to the attraction of the secondary ionization electrons and holes as the charge clusters were pulled apart by the drift field. However, that effect was larger for the compact clusters from the X-ray and dominated the effect from the drift distance spreads.

xiv. 3D VERSUS PLANAR: THE SELECTION OF TECHNOLOGY FOR THE ATLAS INSERTABLE B LAYER

While visiting CERN for work on the Microplex chip, I would also, on occasion, drop over to see Claus Goessling, who was working on a non–high-precision silicon vertex detector at the Intersecting Storage Rings (ISR) at CERN. Years later, when we were working on the proposal for a 3D pixel sensor for the new insert able B-layer, a group where Claus was working was proposing to use planar sensors—such as the planar sensors we used years before. With the planar elements short enough to make pixels, we would both connect our sensors to readout circuitry using bump bonds. Claus's main point was that their planar sensors would require extremely large bias voltage, and he thought we would not be able to make enough working 3D sensors to cover the required area. Their desired bias voltage at the maximum of radiation damage—1500 volts—was too large to be accepted to ATLAS, so he settled for 1000 volts. Our 3D sensor for comparable level of radiation damage would need only a few hundred volts.

The choice between the two approaches was sent to an ATLAS committee. The committee recommended that ATLAS use both planar and 3D on a 75%—25% area basis. Planar would be used in the central region, where less radiation damage was expected, and 3D on the outer 25%, where it was expected to be far more severe. At a private meeting of the two groups, Claus told other members of our group that we would have to accept 75% planar, 25% 3D, or else they "would vote us down to zero." This was a realistic threat, as each group got one vote and there were more German groups, so if they voted as a block, they probably could have voted us down. So we accepted. After they came out, in the open meeting, Claus, at the front of the room speaking to a German colleague further back in the room said, in English, "Order for 100%"—in other words, they did not believe the 25% 3Ds would have been ready in time, being the very first production. This is not what happened: 3Ds were installed, on schedule, together with planar, in the IBL.

REFERENCES

1. C. Adolphsen, R. Jacobsen, V. Luth, G. Gratta, L. Labarga, A. Litke, A. Schwarz, M. Turala, C. Zaccardelli, A. Breakstone, C. J. Kenney, S. I. Parker, B. Barnett, P. Dauncey, D. Drewer, and J. Matthews, "The Mark II silicon strip vertex detector," Nuclear Instruments and Methods A, vol. 313, pp. 63–102, 1992.
2. Pieter Jacobus Van Heerden, The crystalcounter: A new instrument in nuclear physics, University Math Naturwiss, Fak (1945).
3. S. E. Derenzo, R. A. Muller, R. G. Smits, and L. W. Alvarez, "The prospect of high spatial resolution for counter experiments at NAL: A new particle detector using electron multiplication in liquid argon," Lawrence Radiation Laboratory (1969).
4. S. Parker, F. Harris, I. Karliner, D. Yount, R. Ely, R. Hamilton, T. Pun, W. Guryn, D. Miller, and R. Fries, "The PEP Quark Search Proportional Chambers," Physica Scripta, vol. 23, pp. 658–661, 1981.
5. T. C. Madden and W. M. Gibson, "Silicon dioxide passivation of p-n junction particle detectors," Review of Scientific Instruments, vol. 34, pp. 50–55, 1963.
6. J. Kemmer, "Fabrication of low noise silicon radiation detectors by the planar process," Nuclear Instruments and Methods, vol. 169, pp. 499–502, 1980.
7. J. Kemmer, "Improvement of detector fabrication by the planar process," Nuclear Instruments and Methods, vol. 226, pp. 89–93, 1984.
8. Laurence Nagel, 1975. SPICE2: A Computer Program to Simulate Semiconductor Circuits. University of California, Berkeley, Electronics Research Laboratory, Memorandum No. ERL-M520.
9. J. Walker, S. I. Parker, B. Hyams, S. Shapiro, "Development of high density readout for silicon strip detectors," Nuclear Instruments and Methods, vol. 226, pp. 200–203, 1984.
10. G. Anzivino, R. Horisberger, L. Hubbeling, B. Hyams, S. I. Parker, A. Breakstone, A. Litke, J. Walker, N. Bingefors, "First results from a silicon-strip detector with VLSI readout," Nuclear Instruments and Methods A, vol. 243, pp. 153–158, 1986.
11. G. Trilling, G. J. Feldman, J. M. Dorfan, W. B. Atwood, H. C. DeStaebler, R. Pitthan, L. S. Rochester, A. Boyarski, F. Bulos, R. J. Hollebeek, A. J. Lankford, Rudolf R. Larsen, V. Luth, D. L. Burke, G. Hanson, Walter R. Innes, J. Jaros, M. L. Perl, C. Hoard, D. Hutchinson, L. Paffrath, D. I. Porat, C. Akerlof, D. F. Nitz, G. S. Abrams, G. Gidal, G. Goldhaber, D. Herrup, J. A. Kadyk, K. Lee, M. Nakamura, B. C. Barish, G. Fox, T. Gottschalk, R. Stroynowski, D. E. Dorfan, C. A. Heusch, H. F. W. Sadrozinski, T. Schalk, A. Seiden, S. I. Parker, D. Yount, "Proposal for the Mark II at SLC," Caltech-LBL-U.C. Santa Cruz, University of Hawaii, University of Michigan, SLAC Collaboration (1983).
12. A. Litke, C. Adolphsen, A. Schwarz, M. Turala, A. Steiner, A. Breakstone, S. Parker, V. Luth, B. Barnett, P. Dauncey, D. Drewer, "A silicon strip vertex detector for the Mark II experiment at the SLAC linear collider," Nuclear Instruments and Methods A, vol. 265, pp. 93–98, 1988.
13. C. Adolphsen, G. Gratta, L. Labarga, A. Litke, A. Schwarz, M. Turala, C. Zaccardelli, A. Breakstone, S. Parker, B. Barnett, et al., "An alignment method for the Mark II silicon strip vertex detector using an X-ray beam," Nuclear Instruments and Methods A, vol. 288, pp. 257–264, 1990.

14. L. LaBarga, C. Adolphsen, G. Gratta, A. Litke, M. Turala, C. Zaccardelli, A. Breakstone, S. Parker, B. Barnett, P. Dauncey, D. Drewer, J. Matthews, R. Jacobsen, V. Luth, "The Mark II silicon strip vertex detector and performance of a silicon detector telescope in the Mark II detector at the SLC," IEEE Transactions on Nuclear Science, vol. 38, pp. 25–29, 1991.
15. C. Adolphsen, G. Gratta, A. Litke, A. Schwarz, M. Turala, A. Breakstone, S. Parker, B. Barnett, P. Dauncey, D. Drewer, R. Jacobsen, V. Luth, "Status of the Silicon Strip Vertex Detector for the Mark II Experiment at the SLC," IEEE Transactions on Nuclear Science, vol. 35, pp. 424–427, 1988.
16. P. Dauncey, B. A. Barnett, D. Drewer, J. A. J. Mathews, A. Breakstone, S. Parker, C. Adolphsen, G. Gratta, A. Litke, A. S. Schwarz, M. Turala, R. Jacobsen, V. Luth, "Radiation Damage Studies of a Custom-Designed VLSI Readout Chip," IEEE Transactions on Nuclear Science, vol. 35, pp. 166–170, 1988.
17. Mark II Author list (167 authors), "The Mark II detector for the SLC," Nuclear Instruments and Methods A, vol. 281, pp. 55–80, 1989.
18. C. Adolphsen, A. Litke, A. Schwarz, M. Turala, G. Anzivino, R. Horisberger, L. Hubbeling, B. Hyams, A. Breakstone, R. Cence, S. I. Parker, J. Walker, "Initial beam test results from a silicon-strip detector with VLSI readout," IEEE Transactions on Nuclear Science, vol. 33, pp. 57–59, 1986.
19. Walter Snoeys et al. First beam test results from a monolithic silicon pixel detector Nuclear Instruments and Methods in Physics Research A326 (1993), pp. 144–149.
20. T. Kamins, Polycrystalline silicon for integrated circuit applications, Boston, MA: Kluwer, 1988.
21. S. Parker, C. J. Kenney, J. Segal, "3D—A proposed new architecture for solid-state radiation detectors," Nuclear Instruments and Methods A, vol. 395, pp. 328–343, 1997.
22. C. J. Kenney, S. I. Parker, and E. Walckiers, "Results from 3D sensors with wall electrodes: near-cell-edge sensitivity measurements as a preview of active-edge sensors," IEEE Transactions on Nuclear Science, vol. 48, pp. 2405–2410, 2001.
23. C. Morse, C. Kenney, E. Westbrook, I. Naday, S. Parker, "3dx: Micromachined silicon crystallographic x-ray detector," Proceedings of SPIE, vol. 4784, pp. 365–374, 2002.
24. Angela Kok, "Signal formation and active edge studies of 3D silicon detector technology," Doctoral Dissertation, Brunel University, Middlessex, West London (2005).
25. S. Parker, A. Kok, C. Kenney, P. Jarron, J. Hasi, M. Despeisse, C. Da Vià, G. Anelli, "Increased speed: 3D silicon sensors; fast current amplifiers," IEEE Transactions on Nuclear science, vol. 58, pp. 404–417, 2011.

About the Appendix Author

Sherwood Parker died of amyotrophic lateral sclerosis (ALS) on March 9, 2018, just a few weeks before his 86th birthday. As an author of this book, during his illness he agreed to contribute a recollection of his memories of his work on silicon detectors, including 3Ds, an extract of which is presented in this section.

Sherwood Parker

Index

D

S

T

V

W

X

Y